W9-APE-650

PROVING
EINSTEIN
RIGHT

S. JAMES GATES, JR. AND CATHIE PELLETIER

PROVING
EINSTEIN
RIGHT

THE DARING EXPEDITIONS THAT CHANGED

HOW WE LOOK AT THE UNIVERSE

PUBLICAFFAIRS
New York

Copyright © 2019 by S. James Gates, Jr. and Cathie Pelletier

Cover design by Pete Garceau
Cover photographs: Solar Eclipse © SSPL / Getty Images; Solar Eclipse Telescope, Courtesy Special Collections, University Library, University of California Santa Cruz. Lick Observatory Records.
Cover copyright © 2019 Hachette Book Group, Inc.

Hachette Book Group supports the right to free expression and the value of copyright. The purpose of copyright is to encourage writers and artists to produce the creative works that enrich our culture.

The scanning, uploading, and distribution of this book without permission is a theft of the author's intellectual property. If you would like permission to use material from the book (other than for review purposes), please contact permissions@hbgusa.com. Thank you for your support of the author's rights.

PublicAffairs
Hachette Book Group
1290 Avenue of the Americas, New York, NY 10104
www.publicaffairsbooks.com
@Public_Affairs

Printed in the United States of America

First Edition: September 2019

Published by PublicAffairs, an imprint of Perseus Books, LLC, a subsidiary of Hachette Book Group, Inc. The PublicAffairs name and logo is a trademark of the Hachette Book Group.

The publisher is not responsible for websites (or their content) that are not owned by the publisher.

Print book interior design by Amy Quinn.

The Library of Congress has cataloged the hardcover edition as follows:
Names: Gates, S. James (Sylvester James), author. | Pelletier, Cathie, author.
Title: Proving Einstein right : the daring expeditions that changed how we look at the universe / S. James Gates, Jr., and Cathie Pelletier.
Description: New York : PublicAffairs, [2019] | Includes bibliographical references and index.
Identifiers: LCCN 2019019277 (print) | LCCN 2019021752 (ebook) | ISBN 9781541762237 (ebook) | ISBN 9781541762251 (hardcover)
Subjects: LCSH: Solar eclipses—1919. | Total solar eclipses. | Relativity (Physics)
Classification: LCC QB544.19 (ebook) | LCC QB544.19 .G38 2019 (print) | DDC 530.11072/3—dc23
LC record available at https://lccn.loc.gov/2019019277

ISBNs: 978-1-5417-6225-1 (hardcover), 978-1-5417-6223-7 (ebook)

LSC-C

10 9 8 7 6 5 4 3 2 1

Dedicated to the memory

of

William Wallace Campbell

Charles Dillon Perrine

Sir Arthur Stanley Eddington

Sir Frank Watson Dyson

Charles Rundle Davidson

Andrew Claude de la Cherois Crommelin

Erwin Finley-Freundlich

Edwin Turner Cottingham

and

Albert Einstein

CONTENTS

Photo insert located between pages 214 and 215

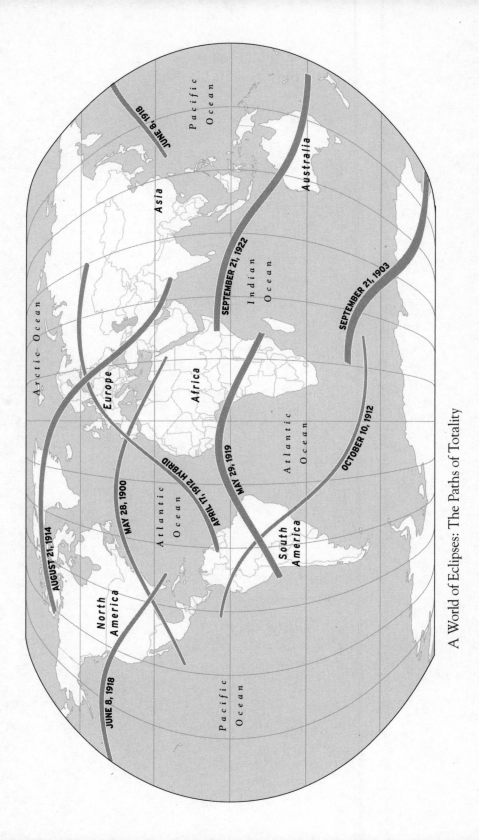

A World of Eclipses: The Paths of Totality

Prologue

I must search in the stars for what is denied me on earth.

—*Albert Einstein*

You are standing in the path of totality, waiting for a total eclipse of the sun. You have never witnessed one before, but each is different. It happens slowly, as if giving your brain time to prepare as the shadow of the new moon speeds forward to devour our closest star. At first contact, it looks as if a dark mouth has taken a bite of yellow from the sun. But this is still your world, the one you have known all your life. The bite grows larger as more of the sun disappears. With ten minutes to second contact, the total eclipse, daylight has gradually slipped away, replaced by a bluish twilight. The once magnificent sun has been eaten down to a thin crescent. Familiar landmarks now exist in a world of monochrome color you do not recognize. Odder things begin to happen. Reflected through the leaves of nearby trees, thousands of small images of the sun's crescent are spilled on the ground around you. Animals have sensed the loss of daylight. Birds flutter in confusion. Cows herd into the barn. Nighttime insects rattle in surprise. Diurnal flowers fold their petals. A shivering dampness flows over you. At your feet, the grass has turned to silver. This is no longer the earth you knew.

Now the sun has become a thin sickle, clinging to its last rays of light. The moon's shadow comes quickly from the western horizon, a massive wall of darkness speeding toward you at over two thousand miles per hour. With

1

totality soon to begin, the crescent of sun breaks into blazing beads of light that flow into each other, like drops of water fusing, until only one bead is left. It glitters in the darkened sky above you like a diamond ring. Jets of red flame burst from behind the black body of the moon before it finally covers the sun, the source of life for your planet. Songbirds are silent. Bats are on the wing. This is when you can look with your unprotected eyes at the spectacle above you. Planets that were lost in the light of daytime are now visible. The brightest stars twinkle. The solar corona, shimmering like a milky halo around the sun's hidden disc, is the color of liquid pearls. Its gray-white streamers, laced with crimson, are spilling backward into space for millions of miles. Your world has been thrown into a dreamlike trance. Distance now has no meaning. The heavens reach down, bringing the universe closer. The vastness of space reminds you of your mortality.

But now the reverie starts to undo itself, slowly reversing its steps. The diamond ring is back, a pulsing bead. It soon blends into a glowing string of pearls. The moon is gradually uncovering the sun, which has not perished after all. There's the crescent again, a blessed slice of yellow. The planets and stars have disappeared. Songbirds begin singing from the trees. The air around you warms as the sky once again lightens. The life-giving sun is on its way back. The world you thought you knew is returning. But it will never be the same one you left minutes earlier. You are now changed. You have been transformed by the magic and the miracle of a total eclipse of the sun.

Although the astonishing splendor of a total solar eclipse has never changed, what *has* are the people who view one. If you live near or within the path of totality, you're just plain lucky. While professional astronomers and experienced amateurs will travel great distances around the world for a total eclipse, even the casually curious can become eclipse chasers, especially within their own country. They catch planes or they drive to that slim track where the view will be perfect, despite the fact that totality may last only a couple of minutes. Days before an eclipse, airports become crowded, car rental companies are besieged, and hotels fill up. Professional tour packages headed by seasoned astronomers are fully booked months in advance by teachers, bank tellers, college students, and doctors.

Eclipse day is filled with news coverage, commentaries, and precautionary advice. Enthusiasts can check up-to-the-minute weather reports and track the path of the sun on their phones and iPads. If the viewing place they chose has clouded over, the more ambitious jump into cars or onto tour buses and speed to another spot, fifty or a hundred miles away, where the skies are cloudless. Wealthy eclipse chasers purchase seats on chartered jets that fly them above any inclement weather as they follow the moon's shadow. In June 1973, a select group of astronomers chased the eclipse path across the Sahara Desert at twice the speed of sound in an early model of the Concorde. In 2017, the National Aeronautics and Space Administration (NASA) sent out two jets with telescopes mounted on their noses, maximizing the minutes they could photograph the eclipse during totality.

But it wasn't always this easy. From the mid-1800s into the first two decades of the 1900s, long-distance travel to eclipse paths was mostly done by professional astronomers and knowledgeable amateurs. These men of science, and the rare women of science, were unwavering in their desire to observe a total eclipse of the sun for those few brief minutes. Lured by this remarkable phenomenon, they planned for months and even years before journeying to exotic parts of the world. Travel by boat, train, wagons, and pack animals was always rigorous and often dangerous. And yet, their best-laid plans could be obliterated in seconds by rain or clouds. The outbreak of regional or national conflict could entirely undo a well-planned expedition. Gone from their families for months at a time to foreign lands and unforgiving climates, these astronomers faced illness and possible death from the bubonic plague, malaria, yellow fever, and the Spanish flu. They protected themselves as best they could from wild animals, poisonous snakes, venomous insects, floods, forest fires, food poisoning, and local superstitions. But through it all, their mission remained clear: advance scientific research to better understand the cosmos.

Until the last years of the nineteenth century, drawings by sketch artists using pencils, charcoals, and pastels had been the dominant method of capturing images during a total eclipse. As photography evolved and became more sophisticated, it soon replaced drawings, which had to be done quickly. Photographic plates were permanent records that could later be reviewed and analyzed. But until 1911, astronomers were focused mostly on solar-related features during eclipses, especially on the corona, that halo of matter surrounding the sun. From this coronal structure, which could only be observed during totality, they would learn much about sunspot cycles, solar flares, and the solar atmosphere itself. Their attention during these eclipses, therefore, was on the sun, our own star. Other stars were studied at night, in observatories and with amateur telescopes around the world. And then along came a young German physicist with a challenging question about photographing other stars during a total eclipse.

Before he turned thirty, Albert Einstein had been working on a geometric theory of gravitation that embraced his 1905 theory of special relativity and expanded Newton's law of universal gravitation. Commonly known as general relativity, the theory provided a provocative and unified description of gravity as a property of space and time, what would become known as *space-time*. The curvature of space-time would be linked to the energy and momentum of any existing matter and radiation. If this idea was correct, the path of light would follow the arc of space-time. Thus, when passing close to a large body or mass, light would bend by an observable amount. But how to prove it? It was obvious to Einstein that his answer lay in those large bodies of the cosmos, a planet or star big enough to attract light waves. A physicist, he now needed the help of astronomers. The only way to test this premise would be to photograph starlight as it passed the gravitational pull of the sun. And the only time that could be achieved was during the brief darkness of a total eclipse.

Over the course of a decade, several esteemed astronomers in four countries would take on the "Einstein problem" in what would become an epic tale of frustration, faith, and ultimate victory. To succeed, at least one of them would need access to the path of totality for an upcoming eclipse. He would need ample funding. He would need the proper instruments. He might have to travel thousands of miles. He would hope for unity in a world that often perched on the brink of war. And he would pray for clear weather during those few fleeting minutes. If all these elements came together, the answer he found in the stars could shake the foundations of physics that had been in place for two centuries. The very concept of gravity, as the world understood it from Sir Isaac Newton, was at stake. Would the apple fall in a new way?

1

A PATH MADE OF MAGIC

The First Expeditions

I shall only say that I have passed a varied and eventful life, that it has been my fortune to see earth, heavens, ocean, and man in most of their aspects; but never have I beheld any spectacle which so plainly manifested the majesty of the Creator, or so forcibly taught the lesson of humility to man as a total eclipse of the sun.

—James Fenimore Cooper, "The Eclipse"

A SOLAR ECLIPSE occurs when the moon passes between the earth and the sun, totally or partially hiding the image of the sun from observers on earth.[1] While our sun is four hundred times the diameter of the moon, it's also four hundred times farther away from earth. That's how the cosmos can perform this dramatic sleight of hand in the first place. A *total* solar eclipse—quite common though rarely viewed in any one place—occurs when the moon appears to cover the entire face of the sun. The moon's shadow, or *umbra*, blocks all direct sunlight, turning daytime into night. Although a *partial* solar eclipse—its shadow is called the *penumbra*—is visible over a region that is thousands of miles wide, a total eclipse is observable only along a narrow path that cuts across the surface of the earth. Called the *path of totality*, it's typically about 10,000 miles long, but only

7

50 to 100 miles wide. The chances of your house sitting in the umbra of a total eclipse are about once every 375 years.[2] To witness the sun totally eclipsed by the moon, you must be standing somewhere within that slender track. And you need the luck of the draw when it comes to clear skies.

Solar eclipses figure prominently in human history. The word *eclipse* comes from the ancient Greek word *ekleipsis*, which means "being abandoned," a rather frightening thought when you consider how important daylight and warmth were to early humans in their struggle for survival. These people had no warning, after all, let alone an explanation. Folk astronomy dates back to the early Paleolithic period, when humans formed small bands for safety and became hunters and gatherers. Solar and lunar eclipses took their place alongside other unexplainable celestial activity, such as meteor showers and lightning storms.

All over the world, myths arose about these terrifying eclipses that no one could understand. It wasn't so many decades ago that shutters were nailed over windows during an eclipse to keep out the sun. Cattle were driven into barns for fear of them becoming blind. Pregnant women were hidden, certain foods couldn't be eaten, candles were lit, and desperate prayers were uttered. Various cultures around the world knew what was happening, of course. Sorcerers were at work. The devil had flicked his tail between the earth and the sun. Fairies were riding on horseback, provoking the moon to steal the sun. Demons were hungry. Or perhaps it was a bear, or a frog, that needed nourishment. In many parts of the world today, folk astronomy still holds sway over science.

Records show that as early as 2500 BCE, the Babylonians and ancient Chinese were adept at predicting solar eclipses. This ability was vital since eclipses were bad omens that threatened the welfare of any ruler. The Babylonians, for example, put substitute kings on the throne, thinking they would fool the eclipse should it attempt to harm the real king. When the eclipse was over, the imposter would step down and the true ruler would again ascend the throne. In China, in particular, it was the astronomer's job to study the skies, track solar and lunar eclipses and the movements of the planets, and then report these findings to the emperor. Because the emperor's prosperity was at stake, these natural phenomena were so important that early astronomers might be killed if they failed to predict their occurrences.

The most famous of these stories concerns the royal astronomers Hsi and Ho. On October 22, 2134 BCE, the two became so inebriated they fell asleep and missed the solar eclipse, the oldest in recorded history. The emperor was not amused. Had they predicted the eclipse, he would have had time to form teams of his men. There would have been arrow shooters to aim at the sun and drum beaters to urge them on—an army that could fight off the invisible dragon that was devouring the sun. Of course, the sun survived anyway on that October day over four thousand years ago. Sadly, Hsi and Ho lost their heads. Although the story is apocryphal, it is said that to this day, no one has ever witnessed a drunk astronomer during an eclipse. And this verse remains as a humorous epitaph:

> *Here lie the bodies of Hsi and Ho,*
> *Whose fate, though sad, was visible:*
> *Being killed because they did not spy*
> *the eclipse, which was invisible.*

It would take centuries of eclipses before early civilizations began to see a pattern in the appearances. Because the Mesopotamians had a habit of recording events by writing them down, they first realized that eclipses didn't just happen randomly. Although astronomy pioneers from many cultures and areas of the globe had tried for years to describe and explain them, it was not until Johannes Kepler observed a total solar eclipse, in 1605, that someone scientifically described the phenomenon. Over a hundred years later, Edmond Halley produced a map that predicted the timing and path of the May 3, 1715, eclipse. It would mark the first prediction using Newton's theory of universal gravitation. Halley's calculations were only four minutes and approximately eighteen miles away from being exact. Since the path of totality for this eclipse passed over London and Cambridge, Halley took the opportunity to inform the public about the nature of the phenomenon they would witness, thus advancing a lay understanding of natural science.[3]

Those early eclipse watchers—prehistoric, Mesopotamian, Chinese, Indian, Greek, Egyptian, Mesoamerican, Persian, and all the rest—whose solitary hard work and observations over the millennia had put in place the celestial road signs and the cosmic maps for modern times, were now mostly lost in the dust of antiquity, if their names had been recorded at all. As startling as the spectacle of totality could be, the phases were now anticipated by informed observers. In advanced parts of the world, the mystery had long been solved. Fairies were not riding on horseback. The devil had not wagged his tail. Nor had a frog eaten the sun.

AMERICA'S FIRST EXPEDITION: OCTOBER 28, 1780

Solar eclipses don't always have good timing. America's first official expedition was undertaken while the colonies were fighting England for independence. It was headed by Samuel Williams, third Hollis Professor at Harvard College. Science was still struggling for a foothold, despite the inroads paved by John Winthrop, considered America's first scientist.[4] At this time, colonial researchers had to deal with a public that was not sympathetic to a worldview grounded in evidence and observation. The witchcraft trials at Salem, twenty miles up the road from Harvard, were less than a century old. Superstition trumped logic in the majority of colonial homes, and science was a perceived attack on religion. But the bigger complication for this expedition was war. With the eclipse path falling over what the current map called "the east side of Longue Island"—this was Penobscot Bay in what would become the state of Maine—Williams would be traveling into enemy territory. He would need a flag of truce to travel safely.

With the help of men like Benjamin Franklin and members of the freshly formed Academy of Arts and Sciences, plans for the expedition went ahead. John Hancock wrote to the commander of the British forces, "Though we are political enemies, yet with regard to Science it is presumable we shall not dissent from the practice of all civilized people in promoting it either in conjunction or separately."[5] The British consented, but with the condition that the entourage leave the area the day after the eclipse, a demand that compelled Williams to feel "retarded and embarrassed by military orders." The Massachusetts Board of War provided

Williams with the *Lincoln*, an old 250-ton galley large enough to transport the passengers, crew, astronomical instruments, and food supplies he requested. The extensive list he turned over would have been coveted by George Washington, three years earlier at Valley Forge. Williams's request included 675 gallons of wine, three dozen smoking pipes, fifty pounds of butter, and seventy-two live chickens. That he decided not to take the kitchen sink could be attributed to the fact that it had not yet been invented.

Williams and his team, which included his ten-year-old son and six of his brightest students from Harvard, were eight days at sea after leaving Boston Harbor. They were allowed to set up camp near a settler's log house, provided they slept in his barn. And yet, this well-equipped expedition missed the track of totality on eclipse day by thirty miles. For two hundred years, American astronomers have been mystified as to why Williams made such a mistake. Was he using erroneous astronomical tables from Europe? Did his map have the wrong latitude for Penobscot Bay? After two centuries, letters written by John Davis, one of his students on the expedition, eventually surfaced to provide an answer. The British had wanted the party to stay on an *island* for greater security measures, denying them access to the path of totality. "Altho from the absurd policy of the military gentlemen in the neighbourhood," Davis wrote, "we were denied the gratification of beholding that uncommon phenomenon a total eclipse of the sun, we were nevertheless favored with appearances that were new, striking, and surprising."[6]

Regardless of its imperfections, this expedition was not an entire failure. For one thing, Williams described in great detail those "appearances" mentioned in the letter of 1780. They would later be known as Baily's beads, which occur just before and after totality, when the sun is a mere crescent of light. The "beads" happen when sunlight streams through craters and other features on the moon. Edmond Halley had written of them as early as 1715, but not until astronomer Francis Baily described them in 1836 did the astronomical world pay close enough attention. Consequently, the phenomenon now bears Baily's name. The history of discovery can be fickle and unforgiving, and it would become more so. Williams's exploration may have been the first time that a war would interfere with an official eclipse expedition. But it wouldn't be the last time. A more famous eclipse, coming in 1914, would greatly affect the astronomers in this story.

THE ADVENT OF THE ECLIPSE EXPEDITION

Total solar eclipses have shaped the lives of many scientists. Astronomers who happened to live within a reasonable distance from the path of totality usually traveled there, as Baily went to Scotland in 1836. Baily was an exception, an adventurer at heart having already toured extensively in America. He and George Biddell Airy, who was then astronomer royal, also observed the eclipse of July 8, 1842, in Italy. Baily set off to the Continent with a 3½-foot telescope to do his viewing at Digne-les-Bains, in southern France. When he reached Lyon and realized that he "had a few days to spare," he changed his route to Pavia, Italy, since he had always wanted to visit Venice. He was mindful enough to follow "along the line of the moon's shadow."

Airy didn't do much better planning than Baily did when he left for Turin. He had with him only a small telescope and tripod. He crossed the Alps and reached Turin on the night of July 5, less than three days before the eclipse. He turned down an offer to view it from the observatory there and decided instead on the Basilica of Superga, a structure set in the mountains eight hundred feet above the Po River. Amazingly, he didn't start up the mountain to view his observational prospects until about two o'clock on the morning of the eclipse. He left Turin and traveled in darkness up the mountainside to arrive at the church at five that morning. These hasty methods of planning would make later astronomers cringe. But during the 1842 eclipse, Airy described on paper for the first time the phenomenon of shadow bands.

The solar eclipse on July 28, 1851, added an important milestone in eclipse history: planning and travel. It would also mark the first time that many astronomers would witness this miracle of the natural world in person. The 1851 track missed the United States altogether since Alaska was still Russian owned. It cut across northernmost Canada and then southern Greenland. It caught the northern tip of Iceland before crossing southern

Norway and Sweden and then heading southeast across Poland and into the Ukraine. Some lucky astronomers of the day could stay home since the path of totality passed right over them. Before this eclipse, no properly exposed photograph of the solar corona—the most dramatic stage of an eclipse—had yet been produced. The timing of the eclipse was good as far as photography was concerned. Just a dozen years earlier, Louis Daguerre had invented a photographic process that created an image on a polished, silver-plated copper sheet sensitized with iodine vapors. This result was the daguerreotype, considered the Polaroid film of its day.

Photographing a total eclipse in those years was problematic in more ways than one. First, the great contrast between the dark shadow of the moon and the corona during totality was as much a challenge for cameras then as it still is today. Add to this the awkward angle at which the photographic equipment had to be mounted to aim at the sun, and eclipse photography became an extra challenge. Despite these drawbacks, one photographer succeeded. Hoping to get the first professional image of the corona, members of the Royal Prussian Observatory in Königsberg (now Kaliningrad, Russia) commissioned Johann Berkowski, one of the most skilled daguerreotypists in the city. Using a 2½-inch refracting telescope, Berkowski captured, in eighty-four seconds, a photograph that is still admired today as much for its beauty as for its place in eclipse history.

Several British astronomers had sailed across the North Sea to Norway and Sweden to observe the same eclipse. It rained heavily in most of Scandinavia the day before, but the sky remained cloudless on eclipse day. In the different towns, villagers had gathered out of curiosity, sitting on the grass nearby to watch these scientists whose language they did not understand. And then the moon began its slow advance on the sun. Even an experienced astronomer such as John Couch Adams, who was set up in Christiana, Norway, had never observed a total solar eclipse. He wrote of it lyrically, as though it were coming from the pen of a novice scientist:

> I now quitted the telescope and looked first at the moon then around on
> the sky. The appearance of the corona, shining with a cold unearthly
> light, made an impression on my mind which can never be effaced, and

an involuntary feeling of loneliness and disquietude came upon me . . . A party of haymakers, who had been laughing and chatting merrily at their work during the early part of the eclipse, were now seated on the ground in a group near the telescope, watching what was taking place with the greatest interest, and preserving a profound silence. A crow was the only animal near me; it seemed quite bewildered, croaking and flying backwards and forwards near the ground in an uncertain manner.

Although few astronomers had gone great distances before 1851, *afterward*, if they could get to India, Egypt, or a necklace of islands in the Pacific Ocean, they organized wisely. They traveled the globe when a good path of totality was predicted. They booked passage on stages, trains, and steamships. They rode on the backs of horses and donkeys. From the mid-1800s onward, they would stand on mountains or rooftops, take boats out to sea, and cross deserts and valleys. Now they were addicted. If the eclipse would not come to them, they would go to the eclipse. Academic reputations that would last a lifetime could be made at a single viewing.

WHEN AN ECLIPSE COMES TO TOWN

As the new century turned, people in all walks of life were now learning more about the moon and that blanket of stars that filled the night skies. Amateur star watchers were growing in numbers, and many were making important contributions, thanks to the encouragement from prominent astronomers at observatories in many countries. The British Astronomical Association (BAA) had formed a decade earlier for the express purpose of supporting amateurs. Turning eyes and telescopes to the skies at night is usually a solitary and quiet task. But when the most famous star of all, our sun, puts on a grand show like a total eclipse, the event tends to be a noisy affair. With seasoned astronomers now accustomed to traveling to remote areas around the globe, the setting up of a campsite became more complicated. They would need local workers and volunteers to assist them. Having a scientific expedition of prominent astronomers visit could set any small town in an uproar. For most laypeople, it would mark the most thrilling event in their lifetimes.

The century's first total eclipse, on May 28, 1900, was a perfect example of the planning that goes into accommodating a visiting expedition. This eclipse shadow crossed the face of the earth like a current of electricity, connecting a string of human beings who waited along a path more than 10,000 miles long. Its track of totality was narrow, only 50 miles wide. But its length was expansive in that it reached many places that would provide good viewing around the globe. As the moon's shadow moved eastward at approximately 2,200 miles per hour, it was observed by scientists in Mexico, the United States, Portugal, Spain, Algeria, Tunisia, Libya, and Egypt before it ended near the Red Sea. In the United States, its path swept across several southern states, stretching 925 miles from New Orleans to the coast of Virginia. There would not be another total eclipse viewable in the country for eighteen years, and the press jumped on it.[7]

Two of the towns in the southern states that lay in the eclipse path were Washington, Georgia, and Wadesboro, North Carolina. Their locations, far from city lights, would provide good sites for expeditions to set up. It was a major event in these sleepy towns when astronomers descended like locusts in 1900. They may not have been the big theatrical stars of the day, but as distinguished scientists, they were close to it. Even before the teams appeared in person, the excitement had begun. Local buildings were given fresh coats of paint. Lawns were manicured. Cafés and shops stocked up on supplies, anticipating a windfall in sales. Town committees and social groups planned afternoon teas and evening lectures for the esteemed guests. There would also be farewell picnics and barbecues when it was all over. This meant gallons of ice tea made, dozens of peach pies baked ahead, and enough skillet cornbread and barbecued pork to feed a small army.

Hotel rooms were reserved, and private residences made available to rent. Journalists and photographers arrived and found lodging. Often, the mountains of astronomical equipment would be shipped ahead, accompanied by a team member to oversee its safety. Local boys and men were then hired to unload the many railroad cars. Crates and boxes would be packed into wagons pulled by horses or mules and transported to the campsites. The folks living in the path of totality prepared as well as they could. Then they waited for the "clippers," as they called them, to finally arrive.

A dozen of the most prestigious American observatories sent teams to these southern locations. The cast of characters would read like a who's who of famed pioneers in the field of astronomy, both the old guard and the new. When the astronomers arrived, the serious work began. Masons would begin erecting the brick piers, and carpenters would build the wooden platforms to hold the telescopes. The canvas huts and awnings that would protect such expensive equipment from rain and sun would need erecting. Volunteers were chosen to act as security guards, standing watch at the huts during the night. Helpers who were needed on eclipse day to assist in scientific tasks such as handling the stopwatch and calling the time might be found among local merchants, blacksmiths, and farmers. They would be rehearsed in their tasks for two or three days before the eclipse. Sketch artists who could expertly capture the image of the corona either came with the expeditions or were found locally. By eclipse day, telescopes forty feet high would be pointing at the sky, like carnival rides at a state fair.

But when eccentric guests from England arrived, speaking with accents that most small-town Southerners had never heard before, the carnival coming to town was not nearly as exhilarating. Hotel chambermaids knocked on the British guests' doors with fresh towels just to hear them speak English. One amateur member of the BAA was the Reverend J. M. Bacon, a colorful character who brought along his photographer-aerialist daughter, Gertrude. Having arrived in New York by steamer, they considered the train trip down to Wadesboro an exciting new adventure, even for the death-defying Bacons. As the train flew through the starry southern night, the reverend witnessed an explosion of glittery sparks that he believed to be a shower of meteors beyond any he had ever seen. And he would have known since he and Gertrude had once observed the Leonid shower from a hot-air balloon before they were nearly lost at sea. "I shouted this intelligence aloud that all the Pullman car might hear, meeting only with a rebuke from our dusky and amused conductor. 'Dem are lightning bugs, Sar!' Of course they were, but I must be held blameless for this was the first time I'd ever seen a firefly."[8]

CONNECTING THE ASTRONOMERS

Nature was kind on May 28, 1900. On the day of the eclipse, there would be clear blue skies along most of the path from New Orleans to the Atlantic Ocean. The streets of Washington, Georgia, were packed tight with people, many reminding each other, "Don't look with your bare eyes!" On out-of-the-way hilltops and in open fields, the astronomers kept close watch on the crescent's length. Their numerous telescopes and cameras were ready. Business owners locked up shop and went with their employees into the streets. The many sketch artists, including "five young ladies from town," stepped up to their boards, excited. Once totality began, some artists would sketch just what they saw with their naked eyes. Others would peer into the eyepieces of telescopes as they drew.

At 8:10 a.m., as the spectators waited, breathless, the moon began its ascent on the sun. As viewed in Washington, the full eclipse lasted for one minute and twenty-five seconds. That was long enough to disturb the natural world. Flocks of purple martins and swallows flew in circles overhead as cicadas and insects rattled in confusion. The *Macon Telegraph* commented on this disruption as the umbra fell over the countryside: "The cows that were being driven in small groups to the pastures near the city stopped in the streets and tried to turn back. Chickens and fowls cackled and cawed, denoting their alarm. The ignorant and superstitious dropped on their knees and prayed to be forgiven." In the town of Washington, the crowd cheered wildly as the darkness receded and the sun began its reappearance.

Traveling at such a phenomenal speed, the eclipse shadow passed over South Carolina and raced on to North Carolina. The streets in Wadesboro were also flooded with spectators, an excursion train from Charlotte having arrived earlier with hundreds of people. Most of the scientists were dispersed, set up and waiting in individual encampments. Despite the early hour, the temperature was almost seventy degrees. Although the "Yanks" had been warned of the southern heat even in late May, several would collapse and need treatment by local doctors. As the full eclipse began, people watched from sidewalks or climbed to the rooftops of buildings, holding in their hands pieces of smoked glass or darkened binoculars. A smattering of small and inexpensive telescopes owned by schoolchildren were pointed at the sky. The total eclipse was over in less than ninety seconds when a sliver of yellow sun began to emerge. The shadow had already sped on to Norfolk, Virginia. There, it would dash across the Atlantic to reach the onlookers eagerly waiting in Europe and Africa.

To view the eclipse on May 28, 1900, members of the BAA were scattered along several thousand miles of the path of totality. They were almost certain that at least one group, if not many groups, would have a successful viewing. Three of the association's members were attending the farewell picnic back in North Carolina when the eclipse reached the USS *Austral*, off the coast of Portugal. An amateur astronomer using nothing more than a piece of smoked glass successfully viewed the full eclipse from the ship. The shadow then made landfall and sped on to Ovar, Portugal, where William H. Christie, the astronomer royal, was waiting with his team from the Royal Observatory, Greenwich.

Many BAA members had gone to sites in Spain to view the eclipse. Astronomers had learned by now that if they could afford more than one expedition, their chances of success with the weather on eclipse day would be greater. Not all expeditions would seek out small towns, such as those in the United States. Three BAA groups had gone to the city of Algiers, one group headed by E. Walter Maunder, a chief astronomer also from the Royal Observatory. The Maunder group was checked into the Hôtel de la Régence since they would view the eclipse from its rooftop. Another BAA group in Algiers had gone to the Hôtel Continental, and a third went to the house of the British vice consul.[9]

On the day of the eclipse, the Maunder team members arranged themselves along the viewing side of the roof in four separate sets, for convenience. As they waited, they cleaned lenses and rubbed lamp black, the soot from oil lamps, into their cameras and telescope tubes to absorb any possible stray light. Behind them, numerous houses climbed the hillside in terraces, each rooftop filled with chattering spectators. Over at the Hôtel Continental, the same preparations were being done to ready the instruments. On the rooftop of the vice consul's villa, the young assistants and artists got ready. They climbed up among the tall chimneys with star charts and sketch pads to wait. On the streets below, the crowd had grown to thirty thousand strong, many of them disbelievers in eclipses. They laughed and jeered as they watched the crazy English scientists, gathered like swallows on the rooftops.

Daughter Edith Maunder would act as timekeeper for her sister Irene, who would take photographs during the eclipse. Edith sat behind a wooden table that had been carried up to the rooftop by hotel staff. Many of the Algerian waiters were allowed to linger on the roof. They had been given a respite from work since the hotel would be empty during the eclipse. Not knowing what to expect, they asked the English scientists questions through an interpreter. Patriotic, they were not pleased to learn that Spain would see the eclipse before Algeria and that it would last longer there. Amused by this, Irene took her place behind a four-inch photographic telescope. A bell would ring three times to indicate the time remaining before the eclipse commenced: five minutes, one minute, and ten seconds. At the third bell, Irene was ready. The sky overhead had turned a deep purple as totality approached. When the moon slid over the sun, her father shouted, "Go!" The crowd on the streets below had gone from wild cheers to wailing at the strange darkness that now enveloped Algiers.

Irene had her own account of this eclipse published in the prestigious journal of the BAA, with her father as editor. Her photographs taken, she wrote the following description

> I shall never forget the crimson glow, and above and below it a milk-like flame stretching its long streamers away into the purple. The darkness, the cold wind, the silent workers around me, and the shouting crowd below all tended to make this strange and glorious sight still more impressive, and I found myself stretching out my arms to that exquisite corona in a perfect ecstasy. Suddenly the moon slipped off the other side of the sun and out he shone in a blaze of light, or so it seemed in comparison with his eclipse. An Englishman cheered. Some Frenchmen clapped. Totality was over![10]

The eclipse picked up speed on the last leg of its journey. It left Algiers behind and made its way to Libya. Tunisia and Egypt were next before the moon finally shook off its shadow and let it drown in the Red Sea.

THE PRINCIPAL PLAYERS GET READY

Scattered along the path of totality to view the eclipse of 1900 were five astronomers whose careers would later be linked to a man named Einstein, who was yet to graduate from the Swiss Polytechnic Institute in Zurich. In Thomaston, another tiny town in Georgia, from the Lick Observatory in California, were William Campbell and Charles Perrine. In Portugal, from the Royal Observatory, were Frank Dyson and Charles Davidson. And standing on the rooftop of the Hôtel de la Régence in Algiers was Andrew Crommelin, also from the Royal Observatory. The last three men who would have an important role to play were Erwin Freundlich, a fifteen-year-old German schoolboy who was dreaming of becoming a shipbuilder; Arthur Eddington, an English college student who had just that year discovered a love for physics; and Edwin Cottingham, who was busy in his clock shop in the tiny village of Thrapston, England.

A dozen years down the road from the 1900 eclipse, Freundlich would send Perrine a letter in which he would refer to a Professor Einstein who had just conceived of a new theory that needed verification. This letter would set the ball rolling, and that theory would change all their lives. It would also mean forsaking the Newtonian physics they had learned as students. But this chain of events was still more than a decade away. It was now a brand-new world that had just entered into a brand-new century. From that day in 1900 to the eclipse in 1919 that would make Einstein world-famous, there would be twelve more total solar eclipses. Their paths of totality would touch on all seven continents and cross all oceans. These marvels of nature still had much to teach about the secrets of the universe. The astronomers were learning. It would be a matter of the right eclipse, the right place, and the right time.

2

EINSTEIN'S VISIONARY YEARS

Thought Experiments and a Streetcar Ride

These are nearly the considerations which drove me to still think about the perturbation of light rays, which as far as I know was not studied by anyone . . . Hopefully no one finds it problematic that I treat a light ray almost as a ponderable body."

—*Johann Georg von Soldner, "On the Deflection of a Light Ray from Its Rectilinear Motion," 1804*

THE CONCEPT OF dwarfs standing on the shoulders of giants—*nanos gigantum humeris insidentes*—is first ascribed to the twelfth-century philosopher and scholar Bernard of Chartres. It was made famous in English by Sir Isaac Newton and later invoked by Einstein when he was referring to his own advances in physics. But Einstein didn't just improve and cultivate the accomplishments of his predecessors; he created dazzlingly new concepts. His manner of thinking appeared to be opposite that of his peers. He perfected the approach of considering experiments in the realm of pure thought. He was not bound to mere observation of an experiment. As a result, his work would produce a mind-wrenching shift in how humanity viewed its place in the universe. He might have needed those

21

gigantic shoulders of Newton and nineteenth-century mathematical physicist James Clerk Maxwell, as well as the broad shoulders of his colleagues
Max Planck, Hendrik Lorentz, Ludwig Boltzmann, and others, so that he
could *see* farther. But he also needed the visual acuity of a prophet.

In 1895, when he was sixteen years old, Einstein began using a plan of
conceptualization that would later rescue him from what he had believed
would be a lackluster academic career. While the world around him was
still comfortable with Victorian railways, ocean liners, hansom cabs, and
steamer trunks, he imagined himself chasing a beam of light as it sped
through the blackness of space, traveling faster than 186,000 miles per
second. Within that early idea conceived by a teenager lay a seed that
would later blossom into his first major scientific breakthrough.

Einstein certainly wasn't the first thinker to use this process, called
a *thought experiment*, a method of exploring complex ideas within the
mind, especially if physical experiments are impossible. Over the ages,
this device of intellectual pondering has been used in many fields, from
law to philosophy. It predates Socrates. But in physics and related sciences,
thought experiments can be traced back to the early 1600s and Galileo's
ship, an imagined vessel set sail to challenge the belief that the earth was
stationary in the heavens. It was Hans Christian Øersted, a Danish physicist and chemist, who more adequately described thought experiments,
in around 1812. Øersted also gave the method a name, using the mixed
Latin-German term *gedankenexperiment*. The thought experiment would
become the kind of speculative device that seemed custom designed for
Einstein.

After Einstein's youthful thought experiment in 1895, more such experiments would follow, involving trains and flashes of lightning, clock
towers, speeding automobiles, falling workers, and plummeting elevators.
But first, real life, outside the confines of his mind, was also happening. If
we consider a young Einstein less than ten years before his first major accomplishment, the special theory of relativity, we could not imagine him
becoming a world figure. In 1896, at the age of seventeen, he enrolled in
the Swiss Polytechnic Institute, in Zurich. When asked by a professor why
he didn't go into a field like medicine or perhaps even law, Einstein was

self-critical. "Because I have even less talent for those subjects," he replied. "Why shouldn't I at least try my luck with physics?"

In the autumn of that same year, he met his future wife, Mileva Marić, a fellow student and the only woman at the institute. An ethnic Serb from a wealthy family whose large acreage lay on the banks of the Danube River, she was almost four years Albert's senior. Petite and humbly attractive, she walked with a limp, having been born with a dislocated hip joint. She had been sent to Switzerland and the institute by her family because women were still not allowed to attend university in the Austro-Hungarian Empire at that time. A year later, Albert had fallen passionately in love with the woman he would refer to as his "Dollie."

If his future academic life seemed shaky, his courtship and forthcoming marriage would be even shakier. For the next few years, Albert penned passionate love letters to Mileva while weathering the wrath that his parents and sister rained down against his future bride. Miss Marić did not live up to the image of the woman they envisioned marrying their Albert. For one thing, overlooking her intelligence, she was plain in her appearance. For another she had the physical disability. And adding to the strikes already against her, Mileva was certainly not German. "My parents weep for me almost as if I had died," Albert wrote to Mileva. "Mama threw herself on the bed, buried her head in the pillow, and wept like a child."

In the summer of 1900, Einstein graduated from the institute with a diploma to teach science. But even with his father's help, he had no luck in procuring a university job. Marcel Grossman, a family friend, suggested he apply to the patent office in Bern, Switzerland. As Einstein awaited word, he accepted a temporary job teaching math in Winterthur, Switzerland. His family's aversion to his relationship with Mileva Marić did not lessen his ardor for her; nor did it help. He was already burdened with the possibility of not being offered the patent office job. With his family bearing down on him about his love life, he became disagreeable with Mileva. In a conciliatory letter written at the end of 1901, he admitted to being short and temperamental. To make it up to her, he decided they should meet at Lake Como.

Situated on the border of Switzerland and Italy, the lake had been immortalized in literature. The hapless lovers in *The Betrothed*, published in 1827 and the most widely read novel in the Italian language, hailed from a village on Lake Como. The poets Percy Shelley and William Wordsworth had fallen in love with the beauty of the region. After a 1790 walking tour, Wordsworth wrote the poem "Lake of Como," in which he refers to it as "a treasure whom the earth keeps to herself." The musical superstar Franz Liszt would often escape there with his mistress. Albert had selected a most romantic setting. Accepting his apology, Mileva went to meet him, arriving by train as Albert waited at the station. They spent the night together in a cozy inn. The next day, a horse and sleigh pulled them through falling snowflakes as they snuggled beneath coats and shawls. This lovers' rendezvous would change Mileva Marić's life forever.

Albert went back to his job teaching math as a substitute teacher. Since the scientific community of the day supported the idea that everything worth discovering in physics had already been discovered, he decided that mathematics would be a good career path, as it had been for Planck, and for the same reason. During this time, Mileva informed him that she was pregnant. Having once planned to earn her doctorate and become a physicist—a rare vocation for any woman then—she realized that becoming a mother would change all those plans. Albert wrote to her that he was working on the "electrodynamics of moving bodies." He mentioned that he would like to have Mileva with him, "in spite of your 'funny figure,'" referring to her pregnancy. Apparently curious about her appearance, he suggested that she "Draw it for me!" It's possible that his mother, Pauline Einstein, had also learned of the pregnancy. Mrs. Einstein would not be interested in any artwork that celebrated it. She wrote to a friend, "That miss Marić gives me the bitterest hours of my life, if it were in my power I would do all I could to ban her from our horizon."

Before their first child arrived, Albert had managed to find a steady job that might support his young family should he and Mileva finally wed. In December, he received word from the patent office in Bern. Offered the job as a third-level junior clerk, he eagerly accepted it. The patent office wasn't the most inspiring work. Among the patents that would cross his desk for evaluation were those for an electromechanical typewriter and a gravel sorter. But the job paid well enough, and he needed an income. "I am an honorable federal ink pisser with a steady salary," he would write to a friend. The hiring was also timely since a baby girl, Lieserl, was born in January 1902.[1]

SECRETS OF A GRAVEL SORTER

That Einstein was much more enigmatic than anyone realized is a detail that almost escaped the world's notice. He kept a universe of personal secrets inside his head, along with all that brilliance. Most amazing about Mileva's premarriage pregnancy or Lieserl's birth is that almost no one knew about them until thirty years after Einstein died. Remarkably, a good deal of his life as a lover and an eventual husband to Mileva Marić was disclosed only when their granddaughter found a shoebox filled with old letters. The packet had been bound with a worn ribbon. It appears that for almost fifty years, Mileva had saved the letters Albert had written to her.[2]

Given the many turbulent years in Einstein's personal life as he conceptualized ideas that would change the world of physics, it's astonishing that he accomplished anything at all. But other scientists before and after Einstein also utilized emotional turmoil to fuel their productivity. And Einstein had it in spades. Whatever happened to Lieserl has been a subject of speculation for years. The letters back and forth between her parents, at least those that still exist, do not say whether the child died or was adopted. They only speak of the birth at her maternal grandparents' home in Novi Sad, Serbia; of Mileva's difficulty in delivering the baby; and the child's illness from scarlet fever. Albert showed concern for his daughter, at least in his letters. "I love her so much and I don't even know her yet! Couldn't she be photographed once you are totally healthy again? Will she soon be able to turn her eyes toward something?" He enjoyed writing about Lieserl, but visiting her seemed less attractive. It's as if she were another thought experiment, better to imagine in his mind.

The most thorough research indicates that Lieserl was born with intellectual and physical disabilities and likely died of scarlet fever at twenty-one months. From his and Mileva's letters, Albert apparently never saw his daughter in person, and her condition may have been the reason. He and Mileva married in January 1903, a year after Lieserl's birth. Mileva moved to Bern to join her husband, but without the baby. Perhaps Albert had his own reasons for this. His extraordinary career was soon to be launched. He was existing on a meager salary at the patent office as it was. And then, what would his family say about his and Mileva's imperfect child had they been told of her existence? The baby seems to have vanished after Albert's last mention of her, in September 1903, when Mileva hurried home to Novi Sad after receiving word that Lieserl had come down with scarlet fever. In that letter, he remarked on his daughter's illness. "I am very sorry about what has happened to Lieserl. Scarlet fever often leaves some lasting trace behind." He then assured Mileva that she would get "a new Lieserl."

Years before Albert Einstein began thinking of eclipses, there would be one that would add a footnote to his personal history, although he would never know it. A total solar eclipse occurred on September 21, 1903, the same month he last mentioned their daughter in a letter to Mileva. The moon's shadow began in water, halfway between Africa and Antarctica, where the cold waves of the Atlantic meet the warmer waves of the Indian Ocean. This eclipse was mostly ignored by science for its unreachability. Its path of totality traveled down into the icy Southern Ocean and cut across Antarctica, ending very near Ross Island in a world of subfreezing temperatures and summers of constant daylight and winters of complete darkness. Probably the only witnesses to near totality, except for seals and penguins, was Captain Robert Falcon Scott and his crew, whose ship, the *Discovery*, was anchored in the waters of McMurdo Sound and frozen for two years into the sea ice. Even if Einstein had been interested in solar eclipses as early as 1903—and he wasn't—this one would be too insignificant for his attention. But it is likely that it occurred the same day Lieserl died.[3]

Albert would keep that promise to his wife. Nine months after she returned from Novi Sad and her last visit with Lieserl, she gave birth to Hans Albert, their first son.

A BREAKING STORM

Although Albert Einstein was not promoted at the patent office, because he had "not fully mastered machine technology," at least his job was made permanent. By 1905, he and Mileva, with their young son, had settled into married life in Bern. They rented a third-floor apartment in the Kramgasse ("grocer's alley") in the urban center of the city and a close walk to his work. In a way, the patent office was a place free of classrooms and laboratories, where he could think unimpeded. Nor did he have to worry about the curse of funding as he went about his thought experiments, since he was salaried there.

That summer, Einstein published two groundbreaking papers in Berlin's highly respected physics journal *Annalen der Physik*. The first paper was on the photoelectric effect, in which he showed that electrons, knocked out of metals when light was shone on a metallic surface, carry the same discrete amount of energy that particles of light, or *photons*, deliver to the electrons.[4] His second paper, on Brownian motion, marked the first time a human being had ever understood the size of atoms.[5] This man, still twenty-six years old, was not the wild-haired scientist in the baggy sweaters we later came to expect. He was a young husband and the father of a toddler. In photographs, he is dapper and well-groomed, even handsome, a hopeful scientist with a love for the violin. Nonetheless, he was mired in a mundane job at the patent office, possibly for life, he thought, while he yearned for a university post. But a night in May of that same year—1905 would become known as his annus mirabilis, or miracle year—would change the course of that lackluster career to which Einstein felt doomed.

Michele Besso, several years older, had attended the Swiss Polytechnic Institute when Einstein did and now was a coworker at the patent office in Bern. A friend and colleague, he was also a trusted confidant. While Besso was quite brilliant and intuitive, he unfortunately lacked ambition and focus. He was a good influence on Einstein, however. He had already introduced the younger man to the works of Ernst Mach, who would greatly influence Albert's approach to physics. Einstein would later call Besso "the best sounding board in Europe." On that spring evening, he went to Besso's home to discuss a problem that had plagued his thoughts for a decade: the two mainstays of physics—Newtonian mechanics and James Clerk Maxwell's equations—were discordant. In Newton's worldview, all velocities, including those of light, could be added or subtracted. But according to Maxwell, in a view backed up by his equations, the velocity of light is always constant. If one theory were proven to be correct over the other, then the result would mean that all of physics would need to be restructured.

After hours of discussion, frustrated and discouraged, Einstein gave up. He told Besso good night and left for home. What happened next is the most famous streetcar ride in history. As the story is often told, Albert Einstein caught a streetcar home to Kramgasse No. 49, the trolley rumbling eastward over a cobblestoned street toward the apartment he shared with his wife and son. It was a short ride—the entire street is only a thousand feet long—and yet the trip would change the foundation of modern physics. As he stared back at the centuries-old clock tower that stood in the heart of the city, an extraordinary thought experiment occurred to him. He remembered his youthful fantasy of chasing a beam of light through space. What if the streetcar he was riding on should suddenly race away from the clock tower at the speed of light?

Whether Einstein actually caught the streetcar home or walked the entire distance, here is what we *do* know happened that night. He says it best himself: "A storm broke loose in my mind." After his discussion with Besso, this mental storm stirred up the ghost that had plagued him since his teenage years. And it freed him to put on paper the conceptual framework he had discussed in a letter to Mileva four years earlier. He titled it "On the Electrodynamics of Moving Bodies," but the world would come to know it as the special theory of relativity. Tagged onto the end of the paper's original pages would be a thank you to Besso: "In conclusion I wish to say that in working at the problem here dealt with I have had the loyal assistance of my friend and colleague M. Besso, and that I am indebted to him for several valuable suggestions."

That paper would lead him, three months later, to publish another paper, this one on the equivalence of mass and energy: "Does the Inertia of a Body Depend Upon its Energy Content?" Those slim pages of the latter document would become science's greatest afterthought, a brilliant and elegant footnote that would contain $E = mc^2$, the most famous equation on the planet. The magnum opus of Einstein's miracle year, that equation would link three dissimilar parts of nature: energy, mass, and the speed of light. This mathematical result in physics is perhaps as widely recognized as the first four notes of Beethoven's Fifth Symphony.

The special theory of relativity hypothesized that the speed of light is constant and absolute, and nothing that conveys information can go faster. When objects travel at speeds that approach the speed of light, strange things occur. They will be measured by observers at rest to get shorter in the direction of travel, their mass will increase, and time will pass more slowly for all events on the moving object. Up until this time, and thanks to Newton's laws of physics, it was believed that space had three dimensions and that time had only one. But more importantly in Newtonian physics, the rate at which time passes is universal, a trait called *absolute time*. Einstein put space and time together in a four-dimensional system where they can't be separated or observed independently. Thus, energy and mass are the same, which is the fundamental idea of $E = mc^2$. *Energy equals mass times the speed of light squared.* Therefore, a minuscule amount of matter as small as an atom could produce a tremendous amount of energy. And until a patent office clerk envisioned it, no one understood or even noticed the existence of this accessible energy.[6]

THE BENDING OF LIGHT BEAMS

Given his annus mirabilis of 1905, academia finally took notice of the twenty-six-year-old who had once considered physics a hobby. If Einstein had published nothing more in his lifetime than the special theory of relativity, it would have been amazing enough in the career of any physicist. But he was not yet done. In 1907, he began challenging the classical physics of Sir Isaac Newton. In *Opticks*, his monumental book published in 1704, Newton had predicted that if a ray of light from a distant star grazes the edge of a huge object, the light should bend, depending on the object's mass and, therefore, its gravitational field. "Do not bodies act upon light at a distance, and by their action bend its rays; and is not this action . . . strongest at the least distance?" The bigger the object, the bigger the pull. Calculations based on Newton's prediction had determined that light would bend by 0.87 arc seconds. Given his place in time, Newton would have been unable to carry out any tests.[7]

Newton and Einstein were not the only ones to speculate on the gravitational bending of light. Johann Georg von Soldner, a German physicist, mathematician, and astronomer, also addressed Newton's question. The son of a farmer and primarily self-taught, Soldner wondered if the bending of light rays might require an adjustment of certain astronomical observations. His calculations, which were published in a paper in 1804, predicted that light would bend by 0.84 arc seconds, incredibly close to Newton's own numbers a hundred years earlier. But as there was still no practical way to make an observation that could either confirm or deny the calculations, Soldner's paper roused even less attention from his peers than Einstein's did when it initially was released.[8]

In 1909, Einstein was finally able to leave the patent office behind for an academic post at the University of Zurich. He barely had time to settle his family into a new home when a better offer arrived in March 1910. He was asked to consider a full professorship at the University of Prague. It was definitely a step up the academic ladder, even though Mileva, then five months pregnant, was not anxious to move to Prague. In late July, the couple welcomed Eduard, their second son, on the heels of two comets, the unexpected Great January Comet in January, and the highly anticipated Halley's Comet in April. Mileva again had a difficult delivery and was ill for weeks after the baby was born.

The next spring, accepting the new position at the university, Einstein moved his family to Prague. While neither he nor Mileva were thrilled with the shabbiness of the city, they nonetheless enjoyed electric lighting for the first time and could afford to hire a housekeeper. Einstein settled into his new office and went to work. Unaware of Soldner's calculations, he returned to the light deflection problem. He had supposed that the effect would be too small to be observed or measured. In this recent calculation, he repeated the problem based on Newtonian theory. Using his own equation $E = mc^2$, he calculated the bending of light and came up with a value. Light grazing the sun's outer edge would bend 0.83 arc seconds. This calculation was close to the great master's thoughts two hundred years earlier.

But it was still just a calculation. How might one measure the effects of gravity on straight beams of light? What source would be large enough to do this? The answer was the sun, since it has 300,000 times more mass than the earth has. One way this theory could be proven would be to observe the positions of stars. Since the sun is our closest source of a strong gravitational field, starlight traveling through space and passing its field would be bent. Therefore, if one could observe a certain star or stars during the day, they should be in a different place than observing them at night, when the light emanating from them to our eyes would not be traveling past the sun. The only way to see stars in the daytime would be during a total solar eclipse.

In June 1911, Einstein completed his paper "On the Influence of Gravity on the Propagation of Light" and again submitted it to the *Annalen der Physik*. Even at this respected journal, which had been in business since 1799, publication rules were still relaxed in the early twentieth century. There was no peer-review process. By 1911, the journal's submissions were overseen by two editors and an associate editor. Planck's position as associate editor since 1895 meant that Einstein could submit papers to his good friend, who published them in the next available issue. Thus, Einstein's eleven-page paper on light deflection was received by the journal on June 21, 1911. It would appear in volume 35, on the first day of September. Since the journal published its papers in German only, the paper would have a limited reading audience in the scientific community. As the ink was drying on his latest calculations, Einstein needed two things: a total believer, and a total eclipse of the sun.

3

THE TWO ECLIPSES OF 1912

The First Attempt: Is Einstein Right?

Nature and nature's laws lay hid in night:
God said, "Let Newton be!" and all was light.

—*Alexander Pope, An Essay on Man*

TWO HUNDRED YEARS before Albert Einstein put the finishing touches on his special relativity theory, the English satirist Jonathan Swift wrote an epigram: "When a true genius appears in the world, you may know him by this sign, that the dunces are all in confederacy against him." The exact opposite seemed true for Einstein. Despite the colossal achievements of his "miracle year," he was only moderately known to his fellow scientists, hardly enough to cause an uprising even if there had been willing dunces. His 1911 paper on light deflection, predicting that light would bend at 0.83 arc seconds, had been submitted to *Annalen der Physik* in June. Knowing that he now needed astronomers, Einstein was eager for any of them to come forward. He appeared more supplicant than genius in that his paper ended with an open invitation for help. "As the stars in the parts of the sky near the sun are visible during total eclipses of the sun, this consequence of the theory may be observed. It would be a most desirable thing if astronomers would take up the question."

In August, Einstein sought advice from an astronomer named Leo Pollak, also at the University of Prague. Pollak then wrote to the Berlin Observatory, informing its members that the theoretical physicist Albert Einstein wished to test a new theory, soon to be published in a paper titled "On the Influence of Gravity on the Propagation of Light." Pollak's letter was read and immediately answered by a young German who had written his thesis and earned a doctorate in mathematics a year earlier. Erwin Finlay-Freundlich had just turned twenty-six. Instead of testing exciting new theories, he was compiling star catalogs and photometric observations. Feeling as suffocated at the Berlin Observatory as Einstein had at the patent office, he wasted no time in replying to Pollak with an eagerness to help.

Einstein gave Pollak permission to send Freundlich a proof of his still-unpublished paper.[1] Thus, Freundlich was the first to have shown a spirited interest in light deflection and general relativity. That Einstein now had a fervent admirer of his theory must have been reassuring. He was barely thirty-two years old himself. Having found his devotee, he was ready to roll. "Highly esteemed Colleague!" he wrote to Freundlich on September 1, 1911, the same day *Annalen der Physik* would officially publish his paper. This salutation was most complimentary, given that the writer was the father of $E = mc^2$. "Thank you so much for your letter, which naturally interested me greatly. I would be personally very pleased if you took on this interesting question."

Freundlich was certainly qualified. After spending six months as a teenager working in a shipyard in Poland, he had first imagined himself studying shipbuilding as a career when he entered the Charlottenburg Polytechnic in Berlin. By 1905, he had abandoned those plans for the study of mathematics at Göttingen University, with mathematician Felix Klein. He would also study astronomy with the brilliant Karl Schwarzschild, a physicist and astronomer who would later greatly influence Einstein's work. Five years later, Freundlich had earned his doctorate and was an assistant at the Berlin Observatory. In his reply to Pollak, he suggested that an attempt be made to detect the bending near Jupiter to eliminate refraction problems from the solar atmosphere. Jupiter might be a Goliath compared with the earth, but next to the sun, the planet was a peanut, roughly one-tenth the star's diameter. Einstein wisely predicted that Jupiter's gravitational pull on light waves would be too small to detect. "If only we had a truly larger planet than Jupiter!" he wrote. "But Nature did not deem it her business to make the discovery of her laws easy for us." Instead, he recommended that Freundlich contact the Hamburg Observatory for photographic plates taken during past eclipses.

Arnold Schwassmann, chief observer at Hamburg, had no suitable plates. Instead, he advised that Freundlich contact Charles Dillon Perrine, who had taken photographs during several solar eclipses for the Lick Observatory in California. On September 11, ten days after the paper's publication, Freundlich wrote to Perrine, who by then had become the director of the Argentine National Observatory, in Córdoba. "Dear Sir," he wrote. "Although I am not acquainted with you, I nevertheless am so bold to apply to you about a question of truly high scientific interest as I am sure you will be able to give me valuable advice." The letter was in English. "The known physicist Prof Einstein has derived in a paper he intends to publish in the course of the next month an effect, that any field of gravitation produces upon electro-magnetic phenomena for instance an effect of the gravitation of the sun upon the light of a star passing near to the sun." Freundlich then added the value Einstein had currently arrived at, 0.83 arc seconds of deflection, and gave more detailed information.

He soon discovered that European observatories had a dearth of the photographic plates he had hoped to find. As he waited to hear back from Perrine, he reached out to a wider community. He wrote to several American observatories, including Harvard College Observatory, where Edward C. Pickering was director. Freundlich confessed that he was soliciting the "support of astronomers who have eclipse plates." Pickering answered quickly and proudly mentioned that his brother, William Pickering, had developed the best method for "photographing the eclipse of the sun and stars on the same plate." William had taken the photos his brother alluded to during the eclipse of May 28, 1900, which he had observed in tiny Washington, Georgia. But on those plates, the images of the fainter stars had been obliterated. Edward Pickering suggested that

Freundlich try other observatories, one being the Lick Observatory, whose director was William Wallace Campbell. The threads were slowly weaving together.

ASTRONOMY TO THE RESCUE

An American born in Ohio, the self-taught Charles Dillon Perrine had been at the Lick Observatory, near San Jose, California, for over fifteen years before he accepted the position at Córdoba, in 1909. In 1897, he had been awarded the Lalande Prize by the French Academy of Sciences for advancements made in the field of astronomy. He and Campbell, by then Lick's director, were close colleagues during the many years they had spent together. Also born in Ohio, Campbell was a pioneer of astronomical spectroscopy and was highly respected for his study of solar eclipses.[2] He often referred to Perrine as his "right-hand man."

While at Lick, the two men had joined other astronomers in the search for a theoretical planet named Vulcan, whose existence had been suggested by nineteenth-century mathematician Urbain Le Verrier to account for the irregular behavior of Mercury's orbit around the sun. Because Le Verrier had discovered the existence of Neptune using only mathematics—he was known as "the man who discovered a planet with the point of his pen"—many astronomers put faith in his theory. In their own search for "the ghost planet," Campbell and Perrine had made numerous photographic plates during several eclipses, including their expedition to Thomaston, Georgia, in 1900. After the January 1908 eclipse to desolate Flint Island, in the middle of the Pacific Ocean, they had finally concluded that Vulcan didn't exist. But those were the plates that interested Freundlich.

Perrine already had a foothold in the history of astronomy for having discovered two moons of Jupiter and nine comets. He was also well respected for his promotion of astrophysics in Argentina. Having gone on several expeditions, he had headed the Lick trip to Sumatra in 1901. The Lick Observatory had been given a large, thirty-six-inch reflecting telescope, one of the first large reflectors of the day, by a benefactor in England named Edward Crossley. Thus, Perrine also had extensive experience with large reflectors. And the Lick team did indeed have plenty of photographic plates. But Perrine was on a steamer bound for Europe by the time Freundlich's letter arrived in Argentina.

In October 1911, Perrine attended a meeting of the Carte du Ciel committee in Paris.[3] Once the conference ended, he and Johan Oskar Backlund dropped by the observatories in Bonn and Berlin on their way to Saint Petersburg, Russia, where the Swedish Backlund was director of the observatory there. Imagine Freundlich's surprise at meeting in person the man to whom he had just mailed a letter a few weeks earlier. He came by Perrine's hotel room to discuss the Vulcan plates. Perrine wrote an account of that brief stopover in Berlin. "Dr. Freundlich was then much interested in the theory of relativity and came especially to enquire of the suitability of the eclipse photographs taken for other problems by the Lick Observatory, for testing the displacement of the stars called for by that theory." Perrine told Freundlich that the images on the Vulcan plates were "not suitable for such delicate measurements as that problem required." Nonetheless, he advised Freundlich to contact the Lick and other observatories for a careful examination of their photographic plates to be certain.

Campbell never received his copy of the general letter Freundlich had sent out to observatories. But on March 13, 1912, the American answered Freundlich's second letter, informing him that he had seen Einstein's paper and would "be glad to assist you as far as possible in testing the question." Campbell then wrote to his former colleague Perrine, telling him that he had sent Freundlich glass "positives" of the Lick's eclipse plates. Given that the photographs taken in their search for the illusive Vulcan did not have the sun in the center of the plates and that the exposures had been short, Campbell also suspected that the results would be inadequate. As though being bitten by the light-deflection bug himself, he asked if Perrine would be willing to photograph the stars during the next eclipse coming in October, with its path of totality cutting across Brazil. Perrine replied that he had already discussed this possibility when he and Freundlich had met in Berlin the autumn before. And he was up to the challenge.

It was agreed that the Lick Observatory would ship down to Córdoba two of the photographic lenses that Campbell and Perrine had used in their intra-mercurial search for Vulcan. Einstein was grateful for Freundlich's continued interest. "I am extremely pleased that you have taken up the question of the bending of light with so much zeal." And now there were two American astronomers on his team, one in Argentina and the other in California. That his German colleagues, other than Freundlich, weren't rallying in support did not seem to deter him. Jonathan Swift's "true genius" was more interested in getting results than dwelling upon the "confederacy of dunces," some quite brilliant, who were now slowly gathering to question his new theory. Einstein knew the odds. In his very first letter to Freundlich, he had reminded the astronomer of one certainty. "If such a deflection does not exist, then the assumptions of the theory are not correct. For one must keep in mind that, even though plausible, these assumptions are quite daring."

On the home front, meanwhile, Einstein's marriage was falling apart, and he considered it irreparable. He was no longer in love with the woman he had once referred to in early letters as his "little everything," his "Dollie," for whom he felt such desire that his "pillow catches fire." While Mileva was becoming more despondent by the day, her husband's reputation was growing. He was busy giving lectures around Europe as she stayed home with their children, missing out on the intellectual stimulation she had once enjoyed as well as a career of her own.

In Berlin during the Easter holiday, Albert met up with his first cousin Elsa Einstein, whom he had known well in childhood. At thirty-six, Elsa was now divorced from her husband, Max Löwenthal, a textile trader with whom she had had three children. She was three years Albert's senior and as opposite from the intellectual and increasingly brooding Mileva as two women can be. It seems that Einstein had arrived at a place in his life where he now preferred his pillow to be fluffed, rather than set on fire, at least where spouses were concerned. Elsa, more motherly than collegiate, seemed the perfect woman to do that. But the physicist still needed a total solar eclipse more than he needed a new wife.

THE HYBRID ECLIPSE OF 1912

There would be two solar eclipses in 1912, one in the spring and one in the autumn. The one on April 17 was a *hybrid* eclipse, which begins as an annular, changes into a total eclipse along the track, and then moves back to an annular before the end of the path. Even if Freundlich had been ready with funding and preparation—and he certainly wasn't—the spring event was not the eclipse for him. With totality lasting a mere two seconds in the ocean west of Portugal, the day became more of a social event than a scientific study. Two days earlier, before midnight and under the darkness of a new moon, the RMS *Titanic* had rammed into an iceberg off the coast of Newfoundland and went to rest in 12,500 feet of water.

When tragedies coincided with natural phenomena, excitement often mixed with fear, a primitive instinct to link the heavens to earthly events. The annular shadow of this hybrid, soon to be nicknamed "the Titanic Eclipse," began in Venezuela and eventually fell across a narrow chunk of Portugal, France, Belgium, the Netherlands, Germany, and the Russian Empire. With search and recovery still going on where the *Titanic* had disappeared into the dark waters, the public turned out in masses to view the eclipse, convinced that the two events were cosmically connected.[4]

Several official astronomers had traveled to France because the annular path would pass very close to Paris, in the northwest suburbs. But there was no need to send out highly equipped expeditions. Andrew Crommelin did a viewing in Saint-Germain, as did the American astronomer G. B. Harrison, of the Maria Mitchell Observatory in Nantucket, Massachusetts. Harrison settled at a spot overlooking the Seine, about twelve miles from Paris. He was shocked to find that a horde of ordinary Parisians had left the city to witness the eclipse, "an animated crowd numbering some thousands." The Observatory of Paris had sent a balloon aloft carrying the inspector general of the French navy and popular science writer Camille Flammarion, also an astronomer. As the annular shadow moved in and the landscape around him grew darker, Harrison wrote that "the weirdness of the hour was enhanced by the appearance of some dozen balloons which had arisen from different parts of the country and which now hung apparently motionless like spectres in the darkening sky."

THE LIGHTER SIDE OF EXPEDITIONS

For this 1912 hybrid eclipse, Frank Watson Dyson, then astronomer royal—this appointment by the crown was synonymous with that of director of the observatory—went to France with his good friend John

Jepson Atkinson. If you read only scientific papers in academic journals, you would think that these Victorian astronomers were concerned only with hard work. But private letters and accounts sometimes suggest otherwise, especially if a colorful British character named J. J. Atkinson came along on an expedition. Dyson had met the older man at one of the Royal Astronomical Society meetings and dinners. The RAS was an association where amateurs and academics could comingle. A county squire, an inventor, a horse racer, and an amateur athlete in his youth—he was now sixty-eight—Atkinson stood six feet tall and weighed in at 18 stones, or just over 250 pounds. A bit of an eclipse groupie, he had paid the thirty guineas fee to join the BAA's first expedition, to Norway, in 1896. He was something of a generous and intelligent Falstaff, and Dyson liked him immediately. That's why he had invited Atkinson to join the 1900 expedition to Ovar, Portugal.

Called "Atky" by his friends, Atkinson soon became a fixture on British eclipse expeditions and not just for helping to unpack and install the instruments. He always brought along his magnanimous personality and appreciation for the fruits of the vine. He was forever encouraging Dyson to take a sip from his trusty bottle of Plymouth sloe gin, which Dyson declined for his usual soft drinks. Atky was first made famous among British astronomers during that 1900 eclipse, when he smuggled out of Portugal a cask of a certain wine he had come to appreciate. Since it was produced for native consumption only and export was illegal, he shoved the barrel in among the packed astronomical instruments. As the crates went through customs one at a time, William H. Christie would announce what was in each. "Telescopes. Mirrors. Photographic plates." When Atky's barrel arrived, a customs agent asked Atkinson what was in it. "A special instrument used to study double stars," he said, and winked at his companions.

Now a welcome guest at RAS dinners and parties, Atkinson had already been on two more eclipse expeditions since the Portugal event in 1900. He and Dyson had journeyed to Sumatra in 1901. They camped on a small island a few miles from the capital city Padang, hoping their photographic plates would solve the mystery of Vulcan. Atkinson had gone again with Dyson and Charles Davidson to Sfax, Tunisia, in 1905, and again made his presence known after the hard work was done. Dyson recalled this event: "a festival dinner with tables set in the open street, the company of British, French, and Italian astronomers, and Atkinson pouring out a tremendous fire of chaff, and asking in vain for the French to 'Keep the ball rolling.' But while he entered into that game like the splendid old boy he was, and though he never became an astronomer in any technical sense, the science meant a real interest to him."

By the time of the hybrid eclipse, in April 1912, plans had already been made for Atkinson to go on the upcoming expedition to South America that autumn, with Arthur Eddington and Davidson. Atkinson's participation in the expedition would make him a minor footnote in the story of general relativity. Brazil was too distant for Dyson to travel. As astronomer royal, he could not be gone from the observatory for that many weeks. Perhaps that's why this April excursion with Atkinson was important to Dyson, even if the eclipse itself was not.

"In the spring of 1912," Dyson wrote, "I went at his invitation to see another eclipse. He took his two-seater car from Southampton to Havre." Atkinson brought his thirty-horsepower automobile, a Gobron-Brillié, on the steamer with them so that the two friends could enjoy the countryside post-eclipse. They attended a lavish lunch in Paris and then a birthday party for twelve-year-old Ruth Turner, daughter of astronomer H. H. Turner. They presented the girl with a box of chocolate bells. It was just past Easter, and as chocolate "flying bells," or *cloches volantes*, were an integral part of the holiday tradition in France, it's a given that shop windows were still filled with the candy. One can only imagine the astronomer royal and the comical Atky poring over the candies at some tiny boutique, the Gobron-Brilliée parked out front. "We went leisurely up the Seine in time to see the short eclipse at St. Germaine," Dyson added, "and back again via Dieppe, and along the coast to Havre taking a trip by boat to see the Bayeux tapestry."

THE BRAZILIAN ECLIPSE OF 1912

Before the turn of the century, expeditions to South America had mostly been mounted for their naturalist and botanical interest. The Portuguese had already been in Brazil for over two hundred years, and the British had been invested in mining there for several decades. While Lick had sent an expedition to Chile in 1893, this eclipse would mark an upcoming awareness of the Southern Hemisphere as a good place to send future astronomical expeditions.[5] The eclipse's path was expansive. It began off the coast of northern Ecuador, caught a bit of southern Colombia, swept across the heart of Brazil, and then crossed several thousand miles of ocean, ending in the South Atlantic. Five countries would organize expeditions to set up camps in Brazil.

From England was the Greenwich team championed by Dyson and headed by Arthur Eddington. This would be Eddington's first eclipse, and his agenda was to photograph the solar corona. France sent a one-man team from the Bureau des Longitudes in Paris, and Chile a team from its national observatory. Argentina sent two expeditions, one from the observatory in Córdoba and the other from La Plata. With totality occurring within the country, two Brazilian expeditions were organized. The larger one was led by Henrique Morize, from the nearby Rio de Janeiro Observatory. Morize had plans to travel inland with the British team. He would figure prominently in the upcoming 1919 eclipse in that country, again in partnership with the British.

The most historically important expedition, at least in scientific hindsight, was headed by Perrine, who had now been director for two years at the National Observatory of Argentina, at Córdoba. Having given up on Vulcan four years earlier despite the still-unexplained wobble in Mercury's orbit, Perrine was ready for this new mission at hand. He would photograph the stars for Einstein, whose newest calculations predicted that light grazing the sun's outer limb would bend at 0.85 arc seconds, even closer to the Newtonian value.

On September 5, Perrine had sent an official announcement about his expeditionary plans to the *La Voz del Interior*, the largest newspaper in Córdoba. This statement would be the first newsprint story that alluded to Einstein's theory, although the physicist is never mentioned by name. "At the request of the Royal Observatory in Berlin, photographs of as many stars as possible near the sun will be taken on a large scale to determine, if possible, if there is any diffraction of the star's light as it passes by the sun, as the theory shows must exist." Few astronomers of the day were more qualified than Perrine. His former colleague William Campbell described him glowingly: "A veteran of five eclipses, he has no superior as an eclipse observer." Einstein would have been gratified to know that such an experienced astronomer had taken on the problem.

THE GREENWICH TEAM

The English expedition left Southampton on August 30, aboard the newly launched ocean liner RMS *Arlanza*, which had been constructed in Ireland expressly for travel from England to Brazil and Argentina.[6] Astronomers were considered celebrities of the day, and those whose observatories could afford it were now traveling in this first-class style. In deference to science, the shipping company even allowed the scientists' many bags and crates of instruments to be transported free of charge. The team consisted of Eddington, who was Dyson's assistant and the newly elected secretary of the RAS; Davidson, a computer known for his skill with astronomical instruments; and the ever-jovial Atkinson, who was suffering from flare-ups of gout but who probably had a bottle of sloe gin in his coat pocket. After a four-hour stop in Madeira, the steamer headed west, across the Atlantic.

It was after dusk on Sunday evening, September 15, with the lights of Rio de Janeiro glittering in welcome, when the *Arlanza* sailed into the harbor with its many small islands. Earlier that day, Eddington had received a marconigram, or a radio telegraph message, from Morize, advising him to stay on board for the night. He would come for the three men on Monday morning. Morize arrived at eight o'clock with an Englishman named Theophilus Lee, a chemist assigned to the Geological and Mineralogical Service of Brazil. Lee was to assist the astronomers during the eclipse and to act as interpreter. His language skills were helpful, because Eddington would soon learn that Morize, congenial though he was, spoke "the worse English I have ever heard."

The British astronomers would be treated generously. All lodging, food, and transportation was at the expense of the host country. Any telegraphing that needed to be done was also gratis. After Morize arranged for the mountain of baggage to be taken to the train station, the party boarded a motorized launch that swept them around the harbor, past Rio's well-known Sugarloaf Mountain rising up from its peninsula—the famous Christ the Redeemer statue was still two decades in the future—to where a motorcar was waiting, as were the local newspapers. Once photographs were taken by the press, the party was off again in the car, down avenues lined with palm trees to the Hotel dos Estrangeiros, the "hotel of foreigners." This was an older restored building at the edge of the city, famous for the political meetings held there. Eddington commented in a letter home that it was "the swagger hotel" but that it was not up to English standards, despite its brochure promise of "disinfectants in the water closets, and drinking water filtered by the Pasteur system."

CHARLES PERRINE, FROM CÓRDOBA

Perrine and his team reached Buenos Aires by rail, a thirteen-hour journey. From there, they sailed on the RMS *Aragon*, a sister luxury ship to the *Arlanza*, which was on its return voyage to Southampton. After six days at sea, they docked at Rio de Janeiro on Wednesday, September 18. They were put up at a hotel with even more swagger than Eddington's, the newly opened and majestic Hotel Avenida that dominated Central Avenue with its belle époque architecture.

The next day, Perrine went off to find the Englishmen. He arrived just in time, for Eddington was in a fluster over one of his cases having gone missing. Charles Davidson was always overly protective of the photographic plates, especially *after* an eclipse, never letting the hampers out of his sight. But the rest of the expedition's baggage was to have been taken separately off the ship and hurried through customs. Instead, it had been sent to the customs house with all the other *Arlanza* passenger baggage. Eddington had spent desperate hours that week trying to locate the lost case, with the aid of a "Portuguese gentleman" who spoke no English. It was to Eddington's great relief when the Portuguese assistant knocked on his hotel room door and had with him the American-born Charles Perrine.

The tall and slender Eddington and the quietly elegant Perrine, with his distinct mustache and light blue eyes, had already met briefly at Greenwich. Now they spent the entire Thursday together in Rio de Janeiro. They went first to customs, and after sorting through the piles of baggage, they located the missing case. By then it was time for lunch, which Eddington was set to have at the residence of Sir William Haggard, British envoy and minister to Brazil and brother to author H. Rider Haggard. Edwin Vernon Morgan, whom President Taft had newly appointed as American ambassador to Brazil, had come to meet the astronomers. Also attending the lunch was an amateur astronomer from England, a young man from a wealthy family. Now a BAA member, James Henry Worthington was at the end of a three-year trip around the world to visit the best observatories.

After lunch, and more photographs of the Greenwich team, English celebrities that they were, the men set off as tourists to do some sightseeing. Their first stop was the famous Rio de Janeiro Botanical Garden, which had been founded a hundred years earlier by the king of Portugal to acclimatize spices from the West Indies like pepper, nutmeg, and cinnamon to Brazil. Now the garden was open to the public, and for two hours, the astronomers marveled at the avenues of royal palm trees, the tropical shrubs, and carnivorous plants. It had been a perfect day for Eddington and Perrine to become reacquainted.

Later that evening, both teams—Perrine and his group and Eddington, Davidson, and Atkinson—gathered for dinner at the Hotel dos Estrangeiros. With Perrine was Enrique Chaudet, his assistant astronomer at Córdoba; James Mulvey, also American, who had designed and built most of their equipment; and a photographer. With no language barrier, the conversation no doubt touched on their upcoming plans. By this time, the British team members had decided on Passa Quatro to do their viewing, about 180 miles from Rio, instead of Cristina as previously planned. Perrine and his team kept their plans to set up in Cristina, fifty miles farther inland than Passa Quatro.[7]

TO THE CAMPSITES

Charles Perrine and his team left Rio the next morning, Friday, September 20, and arrived at Cristina to begin setting up their viewing camp. They found a place behind a church at the edge of town, a picturesque spot that overlooked houses dotting the valley below. Suffering from severe asthma for most of his life, Perrine was lucky that this eclipse would occur in October, or springtime in the Southern Hemisphere. Winters brought on the most severe attacks and kept him house-ridden much of the time. This was a propitious moment in his career. If Albert Einstein was correct in his calculations, it would mark the first time in human history that a test could be revealed to prove that light had weight. The Córdoba expedition under Perrine's direction, from the National Observatory in Argentina, was the only one interested in testing for light deflection in 1912. Charles Perrine hoped for clear skies on eclipse day.

His team consisted of three highly skilled men, but especially James Oliver Mulvey. In Mulvey, he had found a brilliant and self-taught mechanical engineer. Perrine had met him in 1909, at the Armour Institute in Chicago, where Mulvey was attached to the mechanics department. Mulvey was also involved in business with Albert B. Porter at his then-famous Scientific Shop on Dearborn Avenue. At the institute where Porter had been a physics professor, Mulvey had picked up on his own a considerable knowledge of physics, and especially of optics. When they first met, Perrine was on his way to Córdoba to become the new director of the observatory. Having had experience with the Crossley reflecting telescope at Lick, he was interested in installing a large reflector at Córdoba and had discussed this hope with Porter and Mulvey.

A few months after arriving in Argentina, Perrine received a letter from Mulvey. Porter had passed away, and was Perrine interested in Mulvey's services at Córdoba? It was a lucky opportunity for Perrine. For this expedition, Mulvey had designed and built the cameras, as well as the mounts and frames, which were made entirely of wood. Perrine had concluded that wooden frames would be more stable than metal during the drastic temperature changes that occur during those chilly moments of totality.

The many crated instruments and cases belonging to the Englishmen left Rio at midnight on Friday, aboard a cargo train. The three men themselves departed the following night on the São Paulo Express, with a change of trains at Cruzeiro, which they reached at 2:30 a.m. Traveling with them were Theophilus Lee and the amateur astronomer James Worthington. The plan was to sleep in the waiting room of the Cruzeiro station so they could claim their instruments at daylight. With the room packed, Eddington opted to sleep on his trunk out on the station's platform, which amused the porters. It must have been a tense night anyway. The Greenwich team had begun to dislike Lee back in Rio, when he had shown no concern for Eddington's lost case. Eddington's opinion was that Lee hoped to advance his position in Brazil. "I think he had taken this up as a sort of lever to advertise himself and get in with important people; he had somehow got round the British Consul who recommended him to us. We had rather a bad time from him at first, but had the satisfaction of seeing him completely checkmated."

The astronomers had been frustrated to learn that the instruments were delayed and would not reach their destination until a day later. The next morning, after a breakfast of bread, bananas, and black coffee, they boarded a cargo train that would take them the rest of the way. The authorities had seen fit to add a saloon car for their convenience where meals could be served. Passa Quatro was only twenty miles distant, and yet it took the train three hours to climb the steep incline, its track winding up through mountains and a countryside that Eddington thought beautiful and wild. With Lee and Worthington onboard, it's likely the trip seemed twice as long. Now the team had begun to dislike the wealthy young Englishman even more than they disliked Lee. Did Worthington brag about just ending his three-year world tour that carried him to three total solar eclipses—this one would be the fourth—as well as a viewing of Halley's comet? Or did he mention that his family's estate in Somerset had more servants than all the teams at Passa Quatro combined? At least when the haggard Englishmen arrived, they were put up at a family-owned inn that was "clean and comfortable."

Passa Quatro is in the state of Minas Gerais, which translates to "general mines" in Portuguese. This was where the Brazilian gold rush began over two hundred years earlier, when it was discovered in the mountains by early Portuguese settlers. News of the discovery had attracted more

Portuguese by the hordes, over four hundred thousand in all, and with them they brought over half a million enslaved African workers to toil in the mines. The British now controlled the largest mine in Latin America, which was opened in 1830. They later built the Minas and Rio Railway to bring the gold from the mines to waiting ships. With skilled miners imported from Cornwall, England, and with the brutally enforced labor of the black workers, this single mine would produce gold for over 150 years, long after all the humans who came there to see the 1912 eclipse, and even the 1919 eclipse later on to the north, had lived out their lives and disappeared.

Morize finally arrived from Rio with the rest of his Brazilian team, as did the Greenwich baggage and instruments the Englishmen were anxiously awaiting. Their viewing site would be a mile away at the edge of a farm, or *fazenda*, owned by a man with the improbable name of Rudolpho Hess. This location would provide them with privacy and yet be close to the railway tracks, which Eddington believed would be convenient. But to his irritation, all their astronomical instruments had been dropped off at the side of the rails. It took considerable work and carts pulled by six oxen to get the Greenwich equipment loaded and transported to the observation site.

Always accommodating, the Brazilian government had provided the astronomers with a special railway car that would carry them each morning from the inn to the farm. At 11 a.m. the car would bring them back for a two-hour lunch and then fetch them back again to the inn each evening for dinner. "You would be amused to see us all riding down to the Fazenda (eclipse camp) on an engine," Eddington wrote. "There were about 20 of us today clinging on in various places—the cow-catcher is the best seat." The two-hour dinners each night were "rather French in style" for an Englishman and went on far too long, as did the midday lunches. "Dinner occupies most of the evening lasting from 7 to 9. It is a terribly complicated affair of about 12 courses, chiefly meats of various kinds." At least a wood fire was built each day at the farm so that the Englishmen could enjoy their cup of tea.

Despite the hard work in setting up camps and getting the instruments in place, it seems these men of science still had time for a slice of high school drama. As the expeditions at Passa Quatro made ready for the eclipse, the consensus among the many team members and volunteers representing three nations—England, France, and Brazil—was that they *all* hated James H. Worthington and Theophilus Lee. "They are all very nice people," Eddington wrote, "but we very much dislike Lee and Worthington (especially the latter) and it is hard work to avoid a regular rumpus."

WAITING FOR THE ECLIPSE

Little is known about the activity at Perrine's camp in Cristina. But most expedition camps had similar preparations. At Eddington's camp, the hard job of setting up was over. Local workers had built the brick piers that would hold the coelostats and the heliostat. Wood-framed huts covered with waterproof Willesden canvas had been erected to house the instruments. The weather was holding up well, with warm, clear days and cold nights. "I find my helmet very useful but have not worn my drill suit," Eddington wrote. "It is really wonderfully cool weather and one could hardly imagine we are in the tropics." There were no mosquitoes or cockroaches, and the only signs of snakes were cast-off skins. It was also too early in the spring for an abundance of dazzling Brazilian butterflies. There were, however, plenty of flying ants and termites about, with the termites building mounds that were taller than the men.

With the eclipse coming on Thursday, the tenth, each team went through two days of rehearsals, making sure all the instruments were working, loading the photographic plates, and practice-counting the seconds to totality. This careful preparedness could make up for the less-than-two-minute duration of totality. For the English team, Eddington was in charge of the coronagraph. Davidson would oversee the spectrograph, with assistance from Atkinson. The British team had met two young Brazilian men on board the *Arlanza*, one studying engineering in England and the other with English-born parents. They had come to Passa Quatro as volunteers and proved most capable, one handling the six-inch refractor, and the other the six-inch triplet. Another local volunteer would count the seconds during totality.

Unfortunately, the excellent weather conditions began to deteriorate as the hours counted down to the eclipse. From Cristina to Passa Quatro, and for most of the country, clouds moved in on Wednesday morning. By noon, the rain began. It might have been a good omen since rainstorms in that climate were often brief and helped clear the air. But by nightfall, the rain hadn't stopped. The waterproof canvas tops and sides of the huts at least protected the equipment, but the mood of the men was quiet despair.

On eclipse morning the rain was still falling when the teams in the different towns rose before dawn to get ready. At 7:30, a special train arrived in Passa Quatro from Rio de Janeiro, much to the excitement of the townsfolk. They turned out with a band and firecrackers to greet it at the station. Aboard was a huge entourage that included the president of Brazil, Hermes da Fonseca, his wife, and many of his ministers and deputies.[8] Ambassador Morgan, who had attended the luncheon three weeks earlier, was along for the excitement. Also on the train were many students from the Polytechnic Institute of Rio. The train continued on to the farm where the viewing camps were. "Most of the people came and looked round the camp but it was too wet for the President," Eddington wrote. "The American Ambassador sheltered in our shed where our cases are; we like him very much."

The partial eclipse began a few seconds after 8:56 a.m., with the rain still beating down on the huts. Totality started at 10:15 a.m. and ended a minute and forty-nine seconds later. Rain didn't prevent the natural world from responding to the startling loss of daylight. The insects and birds chirped, buzzed, and fluttered as the scene around them grew dark and then light again. One can only imagine the disappointment and heavy hearts of the expeditionary teams, and yet a banquet had been planned by the owner of the farm at Passa Quatro for the important visitors. President Hermes left the train to attend it. His wife, and the wives of all the dignitaries, also stepped down from the train in their long dresses, umbrellas open, as they dodged the many mud puddles. While they ate, the storm intensified, with most of the continent now covered in cloud and lashed by rain. Despite the weather, journalists arranged the astronomers in a group outside the inn for photographs.

A failed eclipse observation didn't affect the prestige of astronomers in the public eye. A dinner was planned at a local hotel in Passa Quatro for a group of about forty. At least there was one bright spot amid all the bad luck. "Lee & Worthington were detested by everyone," Eddington wrote, "and their departure was a great relief." There would be no packing of the instruments until the rain stopped; it continued coming down until the next afternoon. Atkinson had kept the others well entertained. But thanks to the evening dinners with a dozen meat courses and his likely indulgence of more than a case of red wine or sloe gin during his stay, a painful gout attack got the best of him. He left Passa Quatro three days after the eclipse for the privacy of a hotel room in Rio. Because of the intermittent rainfall, the party did not finish packing up the instruments until six days later, even as the flying ants swarmed the canvas huts. By this time, the enormous butterflies and moths were out, so Eddington and Davidson managed more sightseeing, which included a mule ride in moonlight. Caught in another rainstorm, they stood watching as the valley below was set ablaze with the tiny lanterns of fireflies. It was time to go home to England.

On October 23, staff from the Rio observatory drove the Englishmen from their hotel down to the quay in a motorcar. Morize was waiting to wish them bon voyage. They boarded the steamer *Danube* and began the 2½-week journey back to Southampton.

On the day of the eclipse, a disappointed Charles Perrine sent a telegram from Cristina to the United States to announce the results. Then he and his three-man team returned to Rio and sailed back to Buenos Aires. After the thirteen-hour train ride, they were once again in Córdoba. The telegram had gone to Edward Pickering, director of the Harvard College Observatory. It was a one-word message: "Rain." Perrine had just missed the opportunity to be the first man to observe that light has weight.

Pickering immediately cabled the news on to William Campbell, at the Lick Observatory. "Perrine cables from Brazil, rain." Five days later, on October 15—and with no mention of the planning, the travel, the money, the inconveniences, the homesickness, the dashed hopes—the news appeared in bulletin 503 in the *Harvard College Observatory*, mailed out by Pickering: "A cable has been received from Professor Perrine, Director of the Córdoba Observatory, Christiana [sic], Minas Gereas, Brazil, 'Rain.' This appears to indicate that observations of the Eclipse on the Sun on October 9-10 were prevented by bad weather."[9]

During all the time Eddington spent with Perrine and the Córdoba team, he must have learned that they were planning to photograph stars for light deflection in regard to Einstein's new theory. As letters and papers often reveal, these scientists and technicians were eager to discuss their programs and share their expertise. What else would have been the topic of conversation at the dinner table that night in Rio? But unfortunately, no historical evidence has yet been found that Eddington knew of Perrine's plans. And it may never be known. It seems highly unlikely, however, that the conversation would not have included each team's agenda for the 1912 eclipse. Eddington's was the solar corona. Perrine's was Albert Einstein. And yet it has been widely written and believed that Arthur Eddington first learned of Einstein's theory when Willem de Sitter sent him a copy of the paper in late 1916, during wartime.

Albert Einstein seems to have already understood the disappointments that come hand in hand with scientific experimentation when he wrote to Erwin Freundlich. "Nature did not deem it her business to make the discovery of her laws easy for us." But there was another eclipse coming for Charles Perrine to try again.

4

EINSTEIN'S ENTREATY

Astronomers to the Challenge

Nothing more can be done by the theorists. In this matter it is only you, the astronomers, who can next year perform a simply invaluable service to theoretical physics.

—*Albert Einstein, to Erwin Freundlich, 1913*

THE WORD PHYSICS derives from Ancient Greek and translates quite accurately to "knowledge of nature." One of the most fundamental of the scientific disciplines, physics aims to understand how the universe conducts itself, how it behaves. Mathematics is a physicist's third eye. It enables him or her to see beyond the restrictions of ordinary sight when it comes to revolutionary ideas about the cosmos. It opens that gate to an inner and higher perception. Isaac Newton didn't just see light traveling along a straight path. He envisioned it *bending* through an angle of 0.87 arc seconds as it passed near the sun, or so his mathematical calculations informed him. This was the famous question he had asked in *Opticks*. Didn't bodies act on light at a distance? But when it came to answering this question, Newton was locked into his place in time, third eye or not, in the latter half of the seventeenth century. This intellectual barrier must have been frustrating for someone with his enormous brainpower. He was most capable at *describing* gravity. He wrote a useful equation to explain how gravity *acted*, but he had no idea of the mechanism behind it. When Newton died in 1727, his question of light deflection was left in limbo.

As though they were tag-team members from different ages, Einstein had now taken up Newton's question. But how to find the answer? The science of physics depends on a series of steps before arriving at its conclusions:

1. Using the ability to conceptualize.
2. Encoding this concept in rigorous mathematics.
3. Using the mathematics to make error-free calculations that lead to predictions.
4. Using the senses and devices to make measurements and observations as accurately as possible.
5. Comparing the predictions to what is measured and observed.

In 1704, when Newton published his revolutionary work predicting that the path of light would bend by a measurable amount when passing a large body, he had at his disposal only the first three steps, as did Johann Soldner, in 1804. With his own calculations so close, Soldner essentially performed an independent check on Newton's work, discovering that Newton was not only a genius for coming up with his concepts, but he could also correctly calculate with them. That's a skill not always true of a genius. Given Einstein's place in time, two hundred years down the road from the publication of *Opticks*, the German physicist could take advantage of all five steps listed above. That a more advanced technology was now at his fingertips was a gift from the gods.

Einstein was already superbly gifted when it came to the first step, that third-eye ability to conceptualize, which he had exhibited from his teenage years when he "rode" a light beam through the vastness of space. For the rigorous math of step 2, he relied on his good friend Marcel Grossman to help him understand the work of four pioneering mathematicians who had preceded them in time.[1] For step 3, when he first published his paper, he was simply wrong in those early stages of his calculations. Nature might be infallible, but physicists are not, and even intellects like Newton and Einstein were no exceptions. Newton had arrived at a calculation of 0.87 arc seconds. Even if it were proven wrong, the brilliance it required to envision the bending of light in 1704 can't be anything but *right*.

Step 4 is trickier since it depends not just on human senses, but also on current technology. As Charles Perrine and Arthur Eddington stood in their separate huts watching it rain in Brazil in 1912, the art of solar photography had advanced from the day in 1851, when a Prussian daguerreotypist pointed his small telescope at the sun to capture the first photo of the corona. New photographic lenses and filters had been invented over the decades and were now in use. The dry photographic plate was introduced in 1871, replacing the cumbersome wet plate, which had replaced the daguerreotype. The early 1900s had seen the best research telescopes move away from "great refractors" with their large glass lenses to even larger reflectors installed with glass mirrors. Technology was on their side in 1912, even if the weather wasn't. But it was a dicey endeavor to begin with when you consider that the most fortunate astronomers from the best observatories would likely be blessed with only forty minutes or less to photograph the sun during total eclipses in their *entire lifetimes*. And that's if everything went according to plan on an average of a dozen expeditions.

For that last step—comparing the predictions to what is measured and observed—Einstein needed some help. To prove light deflection according to his new theory, he depended on astronomy, a scientific field he had once criticized for its "pedantic accuracy."

In his enthusiasm for the general theory, Erwin Finlay-Freundlich had become its first great proclaimer, attracting attention to it with his general appeal to observatories in Europe and America. He already had William Campbell and the Lick Observatory aboard, as well as Charles Perrine in Córdoba, Argentina. There was a total eclipse due on April 23, 1913, but no observatory had even considered it. Its path began in the Pacific Ocean south of Japan, cut across the Northwest Territories of Canada and Baffin Island, before ending on the west coast of Greenland. None of these places were easily accessible even if astronomers could set up delicate instruments in such freezing temperatures. But an eclipse coming in the summer of 1914 would lay its path across Norway, Sweden, and a huge chunk of the Russian Empire. Freundlich wanted to observe this eclipse. If he could manage funding, it would be his first. But he felt he needed backup. Astronomers had learned the hard way that the more viewing stations along the path, the better the odds.

In February 1913, Freundlich wrote to Frank Watson Dyson at the Royal Observatory, asking for his participation. Answering quickly, Dyson replied that light deflection was not an incentive for the observatory, despite the British intention to send out several expeditions for that eclipse. "It would be an extremely delicate research to undertake at an eclipse, if not quite beyond present possibilities." The astronomer royal did refer the request to the Joint Permanent Eclipse Committee, which agreed with him, adding that the members didn't have the special equipment needed.

Undeterred, Freundlich kept busy. Correspondence fluttered back and forth as he and Einstein discussed plans to have stars photographed during an eclipse. In his reply to Pollok back in September 1911, along with suggesting Jupiter as the large body mass to use, Freundlich had also speculated on photographing stars during *the daytime*. Einstein was intrigued by this suggestion back then. "Is it really possible to observe stars near the sun in full daylight, i.e. in the absence of a solar eclipse, by means of currently available instruments?" he asked Freundlich. "If this can be achieved, then you shall surely be successful in determining whether the theory is valid."

By August 1913, with Perrine's failed attempt in Brazil nearly a year old, Einstein was growing impatient. He had just accepted, in July, a most generous financial offer to move to Berlin and become a professor at the university, as well as the youngest member of the Prussian Academy of Sciences. As an added bonus, he would be made director of the future Kaiser Wilhelm Institute for Physical Research, which was still in the planning stages. He was thirty-four years old. He would not have to teach or carry out any administrative duties. In other words, the position was all riding on his reputation. "The Germans are gambling on me as they would a prize-winning hen," he had jokingly told a friend. Then he added, "But I don't know if I can still lay eggs." This self-doubt may have been his impetus to try *anything*, even photographing the sun in the daytime, in order to prove himself once more.

A MEETING OF MINDS

August 1913 was an eventful month for Freundlich. In Bonn, Germany, attending the International Union for Cooperation in Solar Research was none other than William Campbell of the Lick Observatory in California.[2] Other participants included Campbell's good friend Frank Watson Dyson from the Royal Observatory, Edward C. Pickering from Harvard, and Arthur Stanley Eddington from Cambridge. After the conference was over, Campbell traveled to Hamburg for a few days and then to Berlin. Since Freundlich was too junior in rank at the Berlin Observatory to be sent to the conference, the Lick director graciously came to him. They discussed the upcoming eclipse preparations for the following year. On the heels of such a successful conference of scientific goodwill, Campbell was pleased with the American-German alliance in testing for the general theory. He would be happy to send Freundlich any successful photographs he would take during the eclipse, provided the Lick Observatory Bulletin would publish Freundlich's results.[3]

There were more big changes in the air that August for Freundlich. The last structure of the Berlin Observatory, where he had worked for three years, was being torn down. The new observatory, Berlin-Babelsberg, would open its doors the following spring near Potsdam, where there would be less light pollution. The happiest news was that Freundlich married his fiancée Käte Kirschberg, and the couple planned a honeymoon in the mountains near Zurich. Toward the end of the month, Einstein, who still lived in Zurich until the following spring, wrote again to Freundlich. The astronomer had evidently informed the physicist of Campbell's visit to Berlin and their discussion of testing for light deflection during the next eclipse, because Einstein responded to Freundlich enthusiastically in a letter:

I am very pleased that you have managed to arouse such great interest in our question regarding a bending of light rays. From a theoretical point of view, the matter is now settled more or less. Privately, I am

quite certain that light rays do undergo bending. I am extremely interested in your plan to observe the stars near the sun during daylight. This
should be possible unless suspended particles of the order of magnitude
of optical wavelengths, which deflect the light only slightly, are present
throughout the atmosphere. I am afraid that your plan could founder on
that. But you certainly know more about these things than I do.[4]

At the end of his letter, Einstein congratulated Freundlich on his
marriage and added, "I am delighted that I will see you on your honeymoon." What Käte thought of a no-doubt similar conversation with
Einstein on her honeymoon will be left to posterity's imagination. After the newlyweds met with him in Zurich, they attended a speech that
Einstein was scheduled to give later that day about his work in progress.
During his talk, he nodded at the tall gentleman seated in the audience
next to his petite bride. With his abundant dark hair and wire-rimmed
glasses, Freundlich was impressive in stature and demeanor. His being
singled out by Einstein that day must have been most gratifying to the
young couple. Einstein referred to him as "the man who will be testing
the theory next year."

In October 1913, a restless Einstein, having already experienced the
capriciousness of a total eclipse, wrote to astronomer George Ellery Hale.
Hale was not just a brilliant solar astronomer known for his study of sun-
spots, but also a great believer in international cooperation when it came
to research. He was then at the Mount Wilson Observatory, which he had
founded, outside Los Angeles. Few men knew more about the sun than
Hale did. Einstein's one-page letter included calculations showing light
then bending at 0.84 arc seconds.

Three weeks later, Hale replied to Professor Einstein with an apology for taking so long, explaining that he had written to consult William
Campbell, director of the Lick Observatory, knowing that Campbell was
already interested in the question of light deflection. "He writes me that
he has undertaken to secure eclipse photographs of stars near the sun for
Doctor Freundlich of the Berlin Observatory, who will measure them in
the hope of detecting differential deflections." Campbell was referring to

the eclipse coming in 1914. In his letter, Hale noted that he had recommended that Campbell contact Einstein directly with details. But Hale discouraged Einstein from the idea of photographing stars during daylight hours and listed three reasons: the sky increases in brightness near the sun, atmospheric refraction, and the micrometer—an instrument that measures the apparent diameter of celestial bodies—would not have the range to measure the distance of the star from the sun's limb. "The eclipse method, on the contrary, appears to be very promising, as it eliminates all of these difficulties," he wrote.

THE NEXT TOTAL ECLIPSE

The eclipse of August 21, 1914, would lay out an ample path of totality across the globe for viewing, even if the first half was inaccessible. The umbral track began in the Beaufort Sea, at Sachs Harbor, Canada. The harbor was named for the ship *Mary Sachs*, which had run into heavy slush ice less than a year earlier—the vessel was part of the Canadian Arctic Expedition—and had frozen into the sea for the winter. Regardless of that misfortune, the harbor was an impossible place for astronomers to try their luck. Like its predecessor, this eclipse would also cut across Baffin Island and Greenland. It would traverse the Norwegian Sea to reach land at Norway and Sweden, then catch a slice of southern Finland and eastern Estonia before entering western Russia. From there, the path would be most accommodating, stretching down past major Russian cities like Riga, Minsk, Kiev, and Theodosia on the Black Sea. After passing Turkey and the Middle East, it would end at sunset after barely touching India.

A generous track, but with a modest time for viewing. That January, as he often did, Andrew Crommelin had predicted the durations of totality in the *Nautical Almanac Circular*, a periodical published by the Royal Observatory. Maximum duration would occur fifty miles north of Minsk, where totality would last two minutes and fifteen seconds. All other stations, therefore, could count on a few less seconds. Observatories around the world had begun planning many months in advance. Where to find funding? What instruments to take? Where to set up viewing camps? Who would be team members? Numerous countries, including Norway,

Italy, Spain, England, Sweden, France, the United States, and Argentina, began organizing expeditions. Under Perrine's direction, Argentina was the only country in the Southern Hemisphere to take part. Given the convenient path of the eclipse, some European countries were well represented. Russia would send five expeditions in all.

Freundlich and Walter Zurhellen, from the Berlin-Babelsberg Observatory, along with an optical mechanic named R. Mechau, would go to Theodosia on the Crimean Peninsula. William Campbell and the Lick's team had chosen a small town near Kiev. The British teams would go to Minsk, Kiev, and Theodosia. Charles Perrine and his mechanic James Mulvey, from the Argentine National Observatory, would also travel to Theodosia and share a viewing station with the Germans and the British. From the Amherst Observatory in Massachusetts, a colorful scientist named David Todd would travel to Russia with plans to chase the eclipse at 120 miles an hour in an airplane. With him would be his wife, Mabel Todd.[5] It would be a mixed group. Of all these many teams, only Freundlich, Perrine, and Campbell would test for light deflection.

Having spent years hunting for the ghost planet Vulcan, Perrine and Campbell were now apparently in pursuit of an idea conceived by Einstein. As top-notch astronomers, among the best in the world, they were eager to carry out step 5 for the German physicist: *comparing the predictions to what is measured and observed.* Perhaps their enthusiasm had something to do with faith in brilliance. After all, Urbain Le Verrier had discovered Neptune with the point of his pen. And Einstein's pen had written the formula $E = mc^2$.

WILLIAM CAMPBELL ON MOUNT HAMILTON

William Campbell's tenacity was shaped not just by a difficult childhood, but also by the observatory where he learned and practiced his science. No other observatory in the world would have more involvement in, or a longer commitment to, attempting to test light deflection according to Einstein's general theory of relativity than did the Lick Observatory. The idea for an observatory sitting high on a mountaintop began in 1873, when an eccentric and aging bachelor named James Lick decided he wanted a superlative observatory in Northern California. His lack of astronomical knowledge was superseded by his tremendous wealth.[6]

In 1874, Lick set up a trust and endowed it with $700,000, an astonishing figure in those days. It would be $1.2 billion today. He insisted that the observatory would have "a powerful telescope, superior to and more powerful than any telescope yet made, and all the machinery appertaining thereto." He wanted it built atop Mount Hamilton, a mountain peak near San Jose with an altitude of 4,265 feet. Lick then arranged for the county to carve a five-mile road winding up to the top. He died in 1876 as construction was beginning. Crews would blast seventy thousand tons of rock from the mountain's top to create a level place on which to build. When the observatory, with the world's largest telescope installed, was finished in 1888, Lick's coffin was brought from San Francisco to be reinterred at its base. Two years later, a young astronomer named William Wallace Campbell, age twenty-five, volunteered to work as an assistant. The next year, he was invited to stay permanently on the mountaintop and work on spectroscopy.

Born in 1862 and raised on a farm in Hancock County, Ohio, Campbell was four years old when he lost his father. He and his siblings were soon working in the fields and doing odd jobs to help the family survive. A driven young man, he managed to graduate from the University of Michigan with a degree in civil engineering and a new interest in astronomy. His first job was as professor of mathematics at the University of Colorado, but his passion for astronomy called him back to Michigan to begin teaching this scientific discipline at the university. In 1892, a year after he joined the Lick Observatory, he married a pretty college girl—she had been a student in his math class—and brought her to Mount Hamilton to live. Elizabeth Ballard Thompson could have no way of knowing, as she peered down on the Santa Clara valley below, that she would often travel the world at her husband's side as he studied eclipses. By 1899, they had three young sons, and by 1901, Campbell had become the Lick's director.

That unique mountaintop would shape William Campbell for the rest of his life. It was a self-contained community of about fifty people, staff and employees, occupying houses just below the observatory. The director and his family lived in a tiny brick house heated by woodstoves and with a rock-hewn cellar that could store food. A post office, a lumber mill, and a blacksmith shop had been built. At the one-room schoolhouse nestled into the mountainside, a teacher taught the dozen or so students. Church services were held each Sunday in one of the homes, unless a preacher came up the mountain to visit a relative. Services were then moved to the little school. Telephone poles ran up the steep slope, carrying single-grounded wires made of iron to the crank phone on the wall of the director's kitchen. The world's first permanently occupied mountaintop observatory was high enough that it was snowed upon in the winters, turning its huge dome into a white castle overlooking the valley below.[7]

Until 1910, when a gasoline generator was installed in Campbell's house, he read or wrote letters, and his children did their homework, by kerosene lamplight. In the early years, a stage pulled by two horses lumbered daily up the mountain to deliver the mail and groceries, stopping several times to let the animals rest on the seven-hour round trip. A much larger stage came on Saturday evenings, bringing from the hotel down in San Jose the visitors who had paid a fee to ride up to the observatory. Once inside the dome, they were invited to peer through some of the telescopes. Twice, at the turn of the century, the stage was robbed, depriving its passengers of their money and jewelry. And yet, up that mountain trekked famous astronomers and dignitaries from around the world, including Thomas Edison. Frank Watson Dyson, then senior assistant at the Royal Observatory, came up the mountain to meet Campbell in 1901. He was on his way back from the eclipse in Sumatra, with plans to circumnavigate the globe before returning home to England.

Always fond of Americans, Dyson was impressed with Campbell's easy manner and the family's resilience. "We killed a rattlesnake ten yards from Mr. Campbell's front door," he wrote home to his wife, "a big one with nine rattles." With no antivenom available then, all Mount Hamilton boys carried jackknives in their pockets in case of snakebite. "He has three small boys . . . so you can bet he was glad to kill that snake." Even the drive back down the mountain didn't seem to unsettle the Englishman. "The driver was splendid and the way he brought the stage down with his two horses and a precipice on one side and a wall of rock on the other was fine." It was 1910 before an automobile replaced the stage Dyson had ridden on. A visitor could now make the round trip in two hours, instead of seven by horses.

When he wasn't chasing an eclipse, Mount Hamilton was Campbell's world for thirty years. He was a formal man, his speech always proper and refined. He was kind to the mountaintop families, especially the children. He treated the janitor and his family as well as he treated the families of astronomers. But if it was warranted, he could fire an employee on the spot. He was away from the observatory on official business when the 1906 San Francisco earthquake shook his wife and three sons from their beds, knocked bricks from the chimney, and shattered the glass funnel of an oil lamp. With news cut off and with a cloud of gray smoke billowing on the distant horizon, Elizabeth Campbell, Charles Perrine, and others climbed the hill to the observatory and turned a small telescope toward San Francisco and a wall of fire three miles wide.

William Campbell had experience and a reputation for being careful in his research.[8] He had been on four previous eclipse expeditions, to India, the US state of Georgia, Spain, and Flint Island, a dot in the middle of the Pacific Ocean. In 1913, he began putting together a team from the Lick Observatory for the upcoming eclipse. He had supported Perrine, his friend and former colleague, by sending Lick's intra-mercurial camera lenses down to Brazil. He had supported Freundlich and Einstein by offering up his Vulcan plates. A year earlier, he had carefully explained the problem of the plates in a letter to Freundlich: "Your experience with the eclipse plates is about what we had expected: not only is the sun's image near the edge of the plate, but the aberrations of the camera lens at the edge of the plate are unavoidable; and, further, the clock was regulated to follow the sun and not the stars." Now he was ready to take up the challenge himself. He would bring his family with him to Russia. He would "follow the stars."[9]

CHARLES DILLON PERRINE

Charles Perrine was born in Ohio in July 1867. As is often the case for future astronomers, he was given his first telescope at a young age. A few months before he would graduate from high school, his hometown of Steubenville was caught in the Great Ohio Valley Flood of 1884. In some places, the Ohio River rose thirty-four feet higher than it had ever reached before. Over a hundred houses were soon underwater, while others had been washed away and destroyed. Businesses were submerged, railroad tracks torn up for miles, and, as a sad reminder that would last months, thousands of kegs of nails had been scattered like grain across hundreds of acres of once-fertile farmlands, now lying in mud and rust.

For Steubenville's youth, faced with such devastation, their futures may have been determined by that flood. Because it would take many months and even years for the towns to recover, it's hard to imagine what kind of graduation ceremony the Class of '84 could possibly have had. Two years later, at age nineteen, Perrine packed up and left the Ohio Valley for Alameda, California, across the bay from San Francisco. He may have read in one of his astronomy journals about a new observatory being built on a mountain nearby. It's also likely he chose Alameda since his mother's sister was living there at the time. Two years later, in 1888, the Lick Observatory was opened. As it had been for Campbell, on that mountaintop lay Perrine's destiny. He would evolve into a world-class astronomer.

He secured a job as bookkeeper with Armour & Co., a meatpacking firm in San Francisco. To get to work each day, he took the train and then a ferryboat across the bay. Nights traveling home on the ferry, his eyes would search the skies for meteors or comets. When the brand-new Lick Observatory encouraged amateurs to observe and photograph the January 1, 1889, eclipse for possible publication, Perrine knew it was his chance. He and a professional photographer from Oakland planned a trip north to Mendocino County.[10]

That the path of totality for this eclipse was only 150 miles north of San Francisco was a gift to the Lick Observatory. This would be its first expedition. Being so conveniently located, the Lick sent out offers to other observatories in the country and Europe, offering to assist them with preparations. While the professional astronomers gave themselves two weeks or more to travel to the viewing sites they had chosen—in the heavy rains that fell that December, the horse-drawn stages were often sunk wheels-deep in mud—it seems that Perrine and his companion left just one day before the eclipse. The report they sent to the observatory said they had checked their watch with the Merchants Exchange Building clock in San Francisco on New Year's Eve 1888.

Despite the torrential rains before and after, eclipse day was splendid. Using a sixteen-power spyglass, Perrine and his friend succeeded in obtaining eight good negatives. Arriving back in San Francisco on January 2, their watch was now thirty seconds off from the Merchants Exchange clock. They made the time corrections and sent their report and photographs up the mountain to the Lick Observatory. "The corona shone with a soft, silvery light, slightly tinged with blue in the outer streamers," reads the report that Perrine wrote. What must he have felt at that moment? That first eclipse began a career that would carry him on expeditions around the world and astronomical recognition.

Perrine managed to get a meeting with Edward Holden, the observatory's first director, and convinced Holden to hire him as secretary. His job was to keep the books, manage the post office, do inventory of equipment and supplies, read and record the time, answer the single phone line, and type and file correspondence to and from staff members. But at least he was inside the door of one of the world's finest observatories. Nights, he hung out with the astronomers up in the dome, learning all he could from them and forming what would become a long friendship with Campbell. Perrine would learn his craft from a vigilant teacher.

In 1895, the self-taught Perrine—he once said he sold ham to earn a living, referring to his job at the meatpacking plant—became an assistant astronomer at the Lick Observatory. That same year, he discovered his first comet, and then another comet, and then another. When the French Academy of Sciences awarded him the prestigious Lalande Prize for advances in astronomical study just two years later, he had discovered five comets in all and had "rediscovered" two periodic ones. In the spring of 1900, he traveled across the country by train with Campbell to Georgia for his first major eclipse expedition. Within a few more years, he was made president of the Astronomical Society of the Pacific and had discovered two moons of Jupiter. His astrophotography skills were now among the best in the world. In 1905, a few days after Einstein's special relativity theory was published in *Annalen der Physik*, Perrine married Bell Smith on the fourth of July. He was thirty-eight years old. After an expedition to Spain with the Campbells to view the eclipse of August 30, the newlyweds went on a world honeymoon, visiting Italy, Switzerland, Germany, France, and England. On their return to California, Bell went to live in Berkeley so she could finish her last year at the University of California.

Charles went back up the mountain to the observatory. Just after 5:00 a.m. on the morning of April 18, 1906, a razor strap hanging on his wall began to swing back and forth. A bookcase moved forward an inch. Windows rattled, and the door swung wide open. His pendulum clock had stopped at 5:12. Perrine volunteered to go to San Francisco and check on the relatives living there. After the stage ride down the mountain, he caught the train from San Jose to Alameda. Across the bay from the destruction, his parents were safe. And so was Bell, in Berkeley. He telegraphed Mount Hamilton: "Folks are fine." The observatory secretary pinned the message to the bulletin board at the post office, where the mountaintop community could read the good news. The next day, Perrine rode the ferry over to San Francisco. He sent a telegram to Elizabeth Campbell, who was worried about her relatives: "People safely away from danger. Houses are threatened. Greater of San Francisco is gone." Bell's graduating class the next month was the only one in the school's history that didn't have to take finals, because of the devastation of the 1906 earthquake.

Bell Smith Perrine then went to live with her husband on Mount Hamilton, in one of the small houses not far from the Campbells and their sons. Four years later, in 1909, Charles Perrine became director of the Argentine National Observatory, in Córdoba. He and Bell would leave the mountain and move to the star-filled Southern Hemisphere.

FREUNDLICH: THE GERMAN EXPEDITION

Erwin Freundlich was determined to test for light deflection during the 1914 eclipse. But the old problem of funding again reared its head. His colleagues in the scientific community in Germany had no interest in investing in a project they didn't understand or believe in, or both, the project being the theory of general relativity. Einstein felt that if the Prussian Academy of Sciences didn't pitch in—he had by this time accepted its offer to join—that he would dig into his own savings and put up two thousand marks "to start the ball rolling." The Berlin Academy finally offered some funding, which was still not enough to cover expenses. Freundlich was grateful to be able to borrow parts of telescopes from Perrine, at Córdoba. From them, he had assembled four cameras for his mission. But he needed money for photographic plates, travel, and the various expenditures that an expedition to another country requires.

It was Freundlich's good luck that a friend had introduced him to Gustav Krupp von Bohlen und Halbach. He was the husband of Bertha Krupp, who had inherited her family's company at age sixteen when her father, a likely suicide after newspapers exposed him as a homosexual, died in 1902. Kaiser Wilhelm II, a family friend, found a husband for Bertha rather than let the company be run by a woman. Gustav was given the family surname. An old and powerful clan, the Krupp industrial dynasty was known for its manufacture of steel, weapons, and artillery in Europe for over three hundred years. The Krupp company also manufactured U-boats and supplied most of Germany's need for artillery. Freundlich made a good impression on Gustav and Bertha Krupp, who became the major funders for his expedition to Russia. If there was an irony in the pacifist Einstein's innocent or deliberate ignorance of where the money was coming from, it wasn't mentioned.[11]

THE ENGLISH EXPEDITIONS

Although Dyson was not interested in funding an expedition to test for light deflection for Einstein, the Joint Permanent Eclipse Committee prepared to send three British teams to Russia, from the Royal Observatory, Greenwich, the Imperial College in South Kensington, and the Solar Physics Observatory, which had moved in 1911 from South Kensington to Cambridge. The teams' main focus would be tests carried out on the sun's corona and the flash spectrum.[12] As is often the case when expeditions visit a foreign country, resident astronomers and observatories help with the governmental permissions, transportation arrangements, viewing stations, and finding local assistants and volunteers. With the British teams going to Russia, the English had been in touch with Johan Oskar Backlund, of the Pulkovo Observatory in Saint Petersburg, for many months. So had Freundlich, Campbell, and Perrine. Backlund had traveled from Paris with Perrine in the fall of 1911, when they had met Freundlich and first heard of Einstein's new theory.

In January, the august British scientific society known as the Royal Society had asked the Russian government for permission to allow Father Aloysius Cortie and a fellow Jesuit from Stonyhurst College Observatory to be part of the team going to Kiev.[13] Cortie met with the Russian ambassador in London, hoping to demonstrate that he was "neither a dangerous nihilist nor a wily proselytizer masquerading as a scientist," as the *Stonyhurst* magazine put it. But in May, three months before the eclipse, the Russian Foreign Office denied the Jesuits permission. With time running out, they quickly decided on Hernösand, Sweden, even though early weather reports had indicated that August in Sweden would not bode as well as in Russia. Despite this inhospitality, the Russian government nonetheless, in the interest of science, would give free entry to the other foreigners, free railway shipment for their tons of equipment, and a half-rate fee on all passenger tickets.

THE EXPEDITIONS SET SAIL

After long months of planning, the expeditions from various countries started packing up their instruments and preparing for the journeys to their respective stations. First to embark was Heber Doust Curtis, an astronomer and a Campbell colleague, who left the Lick Observatory on June 15. Curtis had been on other Lick expeditions, including the one to Georgia in 1900 as a volunteer. He would travel ahead with the instruments and equipment to oversee a safe arrival in Kiev. Campbell and the remaining entourage would follow in early July. Curtis had written a review article on general relativity as early as 1911. Complete with a bibliography citing all published sources by others, his was the first paper on this topic to appear in an American astronomical journal.[14]

Perrine and Mulvey left Córdoba a day later, on June 16, traveling first by train to Buenos Aires. From there, they would catch a steamer to Genoa, Italy, on the initial leg of their journey. Perrine, now the father of two young children, was already a couple decades into his career by this time, having received numerous awards and an honorary doctorate. Being the first astronomer to attempt to test for light deflection according to Einstein's theory, he was now one of three such astronomers.

Frank Dyson left Liverpool on June 22, on the SS *Ascanius*. As the astronomer royal, he was bound for Australia and the eighty-fourth meeting of the British Association for the Advancement of Science (BAAS). Arthur Eddington, his former assistant at the Royal Observatory, was traveling on the ship with Dyson to attend the meeting. Eddington had been a year now at Cambridge as Plumian Professor of Astronomy and Experimental Philosophy. When Sir George Darwin died the previous December—he was the second son of the famous evolutionist—Eddington succeeded him. Eddington's star was rising. Just three months earlier, he had been appointed director of the observatory and had moved with his mother and sister into the lovely Cambridge apartment that came with the job, along with its gardens and lush grounds.

So important were these annual BAAS meetings among scientists of the world—the gatherings were intellectual pleasure cruises of two or three months for physicists, geologists, mathematicians, astronomers, biologists, and so on—that Australia had begun planning five years earlier.[15] The Australian Commonwealth gave the group a grant that would amount to two million dollars today. It would cover passage and railway fees for all three hundred members. Along with spreading scientific friendship among countries, the members would sail almost completely around Australia, with stops in various cities for lectures and sightseeing. During the summer of 1914, it seemed as if the greatest scientific minds on the planet were at sea, or soon would be. When Dyson and Eddington set sail on that June day, they were leaving behind what the newspapers referred to as the "Irish problem," the "suffragette agitation," and "labour difficulties." They would come home to an England greatly changed.

JUNE 28, 1914: A WRONG TURN INTO HISTORY

It was Sunday morning when Archduke Franz Ferdinand of Austria and his wife Sophie, the duchess of Hohenberg, rode in their 1911 open-style automobile along a street in Sarajevo, then capital of the Austro-Hungarian province of Bosnia and Herzegovina. They had left Vienna by train earlier that morning for a diplomatic visit with the local governor. As the automobile chugged along, a young man stepped out of the crowd lining the street, his arm raised. He was one of a six-member secret cult called the Black Hand, supported by officers in the Serbian military whose current mission was to assassinate the archduke and end Austro-Hungarian rule in Herzegovina and Bosnia. The grenade he tossed at the royal couple bounced off the back of their car and then exploded. The spray of shrapnel injured members of the entourage riding behind them. After arriving at the governor's residence, where they collected themselves, the archduke and duchess were determined to visit the hospital where an injured officer had been taken.

In what has to be the most fatal decision a chauffeur has ever made, the driver turned down a wrong street. Unaware that the itinerary had changed, he followed the lead cars. Forced to back up slowly in a string of autos, gridlocked, the royal car was an easy target. One might question why, given the earlier attempt on their lives, the royals climbed back into an open-style auto. Nonetheless, the bad luck that day was astonishing, for sitting in a café across the street was another member of the Black Hand. Taking a pistol from his pocket, he left the café and strode over to the open car. He shot Sophie first and then the archduke, who was soon begging for his wife to live for the sake of their children. Sophie's white dress and the fallen green plumes from the archduke's hat would be soaked in blood by the time the chauffer finally freed the car and sped away.[16] News of the murders spread quickly around the world. Marconigrams were received by governments, newspaper bureaus, and ships at sea. At this time, the assassinations seemed only a single, tragic event in the eyes of the public. After all, it was just an archduke, and where *was* that little place called Sarajevo?

Telegraphed reports and newspapers gradually began speculating as to what *might* happen because of the assassinations. But it hadn't happened yet, and an eclipse cares little for the woes of civilization. Thus, with Curtis gone on ahead, Campbell left San Francisco in early July. He was bringing with him his entire family: his wife Elizabeth, her mother, their three college-age sons, and the boys' friend Charles Brush Jr., all of whom were skilled in acting as volunteers and operating some of the instruments.[17] Campbell was proud of his sons—Wallace Jr., Douglas, and Kenneth—who had come of age atop Mount Hamilton, jackknives in their pockets and polite smiles for the important visitors. The Lick party would travel to Genoa, Italy, by steamship, then through Austria by railway, to arrive at Kiev, where Curtis would have a temporary home waiting for them.

A cable with news of the assassinations reached the steamer carrying the BAAS members. Disturbing, yes, but so far nothing but worrisome rumblings had followed. Europe was a distant world, far away from deck activities and evening dinners aboard the ship. On July 13, they sailed into port at Cape Town, South Africa, for a visit to the Cape of Good Hope. This was where their friend Sir David Gill had been Her Majesty's astronomer for twenty-seven years. Upon his retirement in 1906, Gill had returned to London and often invited Dyson and Eddington to his home for what he called "star-streaming dinners." When Gill passed away that January, Eddington had been asked to write his obituary for the *Monthly Notices of the Royal Astronomical Society*. He did so as the steamer sailed between England and Cape Town. At night, he paced the deck, searching his thoughts for the right words. In the sky overhead was the Southern Cross, with its four brilliant stars. He had missed an opportunity to see the constellation when he was in Brazil, in 1912, it being the wrong time of year. Now, as he worked on Gill's obituary, Eddington was more appreciative of those southern skies and "the finest stretch of the Milky Way" that flowed over his head. With the archduke's murder now days old, the steamer sailed around the Cape of Good Hope and headed for Perth, Australia.

On July 17, the Royal Observatory team left London. Dyson had put Davidson, the man he often referred to as "the finest instrument in the Royal Observatory," in charge of this expedition to Minsk. The soft-spoken Davidson, known for his dry sense of humor, had become indispensable for his knowledge of the instruments. This was his fourth eclipse expedition, the previous one being with Eddington to Brazil, where they had ridden a mule in the moonlight once the rain had stopped. Accompanying him was Harold Spencer Jones, who had become chief assistant at Greenwich a year earlier when Eddington vacated that position to accept the Plumian Chair at Cambridge. Also along as a volunteer was a colorful British solicitor and amateur astronomer named Patrick Hepburn. Davidson was still not interested in Einstein's general theory at this stage.

The three Englishmen left London aboard the small Russian steamer *Imperator Nikolai II*, which serviced ports in Russia, Germany, and England. The ship saved time by taking the 60-mile-long Kiel Canal that cut across northern Germany to the Baltic Sea, saving some 250 nautical miles of travel. They crossed the Baltic and sailed into the Gulf of Finland to land at Saint Petersburg on July 22. They had planned a quick visit to Moscow before arriving in Minsk. But the decision turned out to be an upsetting, even dangerous one. Below their hotel room windows, on the city streets, angry crowds were forming in demonstrations and demanding war. Uneasy, the Englishmen left Moscow the next morning.

The other two British expeditions also set sail. The Solar Physics Observatory in Cambridge left in mid-July to travel to Theodosia, on the Black Sea. Headed by Hugh Frank Newall, this group would meet up with Freundlich and Perrine in Odessa, Russia, and they would sail together from there. The team from the Imperial College left later than all the others, on July 25. With Alfred Fowler at the helm, they were headed to Kiev, where they would set up camp in the university's botanical gardens.

Perrine and Mulvey had sailed through the Straits of Gibraltar and on into the Mediterranean when they received news of the archduke's murder. Just south of Toulon, France, they watched French warships going through military maneuvers. After reaching Italy, Perrine decided that the ships had been too slow. In Genoa, they let the instruments sail on without them, keeping only the optical components. To reach Odessa, they would have to skirt Serbia, where the royal couple had been murdered.

They boarded a train and rode north to Vienna, which they reached on July 15, and continued on toward Russia. This was almost a month to the day since they had left Argentina.

Police on horseback now patrolled the streets of Vienna, and the railway was heavily guarded by troops. Towns and cities in Poland swept past their windows as the train sped down to the border of Russia. Here, their passports and all reading materials were carefully checked by customs officials. After eating at the station, they changed trains for the journey to Odessa and slept soundly in comfortable berths. As they entered the eastern part of the Russian steppe, they watched from their windows Cossacks putting their horses through drills. In Odessa, they visited the observatory, met with the Argentine general consul, and waited for the Germans and the Englishmen to arrive. From there, they would travel on to Theodosia as a group.

Father Cortie and his fellow Jesuit priest, with a shorter distance now to travel, having been banned from Russia, left Hull, England, on a steamer bound for Gothenburg, Sweden. The men then went by railway to Stockholm and on to Hernösand the next day by boat. The dozen or so expeditions from other observatories in Europe were also on the move or had already arrived at their destinations. With the expeditions focusing on the upcoming eclipse, the world continued to explode around them. Einstein's world was also exploding.

CATASTROPHE AND COLLAPSE

Since the spring of 1914, Einstein had been settled in Berlin with his family. He was still working on his general theory and other papers. Now, he no longer had to mail love letters to his cousin-turned-girlfriend. He could deliver them in person if he wished, since Elsa Einstein Löwenthal also lived in Berlin. Mileva was not happy with the move from Zurich or with her life in general. By mid-July, with the anger and frustration now boiling between her and Albert, she moved with their sons to the home of Fritz Haber, a family friend and chemist at the institute who had been instrumental in bringing Einstein to Berlin.

Albert took this opportunity to draft a stunning document to his wife and the mother of his two sons. It was so devoid of warmth or kindness that calculations on a blackboard would have seemed more affectionate. The document spelled out the conditions under which he would remain living in the same house with Mileva as her husband. Given that it was also written by the same genius whose other papers would change the face of physics, it bears reprinting here:

CONDITIONS:

A. You will make sure:
1. *that my clothes and laundry are kept in good order;*
2. *that I will receive my three meals regularly in my room;*
3. *that my bedroom and study are kept neat, and especially that my desk is left for my use only.*

B. You will renounce all personal relations with me so insofar as they are not completely necessary for social reasons. Specifically, You will forego:
1. *my sitting at home with you;*
2. *my going out or traveling with you.*

C. You will obey the following points in your relations with me:
1. *you will not expect any intimacy from me, nor will you reproach me in any way;*
2. *you will stop talking to me if I request it;*
3. *you will leave my bedroom or study immediately without protest if I request it.*

D. You will undertake not to belittle me in front of our children, either through words or behavior. [18]

This ultimatum may have been Albert Einstein's most unique thought experiment yet. Even after Mileva agreed to his terms, he wanted reassurance that she fully understood what they meant. While he could not be her friend, as he wrote back to her, he could behave as her business partner. They could live in the same house so that he would not lose his sons. "In return, I assure you of proper comportment on my part, such as I would exercise to any woman as a stranger." With this letter, the woman he had loved above his family's fervent disapproval, his "Dollie," his "little witch," finally understood that it was over. By this time, Mileva was walking with difficulty and was battling bouts of depression. It would be impossible to live under the same roof with her husband and endure such unkind terms. She would take the children and leave Berlin.

Albert now had to support another household. He drafted a second document that would give Mileva and the boys a bit less than half his salary. With family friend Michele Besso escorting her, Mileva and her sons left on a train for Zurich. Albert had accompanied his family to the railway station. Once the train pulled away, according to friends, he cried all the way home and all afternoon and evening. Perhaps it was not just for the loss of his boys, but also for the loss of what had been so passionate and overpowering a love story just a dozen years earlier.

The same afternoon that Mileva's train pulled out of Berlin, a love letter, written by a British naval officer to his wife, would suggest an opposite kind of matrimony. "My darling one and beautiful," Winston Churchill wrote to Clementine, who was expecting their second daughter that autumn. "Everything tends toward catastrophe and collapse." He was speaking of world affairs and not their close and loving marriage. Churchill already knew war firsthand, as a soldier and correspondent. Yet the devoted husband welcomed World War I with open arms, even relished it. He was now ready to prove his worth as a military strategist. On the other hand, Einstein, the failed husband whose letter to his own wife seemed to be one of hatred, not love, was against all wars, a pacifist.

Worlds within and without were now colliding, and conditions were being laid down by governments, not just husbands. On July 28, 1914, the day before Mileva's train left Berlin, Austria-Hungary had declared war on the Kingdom of Serbia. Countries in Europe would soon begin mobilizing. That wrong turn by the chauffeur was starting to take its toll. As Churchill waited to see if he would get his fight, Einstein was now poised to get his own wish: "Nothing more can be done by the theorists. In this matter it is only you, the astronomers . . ." His faith was riding on three men: Erwin Freundlich, Charles Perrine, and William Campbell. These stargazers were also mobilizing. As if they were magi, the three teams had traveled far. They were now in Russia, setting up their wares and waiting for one bright star to give them a message.

5

THE 1914 ECLIPSE

The Second Attempt: A Path of Fire

Professor Einstein, the chief developer of the theory of relativity, has been quite anxious that eclipse observations be made to decide the question of the existence or non-existence of such a deflection of light when passing thru a strong gravitational field.

—*William Wallace Campbell, Astronomical Society of the Pacific, 1914*

SAMUEL WILLIAMS MAY have headed the world's first and best-funded eclipse expedition during wartime, but that wasn't the first time a solar eclipse visited the planet during a period of fighting. Wars are too common. It's believed that as early as 585 BCE fighting between the Medes and the Lydians at the Battle of Halys ceased during a total eclipse and a truce was agreed upon. If the 1912 eclipse had appeared as an omen of the *Titanic's* misfortune, this eclipse of 1914 looked as if a heavenly finger had traced a line across the war zone. But in the western countries, most people were not thinking about the upcoming eclipse. They were concentrating on the chess moves by their governments and waiting to see what would happen. Before Germany declared war on Russia, the astronomers who were already in Russia went ahead with their plans. Down along the path of totality, from Minsk to Kiev to Theodosia on the Black Sea, the teams began arriving and setting up their camps.

THE LICK TEAM AT BROVARY, NEAR KIEV

When Heber Doust Curtis, a former Greek and Latin professor turned astronomer, arrived in Kiev on July 11, the war was still a rumor. Having traveled ahead with the Lick instruments, Curtis was frustrated to learn that many officials and local scientists were enjoying their summer vacation and were thus away from the city. He found assistance from the Kiev Circle of Amateur Astronomers, however, and began preparing for Campbell's arrival ten days later.[1] Finding a home for the team meant turning down generous offers of lavish estates and villas, including one from the court chamberlain to the czar, all for free and including travel and labor. It was an honor, after all, to have these astronomers as guests. But the fact that those homes sat several miles from the path of totality meant they were too far away. Campbell wanted his campsite to be at the place of action and to be self-sufficient. Perhaps he had learned a lesson at the 1900 eclipse that he and Charles Perrine had viewed in Georgia. He referred to his landlady as "a terror." Perrine had even come down with a case of food poisoning so serious he bled internally and was barely able to perform his tasks. Campbell had learned the hard way not to rely on outside resources when it came to food and water. His campsites were well stocked and as comfortable as possible.

Kiev was a modern city with half a million people, so the offer of a dacha at Brovary, twelve miles northeast and right on the eclipse track, proved the perfect site. The home sat on eleven enclosed acres with plenty of protective fruit trees and hardwoods. The expansive lawn provided ample room for the tents that would protect the instruments, and there were barns and a cellar if extra storage space was needed. The greenhouse had an underground compartment used in winter; this windowless room was quickly converted into a darkroom for developing the plates. The judge who owned the estate had been transferred to another district, and his wife was leaving shortly. Since the two-story residence would be empty, it was also available to the team. This meant they could dispense with setting up a kitchen and dining tent. Coming with the dacha was a domestic staff to help Elizabeth Campbell oversee household duties, which included boiling all drinking water, another of her husband's expeditionary rules. Also in the deal was a German-born cook who had been in Russia for so many years that her German was rusty.

The place was perfect, and Curtis rented it at once. The 4½ tons of equipment, which included the forty-foot telescope, were delivered on July 18. The Campbell entourage—William, Elizabeth, her mother, their three sons, and the sons' friend Brush—arrived three days later.

TO THE RUSSIAN RIVIERA

The Crimean Peninsula is surrounded by the Black Sea on three sides, with the Sea of Azov bordering its northeast boundary. The Crimean Mountains trail along the southeast coast, just a few miles inland from the water. Before arriving at Theodosia, the steamer from Odessa carrying the British team, the German team, and the Argentinian team stopped at port cities along the coast. At Sevastopol, which was home to Russia's Black Sea Fleet, the astronomers watched as warships went through their maneuvers. Perrine was impressed with the beauty of the place. "It was the best time of the season along the coast of the Crimea, 'the Russian Riviera,'" he wrote. It *was* the Russian Riviera. When railways were laid down in the 1870s, over a hundred thousand tourists began visiting the peninsula annually, either for pleasure or to seek cures for tuberculosis and other ailments.[2] Many came to Yalta, which is where the steamer carrying the teams stopped next. Perrine described it as "the most beautiful and most renowned of all that region, where the Tsar's summer palace is located."[3]

Germany was still a week away from declaring war on Russia when the ship arrived in the early morning hours at Theodosia, on Crimea's southeast coast. As the crow flies, this was about five hundred miles southeast of Campbell's rented dacha at Brovary. The three teams immediately began the search for a perfect area to set up their camps. By noon, they had found one, in a vineyard on the slopes of a hill two miles northwest of the city. Theodosia, with a population of about thirty-five thousand, lay twelve miles from the center of the umbral path and thus would have three seconds less viewing time during totality. The astronomers had decided early on that such a minor forfeiture of time would be worth the conveniences of town. Now that they had their campsite, they looked for a suitable house to rent. When a resident offered them an empty bungalow that stood only a few hundred feet from the vineyard, they spent the next five days furnishing it to suit their needs. A local couple was hired to help with the household chores, to buy food supplies, and to cook daily meals for the group of ten.

THE TEAMS IN THEODOSIA

Many expeditions from Europe had come to the Crimean Peninsula and were set up in or around Theodosia, including ones from Russia, Italy, France, and Spain. From Germany were four teams, including Freundlich's. From England, and sharing the vineyard campsite with Perrine and Freundlich, was the team from the Solar Physics Observatory. With astrophysicist Hugh Frank Newall was his wife, Margaret, an accomplished pianist. His three team members included Frederick Stratton, also an astrophysicist and a temporary officer in Britain's army reserve, Charles Pritchard Butler, their senior observer, and R. Rossi, an Italian volunteer who had joined the group as they came through Trieste. Newall, who had gone to Algiers for the 1900 eclipse, also enjoyed traveling on the same steamer to Sumatra with Dyson and the amusing Atkinson for the 1901 eclipse. He had become director of the Solar Physics Observatory a year earlier, and Stratton was made his assistant director.

All of the instruments that the British party had shipped ahead were safely stored and waiting for them in warehouse sheds at customs. The party had the cargo transferred by horses and wagons up to the vineyard. The teams decided that Perrine and Mulvey would share the bungalow with the British. Freundlich and his two colleagues would find rooms in the city less than a mile away. But they would join the others for "mess," which meant a diverse group around the table for daily meals.

Perrine had brought from Córdoba his capable engineer and designer, the Indiana-born Mulvey. The instruments they shipped from

Córdoba to Crimea included their forty-foot telescope, which had also gone with them to Brazil in 1912. As in previous eclipses, they had plans to study the solar corona. But their most important mission was to photograph star fields in the vicinity of the sun to determine any degree of light deflection. It had been a costly expedition for the National Observatory of Argentina to finance, and coming on the heels of the failure in Brazil, it was not a popular venture back in Córdoba. Much was riding on good weather. But advance climate reports had depicted Theodosia as an excellent spot, with frequently clear skies and the sun higher there on the horizon during the eclipse. The problem was that there was no sign yet of the instruments that had sailed from Genoa without them. As they waited, Perrine and Mulvey concentrated on getting the piers built.

Freundlich must have been jubilant. For almost three years now, he had championed Einstein's groundbreaking work. The astronomer had solicited others in Europe, North America, and South America, hoping to enlist them in the cause. He had remained unwavering during the increasing criticism from German scientists who dismissed Einstein's theory. He had even held fast during the growing disapproval from his own director at the observatory. Hermann Struve believed that Freundlich needed to concentrate on his job. Now here Freundlich was, as prepared as possible and about to witness his first total eclipse. With him to Crimea had come two capable colleagues, Walter Zurhellen, also from the Berlin-Babelsberg Observatory, and R. Mechau, an optical engineer who worked for the Zeiss company.

THE ENGLISH AT MINSK: THE SOLAR CORONA

Charles Davidson and the chief assistant, Harold Spencer Jones, set up camp near Minsk, about nine hundred miles northwest of Crimea. Dr. and Madame Kodis, the chief surgeon of the town and his wife, a volunteer nurse for the Red Cross, had kindly offered the Englishmen their country villa, three miles from the city. Sitting on a swell of hill surrounded by flat land and with a large garden, the villa provided a perfect location. The amateur astronomer along with the team, London solicitor Patrick Hepburn, would assist by taking photographs and changing the plate holders. Davidson's program was the same as it would have been in Brazil two years earlier with Eddington, had they not been rained out. But he was still concerned only with the solar corona and the flash spectrum, not with Albert Einstein.

THE BUILDING STORMS

Within days of his arrival, William Campbell noticed that bad weather seemed to be the norm for that part of Russia in August. It was often cloudy, even rainy, around the eclipse time of day. Mid-totality at Kiev had been estimated at 2:47 p.m. Campbell knew that weather reports could be unreliable, and information from Russia had been difficult to obtain. But he was confounded by just *how unreliable* the reports he had been given were turning out to be. They had predicted that the cloudiness factor would be higher—6.8 on a scale of 1 to 10—in both Sweden and up north at Riga, where the eclipse would first enter Russia.

The farther south one traveled in the country, the better the odds of less cloud cover. Kiev, near where Campbell's dacha sat, was six hundred miles southeast of Riga. Theodosia, on the Crimean Peninsula, was another five hundred miles southeast of Kiev and registered very low, with a 2.3 factor. That's why the peninsula was now crowded with eclipse expeditions from all over Europe. Campbell had chosen Brovary, near Kiev, and yet every day brought rainstorms or cumulus clouds. At sunset, the clouds dispersed and the nights were perfect. As the Campbell team waited for August 21, there was no single day in Brovary with decent weather. But it was still early. Perhaps by midmonth, the days would break clear in the early afternoons and stay that way until the eclipse was over.

Down in the vineyard at Theodosia, the teams headed by Newall, Perrine, and Freundlich had a different kind of problem. Their enemy was the wind. Shortly after arriving, they had noticed that if the wind came from the southwest, it tended to bring cumulous clouds with it. Around noon, the clouds would form on the mountain peak near the small town of Staryi Krym, some fifteen miles distant. This sizable feature, Agarmysh Mountain, lay in almost exactly the same azimuth as would the sun during totality, the same arc of horizon. By afternoon, or eclipse time, the clouds would be blown in broken, gray masses across the sky. Often, they were followed by angry thunderstorms. But if the wind came from another direction on August 21, then the chance of a brilliantly blue sky was almost guaranteed. Two years of planning and considerable expense now rode on the direction of the wind.

Another kind of storm was also building. On July 28, the Austro-Hungarian Empire invaded Serbia. Throughout the Russian cities in the path of totality, all of the individual teams had been noticing signs of growing tension as the country mobilized. At the dacha, the Campbells watched crude wagons creaking past as the carts carried the local husbands and fathers, sons and brothers into Kiev, one of the main mobilization depots. Following along behind were priests with glittering icons in their arms, blessing the soldiers as they went. Stoic and patriotic, the women said goodbye to their husbands as they comforted the children. "It was already a tragedy for these peasant women," Campbell wrote, "left behind to gather in their little crops for the winter. And we could but recall that this same tragedy was doubtless at that time being multiplied ten thousand fold by similar scenes in nearly every square mile of Europe."

In all the cities, the astronomers were impressed with the orderliness of the Russian mobilization. Troop trains now passing through Kiev began to number forty a day. The same was happening at Minsk. At Theodosia, the soldiers sang patriotic songs as they marched to the trains that would carry them away to possible death. War was in the wind, and it was being blown in all directions. The four teams of German astronomers in the country were warned that Russia was not responsible for their safety if they remained to view the eclipse. On the morning of August 1, the German ambassador to Saint Petersburg, representing the German Empire, issued a declaration of war. Germany had regarded Russian mobilization as an act of aggression and had retaliated. Now war was no longer a whispered speculation.[4]

THE BROTHERHOOD OF WAR

That summer, Frank Dyson and Arthur Eddington had sailed on the *Ascanius* to Australia, as part of a seventy-member advance party for the

BAAS meeting. They would give lectures and enjoy the sights until the forum officially commenced. For five years, Australia had planned a grand welcome, with placards going up in shops windows to announce the lectures. Hansoms, taxis, and private autos would meet the ships and fetch the scientists to their hotels or the private homes that welcomed them. The sea voyage from Liverpool via the Suez Canal would be over thirteen thousand miles. By the time the advance party landed at Perth, on the western coast, the two ships carrying the rest of the members were already many days at sea. The usually optimistic Dyson was now apprehensive. The news being transmitted to his steamer had been all about the mess in Europe. He worried about his astronomers, Davidson and Jones, who were in Russia. "The war and rumour of war is very dreadful and makes me wish I was home," he wrote to his wife. "The papers say that the Stock Exchange is closed, and that prices are going up; so I am afraid you are having a very anxious time in England. Perhaps things are exaggerated, but the idea of England and Germany being engaged in a war over some trumpery Austrian and Serbian question seems too ridiculous for words."

At Perth, Eddington gave a lecture titled "The Stars and Their Movements" to an absorbed crowd of four hundred. The next day, August 2 in Australia, Dyson stepped out for an evening stroll. On a placard in front of a newspaper office, he read the words GERMANY DECLARES WAR. On which country? He, Eddington, and the others could only assume it was Russia. The paper would not be printed and available until 6:30 the next morning. The astronomer royal wrote again to his wife that same night. "How you will get on, I don't know, but am afraid you will have difficulties with money and prices. If England declares war, I don't know how I am to get back. . . . This won't reach you until the beginning of September, and no one knows what may happen before then." Dyson was the father of two sons and six daughters by this time, the youngest girl having just turned a year old.

Eddington also wrote home while still in Perth. "We heard definitely of the war between Germany and Russia. Everyone here seems to take it for granted that England will join in. It all seems incredible. We are anxiously awaiting news." They didn't have long to wait. Back in Europe, on August 2, Germany invaded neutral Belgium. Great Britain protested this act as a violation of the 1839 Treaty of London and quickly gave Germany an ultimatum to leave Belgium. When the German imperial chancellor asked the English ambassador why England would go to war over "*ein fetzen papier,*" a scrap of paper, Winston Churchill would get the fight he had hoped for.

As Dyson and Eddington sailed out of Perth, bound for the official opening of the meeting in Adelaide, breaking news was telegraphed to their captain. On August 4, England had "joined in" by declaring war on Germany. This declaration bound all British dominions to follow suit, Australia being one of them. The seven German scientists and one Austrian who were at that moment sailing to Australia to attend the meeting would now be coming ashore in enemy territory. The mood at the meeting, however, was one of friendship, with toasts being raised to science, not war. Sir Oliver Lodge, the group's previous president, gave a conciliatory speech and read the welcoming telegram from the governor-general. While many members were obliged to change their plans and return home, the rest would be shown Australia's hospitality. Among the honorary degrees given out that night were one to a visiting German, the brilliant geographer and geologist Albrecht Penck, and another to Lodge. The aura of camaraderie would not last.[5]

"INTERNATIONAL COMPLICATIONS"

Campbell and Curtis found Russian volunteers who had not been called away to the military. But the automobile the team needed to transport everyone to the dacha for the important drills *had* been called away. Now any civilian needed a permit to ride in a car on Kiev streets. Another impediment was that the 2,500-foot Nicholas Chain Bridge, a suspension bridge over the Dnieper River, was often closed for marching troops and the movement of heavy artillery. Campbell's "self-sufficient" campsite had been disrupted. But he understood that these inconveniences were due to war restrictions. The Russian government had been as hospitable as it could be, otherwise. The local peasants were also kind and unassuming.

Still, the Campbell team members wondered if their American presence, added to a total eclipse that would turn the world to darkness, *and* the outbreak of war, might set the residents in a panic against the foreigners. The local authorities issued brief proclamations explaining the upcoming event, which somewhat eased the team's worries. But the district police chief went as far as to warn that children should be kept inside and cattle not allowed to go to the fields.

The Russian military turned up at the dacha one day because of what Campbell would call "international complications." The soldiers had come for the Berlin-born cook who had been part of the dacha deal and for whom the Campbell crew already had a great deal of fondness. The Americans knew that her German was rusty and that she couldn't read or write *any* language, and they hoped those shortcomings would work in her favor. But she was arrested under the general rule that applied to all subjects born in Germany or Austria. "The arrest of this energetic character left such a gap in the internal economy of the camp," Campbell wrote, "that all our influence and that of our Russian friends was at once exerted, with the result that after four days she was brought back rejoicing to a welcoming camp."

This Lick expedition had started out as an adventure for the Campbell sons and their good friend Charles Brush. Even with the bedrooms in the dacha available to them, the four young college men chose to erect tents on the large lawn and sleep out under the stars, nights being in perfect contrast to the cloudy, rainy days. When the eclipse was over, Campbell planned to show his sons some of Europe's great observatories. He and Curtis put the boys and the volunteers through their drills as they waited for August 21 and hoped for clear skies. As it was, not one day so far had indicated good weather for success. Clouds and war were not what they had come to Russia expecting to find. All they could do was pray.

THE PEACE OF THE VINEYARDS

Shortly after arriving in Theodosia, Perrine, Freundlich, Newall, and the others had watched as ominous red mobilization notices went up along the streets, on lamp posts and in shop windows. The pebbled beaches of the Russian Riviera would see different days as the steamers loaded with tourists slowly vanished. With the country now mobilizing, most of the qualified men in all the viewing cities—the astronomers had expected to find their volunteers from these ranks—had disappeared to become soldiers. Automobiles were seized for military usage, and any available gasoline was commandeered. Even horses that the scientists were depending on to pull the carts, once the instruments arrived, were called into service if the animals were worthy.

But a bigger problem lay in the fact that three of their friends at dinner each night, and beside them in the vineyard each day as they made ready for the eclipse, were German. Two more German groups were expected to arrive soon, from Munich and Potsdam. Another team from the Hamburg Observatory was already set up fifteen miles to the west, in the small town of Staryi Krym, where the mountain liked to catch the clouds. Freundlich had to have known the dangers of going to Russia, given that a possibility of war with Germany was resting on the horizon. He had just turned twenty-nine years old. In his enthusiasm to be personally involved at last with Einstein's theory, he had gone anyway.

Caution had become the daily mantra as the astronomers waited to see what would happen next. Worried residents in Theodosia had advised the teams to move down to the city, where they would be safer. After discussing whether they should listen to these "sage counsellors," the astronomers decided unanimously to stay put. The orderly calm of the Russian troops they saw passing by each day to the trains gave them reassurance. "We followed our own inferences," Newall wrote of that decision, "and holding quietly to our camp, we gained a wonderful experience in the peace of the vineyards." Things were soon to change. The same day of Germany's declaration of war against Russia, the local military authorities sent a notice up the hill to the rented bungalow. Freundlich, Zurhellen, and Mechau—they hadn't even had time to fully set up their instruments—were ordered to leave Russia immediately.

The Russian authorities also sent word to the members of the Hamburg Observatory team, at Staryi Krym. They were to dismantle the equipment and prepare to leave the country. The two other German expeditions, from Munich and Potsdam, had arrived in Theodosia the day before war was declared. By August 5, all the German teams, as well as any German nationals in the country, had reported to customs in Theodosia. Passports were checked and the teams reassured that they were not accused of any wrongdoing. They were simply being repatriated to Germany. Two days later, when their steamship arrived in Odessa, they were declared prisoners of war and detained there. With the fighting now begun in earnest—the Battle of Liège, the first clash of the war, had started on August 5—what had become of Freundlich, Zurhellen, and Mechau would evolve into rumor and speculation among the remaining astronomers as the days wore on. Some reports had them deep in the Volga region, or at an internment camp up north, or sent by train several hundred miles east to Orenburg. Everyone agreed, no matter *where* the Germans were, they were prisoners of war.

On August 4, the day that England declared war on Germany, the British vice consul had come at midnight to the bungalow to alert the team from the Solar Physics Observatory. Hugh and Margaret Newall and the two other team members could carry on if they chose to remain in the country. But Lieutenant-Colonel Stratton, the astrophysicist professor, was being recalled, as were other military reservists in Russia as part of the expeditions. Stratton left a few days later for England.

Still at the bungalow with Newall and the others were Perrine and Mulvey, who had not yet received their instruments. Having carried the optical parts with them in their baggage, the men began to improvise scaffolding and tubes as best they could for possible replacements. Adding to the problem, Freundlich's equipment had been confiscated when the Germans were declared prisoners of war, so it was not at Perrine's disposal. And yet some of it belonged to the Córdoba Observatory. Carrying on with an expedition under the umbrella of war was a stressful undertaking for *all* the astronomers in Russia.

The ninth of August brought good news when the customs office at Theodosia received a telegram that the Córdoba equipment had been shipped from Odessa. When it arrived the next day, Perrine and his volunteers were waiting with horses too unfit for the war effort to cart it in wagons up to the vineyard. But a heavy rainstorm had also arrived, making it impossible for them to navigate through the mud and water. The downpour ceased long enough for the team to transport the equipment up to the tents before the rain started again. This time, the storm had such power that it would last two more days. It was recorded as one of the worst to hit the Crimean Peninsula in many years. The deluge washed out bridges and vineyards and flooded local farm fields.

Some good luck also arrived. Two English merchant ships loaded with grain were anchored in the harbor at Theodosia, prevented from leaving on orders by the Russian government. Both captains and several crew members offered their services for the drills and the photographing during the eclipse. More volunteers were found among local residents. The drills began on August 18, and the sailors proved to be skillful helpers. Now, with the instruments in place, Perrine and Mulvey had the largest and most proficient viewing station in Crimea. What they needed on August 21 was wind coming from anywhere but the southwest, if it blew at all. And they needed two-plus minutes of clear sky during totality. Even half of that would do.

THE DAY OF THE ECLIPSE

Beginning in the cool waters of the Beaufort Sea and advancing with dazzling speed, the eclipse's shadow crossed Baffin Island, then Greenland and the Norwegian Sea before it moved over central Norway and entered Sweden. Cortie, the "Jesuit astronomer" who had been refused permission to join the English team at Kiev, was ready for it in Sweden. Despite having to report daily to the local police station because neutral Sweden had nonetheless mobilized, Cortie's impromptu team of himself, another priest, and their volunteers would be fruitful. Only two days of the three weeks they were in Sweden were not cloudy at totality time, just as the ominous 6.8 cloud factor had indicated. But those two days happened to be August 9 and *the day of the eclipse*. As curious locals gathered respectfully around the roped-off instruments, the two Jesuits captured wonderful images of the solar corona. Capricious as governments, solar eclipses, and the weather can be, this unplanned and rushed expedition to Sweden was highly successful.

REGULUS, AT MINSK

The moon's shadow left Sweden behind, moved across the tip of Finland and eastern Estonia, and then reached land at Russia. With no English team in Riga, the umbra would next visit Davidson, 250 miles down the line at Minsk. He was waiting in the garden of the lovely villa, three miles from town, where his team had set up camp. They had found "four sorry beasts" to pull the carts of equipment when it arrived, and all was ready. Dr. Kodis, the surgeon who owned the estate, and his wife, had been recruited as volunteers. As with the other stations, Davidson had so far witnessed only two days that would be satisfactory for observations. On this day, after a clear-sky morning, the usual cumulous clouds moved in and gathered in earnest.

With the eclipse approaching, the sky was still overcast in patches and the wind was blowing. As clouds moved away from the sun, Jones watched the crescent on the ground glass of the coronagraph. On a sheet of paper, he had marked the lengths for 3 minutes, 2 minutes, 1 minute, 30 seconds, 20 seconds, and 15 seconds. At 3 minutes, Jones shouted, "Get ready!" It was his signal to the volunteers with the spectroscope to start the exposures for the flash spectrum, and for the others to be on their guard. The first plate was inserted into the coronagraph. At 15 seconds, Jones shouted, "Go!"

Madame Kodis, situated at the metronome, began counting down the seconds, but in French, which the astronomers could understand more easily than Russian. Jones saw the crescent of the sun grow smaller and smaller as Madame Kodis counted, "Un, deux, trois, quatre, cinq." Then, as he watched, the last bright bead—the phenomenon named for Francis Baily, not for Samuel Williams—vanished. Totality had begun just as a mass of clouds drifted in. The wind suddenly slacked. It gave one last gust before it died completely. And then, as if by magic, the clouds parted. The world around the villa fell silent for a few seconds before the insects began chirping and buzzing, thinking it nighttime. Dr. Kodis was not as attentive as his wife. Having been given the job of starting the stopwatch at totality, he forgot. In his defense, he had also been handed a spectroscope to manage, not a *stethoscope*. He could be forgiven in seeing for the first time the overwhelming spectacle above his villa and garden.

According to Davidson, the darkness during totality was not intense. Both Mercury and Venus shone brightly, as did the sparkling star Regulus, in the constellation Leo, which was visible through the corona. One of the brightest stars in the northern night sky, Regulus was the one that would command attention from Campbell, Curtis, and Perrine during this eclipse. Relying on Lick's extensive experience with intra-mercurial plates, Curtis had, the year before, published an article that detailed for interested astronomers how to photograph the eclipse during the 1914 event, should they wish to test light deflection according to Einstein's recently revised theory. He had explained that they must let Regulus guide their telescopes. But as Dyson had written to Freundlich in early 1913, the Greenwich team was not interested in "extremely delicate research," which might even have been "quite beyond present possibilities."

Davidson was lucky. Through this break in the clouds, his team captured good photographs. In the town of Minsk, three miles away, a colleague who hoped to sketch the corona had been clouded out. But Davidson was not interested in light deflection. Not yet.

THE GUIDING STAR AT BROVARY

A few minutes and 270 miles farther on, the eclipse shadow reached the dacha near Kiev. The 4½ tons of Lick instruments—Campbell proudly referred to his expedition as "powerfully equipped"—had long been set up on the lawn, and the drills run. The forty-foot telescope, which Campbell would oversee, was mounted and ready to take large-scale photographs of the corona. The polarizing photometers were ready. The one-prism and the three-prism spectrographs were ready. The objective grating spectrograph was ready. The ultraviolet objective spectrograph was ready. The moving-plate spectrograph for the flash spectrum was ready. The Floyd telescope, with its five-inch aperture and seventy-inch focal length for smaller-scale photographs of the corona, was horizontally mounted and ready. The extra-focal photometers were ready. The chronograph, connected to the forty-foot telescope for recording the times of second and third contacts, was ready. Four cameras, each with a three-inch aperture and a focal length of eleven feet, four inches, were also ready.

"As Regulus was very favorably placed, slightly over a degree from the eclipsed Sun," Campbell wrote in his report, "a fifth lens of the same size and focal length was mounted with the other four, having an ocular provided with cross-wires, and set by calculation so as to have Regulus central when the Sun's image was central on the large plate; it was hoped to be able to use Regulus in this way as a guiding star, to insure perfect roundness in the star images." The mounting of the fifth lens was Curtis's responsibility. All the volunteers assigned to the other instruments were set to perform their rehearsed jobs. The signals before totality, and the counting of the seconds, would be given by the Campbell's youngest son, Kenneth.

Of the last eight Lick Observatory expeditions, seven had been major successes. This good record was due to staff decisions to bypass many viewing opportunities if conditions there seemed unfavorable. And the decisions had been proven right over the years. Having planned this trip on unreliable weather reports, Campbell needed luck on his side more than ever. He and his former right-arm, Perrine, and Curtis were known for the experience and skill that had amassed for the Lick Observatory the finest collection of eclipse observations in existence. The comparison plates had been done earlier at Mount Hamilton with the same instrument setup as in Russia and with exposure times of the same length, as needed. The Lick expedition was more than ready.

The Campbell sons and Charles Brush were at their stations. As usual, Elizabeth Campbell was also assisting. As they watched the cumulus clouds gather overhead, they waited nervously. The storm that had swept first over Cortie's team in Sweden and then down to Riga had passed quickly enough that it left behind clear weather for those lucky astronomers. But the tail of that same storm now reached the Brovary dacha at exactly eclipse time. The sky was completely blocked by clouds, a heavy gray blanket. "Nothing could be seen of the Sun," Campbell wrote, his disappointment lacing every word, "and no observations of any sort could be made. Ten minutes later a little might have been secured thru a cloud gap, and one hour later the region about the Sun was beautifully clear."

ON THE CRIMEAN WIND

The forty-foot telescope from Córdoba was pointed like a giant finger at the heavens. Perrine had brought with him an impressive array of instruments, including the camera mounts and frames made of wood, not metal, as they were at other observatories. He was convinced that wood could better tolerate the drastic temperature drop that occurred during totality, and he had been ready to prove it in Brazil in 1912. The brilliant James Mulvey had designed and built the mounts back at Córdoba, with a clockwork system for motion tracking. Mulvey had also invented a new speed-control system. Much preparatory work had been done in Argentina to ready the instruments for shipping and for faster installation once the team arrived in Theodosia.

After the rains stopped, the two men had worked hard in the Crimean heat to set up camp. But at least the climate was such that Perrine did not suffer from asthma attacks. Their camp in the vineyard was close to Newall and his English team from the Solar Physics Observatory. A camaraderie often developed among like-thinking astronomers who traveled the world together. And these two men had eaten their meals at the same table each day, even slept in the same bungalow each night. While Perrine and Newall could not have come from more dissimilar backgrounds— Newall was the son of famous Scottish engineer and astronomer Robert Stirling Newall—they spoke the same language. They were able to discuss their work and their concerns about the war. They hoped that Erwin Freundlich and the others would be safe, wherever they were interned. And they talked about what kind of weather they might expect on eclipse day.

On Friday morning, after an early breakfast, the men from both teams went into the vineyard to make the final adjustments on their instruments. Newall commented on the wind. It was coming from the southwest, exactly the direction they had hoped against. But the day was still clear and the sky blue. Maybe the wind would change. It didn't. At 10:00 a.m., a cluster of gray clouds began to gather around the mountain's peak at Staryi Krym. By noontime, they had broken free and were floating lazily in the direction of the vineyard. The sky overhead was soon filled with them. Perrine and Newall could see that the sky just a mile or two to the north was perfectly open. The eclipse would begin at 3:20 p.m. Newall was still optimistic. "Our only hope lay in the unlikely chance that the total phase would be visible in a small patch of blue sky between the masses of cloud."

Perrine told the volunteers to take as many photographs between cloud patches as they possibly could. Even though a mist seemed to cover the sun's surface, many shots were indeed taken. The photographs would not amount to much value compared with those successful images that Perrine and Campbell had captured at Lick expeditions in the past, or even those taken at other viewing stations in 1914. The Italians, for instance, down in the town and just three miles from the vineyard, were successful. And so was a nearby French team. But these Italian and French astronomers had probably never heard of Albert Einstein. Imperfect though they were, Perrine's photographs nonetheless come attached with historic value. They were the first ones taken in an attempt to verify light deflection according to Einstein's theory of general relativity.

A half hour after totality was finished, so was the wind. The sky over the vineyard was clear and blue and remained that way for the rest of the day. By then, the moon's shadow had already reached its destination off the coast of India. The eclipse of August 21, 1914, was over.

GOING HOME: "GETTING OUT"

Remarkably, many of these post-Edwardian astronomers were world travelers of the highest degree. Once they were done with the questions that lay among the stars, they concentrated on the wonders to be found on earth. They would visit art museums, gothic cathedrals, and ancient ruins. But now, with the outbreak of war, when trains and steamers were commandeered by the military, the teams were only concerned with going home. With forty troop trains a day still running through most cities, civilian trains were rare, especially ones that carried freight. One detail soon became obvious. The valuable instruments, the pride of many world observatories, were not leaving Russia.

Transporting them safely to a place of storage now became a major problem for all the teams. A government official stepped in to secure special passage for the Lick instruments from Kiev to the Pulkovo Observatory in Saint Petersburg. Backlund would safely store them, as he would for many others. But no one knew how long the war would last. It was an expensive cargo to leave behind, and in the future, this restrictive measure would prove costly in more ways than one. The Campbell family had prepurchased their tickets and originally had plans to travel by rail through Berlin and Paris and then by steamer to London. But these plans were now pipe dreams. "For two weeks before we left the observing station," Campbell wrote, "we gave considerable thought to the problem which was always referred to as 'getting out.'" He soon found this would not be easy. "Only one passenger train a day ran from Kiev to Moscow, and neither money nor influence could guarantee a seat on this slow and crowded train."

Yet the Campbells managed to book passage. In Moscow, they heard rumors that even if they reached London, they might not be able to book tickets on an Atlantic passenger steamer, since the vessels were being taken for troop transports. It's an amazing coincidence, given the chaos of travel arrangements and canceled tickets, that in Saint Petersburg, Campbell would meet up with his longtime friend Perrine and the mechanic Mulvey. Also arriving at this frenzied time were Davidson and his team, who had been successful at Minsk.

Perrine and Mulvey had said goodbye to their English roommates at the bungalow and had left Theodosia on August 25, headed north by train to Moscow. It was time to get out of Crimea. They had just received word that a man who had been caught taking photos of a local arsenal had been accused as a spy and killed in the streets while trying to flee. Berths weren't available for half of the nearly three-day trip. The two men slept in the passageway aisle of the train, but at least they were on their way out of Russia. They had a long way to travel before they would reach home. Perrine's firstborn, his son and namesake, had celebrated his third birthday back in Argentina a week before the eclipse. His daughter was eighteen months old. Expeditions were tough on families.

Perrine's disappointment had to have been deep. His hope now was that Campbell had enjoyed better luck near Kiev. But after a few days spent with Backlund in Saint Petersburg, he met up with Campbell and learned the bad news. However, Dyson's team at Minsk, the one from the Royal Observatory, had been successful. The realization now settled on Perrine and Campbell. Of the three teams prepared to test for light deflection, none had done so. They would have to wait for the next eclipse, if the war ever ended. And if they could get out of Europe alive. In Saint Petersburg, Perrine had read in the newspapers that the *Arlanza*, the steamer he was scheduled to take from Southampton back to Buenos Aires, had been sunk off the coast of Brazil.[6]

The astronomers and their teams had determined in Saint Petersburg that the best way back to England would start with a train ride north to Finland. Troop trains loaded with soldiers and weapons seemed to be everywhere. While the visitors managed to secure passage on the rare civilian train, they were not guaranteed berths. But the three teams— Campbell's party, Davidson's party, and Perrine and Mulvey—didn't care. They were happy to be leaving Russia, much as they had been impressed with the quiet confidence of the troops and local citizens. Coincidentally, the same day they left Saint Petersburg, on September 1, the city was officially renamed Petrograd, a more Russian-sounding name.[7] They changed trains in Tammerfors, Finland, for a rail line that would carry them across the country to the west coast. From there, they hoped to book a steamer to Stockholm, which meant traversing the Gulf of Bothnia. With numerous delays and inconveniences—passports had to be checked often by police, and travel permits issued—they were finally escorted from the train in Rauma, Finland, and put aboard a steamer for Stockholm.

Sailing over this body of water rumored to be rife with mines and German U-boats was worrisome. Steamships had their own navigational devices to detect mines, but they also received signals from coastal stations as they steered past the southern tip of Finland to approach Stockholm. Twice, the passengers were called below deck for their own safety.

Perrine commented on the unnerving situation: "While we were passing through these channels and between the islands, we were almost continuously watched by Swedish torpedo boats which appeared between the islands, gave us a look and retreated out of sight again." Only the youngest member of the travelers, Kenneth Campbell, was hoping to see a German warship.

Two days after Perrine and Mulvey left Crimea, the Newall party had emptied the rented bungalow and caught a steamer for Yalta, where they would visit the Simeiz Observatory before heading north to Moscow. As they were passing through customs in Theodosia—they left their many crates and cases there to be stored by agents—they noticed the Hamburg Observatory's instruments stacked in a corner. The information they were given was that the three older German astronomers had been repatriated from Odessa back to Germany. But Freundlich, Zurhellen, and Mechau had been sent by train to an internment camp in Orenburg, or so it was believed. Rumors were now as rampant as the clouds that had floated over the vineyard.

Arriving in Saint Petersburg days after the other teams had already left by train for Finland, Newall would learn of Davidson's successful observation at Minsk and Campbell's disappointing results at Brovary. During Newall's visit to Backlund at the observatory, he saw the amazing photographs that had been secured by the Russian team in Riga. Newall could only acknowledge, "What a superbly interesting corona we had missed." He realized that the big mistake in Crimea had been that the teams were set up too close to each other, even in a restricted area. But what Campbell, Perrine, and Freundlich had missed was so much greater: a chance to be first in determining if light had weight.

THE PRISONERS OF WAR

As Albert Einstein struggled with the remnants of his marriage and further work on his general theory, he wrote to his friend Paul Ehrenfest on August 14 of that same year: "My good old astronomer Freundlich, instead of experiencing a solar eclipse in Russia, will now be experiencing captivity there. I am concerned about him." While the astronomers who had traveled to Russia would not learn the facts until weeks later, and despite even more false rumors, Erwin Freundlich had fared better than was feared. After examining the situation, the governor of Odessa released seven of the older German team members, including one Danish American who had come with the Hamburg expedition. Within a few days, these men were sent back to Germany.

The governor then telegraphed the Academy of Sciences in Saint Petersburg, which had issued letters of recommendation to the astronomers before they came to Russia. The academy quickly renewed the letters at the governor's request, and he allowed Freundlich, Mechau, and another member of the Hamburg team to return to Germany. However, Zurhellen and the remaining three Germans, being army reservists, were sent to Astrakhan, in the Volga district, where they were not released until a year later. Instruments belonging to all the German expeditions were confiscated by the Russian government and stored at a university in Odessa until after the war. According to some reports, Freundlich was back in Berlin as early as September 3. He had beaten all the British and Americans home.

From Stockholm, the three teams still traveling together—Davidson's, Campbell's, and Perrine's—went to Christiana by rail, the city that is now Oslo, Norway. Once there, they booked passage on the Bergen Line, which had just been completed a few years earlier. They boarded the train in Oslo, sped across the Hardanger Plateau, with its alpine climate, and arrived in coastal Bergen, three hundred miles later. They left Bergen aboard a steamer bound for Newcastle upon Tyne, in northern England, where they expected to arrive on September 7. Again, the dangers of now crossing the vastness of the North Sea during wartime were evident. The ship demanded a lights-out at night policy, with port windows boarded and no smoking on decks. The *Titanic* still fresh in their minds, the passengers found it impossible not to count the lifeboats, which were there at the ready.

From Newcastle, it would be another train ride down to London, some three hundred miles to the south. Charles Brush Jr. left the party and took a train across England to catch the steamer he had originally booked for return passage to the states. Douglas and Kenneth Campbell, unable to make their previously planned connection to Rotterdam, went back to the states, traveling in ship's steerage. It had been a world education after all. The remaining four Campbell family members and Curtis; Davidson, Jones, and Hepburn; and Perrine and Mulvey rode south on the train to London, where Campbell cabled Mount Hamilton that all party members were "in good health." The Englishmen were now home, but the Campbells would spend two less-dramatic weeks before catching a steamer to Boston and then a train across country to San Francisco. They would be back on Mount Hamilton in mid-October. Perrine and Mulvey would sail back to Buenos Aires on the *Arlanza* after all—its sinking off Brazil's coast was another "steamer rumor." As Perrine saw it, "Argentina is currently one of the few quiet places in the world."

A week after these three teams reached London, Hugh and Margaret Newall were presenting their passports on the quay at Newcastle upon Tyne. Now, everyone was safely out of Russia, even American aviators David and Mabel Todd, who, according to the August 14 edition of the *New York Times*, were lost there while "chasing the sun." With the outbreak of war—they could find no airplane to do the chase—the Todds left the continent via Sweden and Denmark.[8] Campbell and Perrine managed to visit Dyson at the Royal Observatory. As they waited to say their farewells, they found the city of London serene, except that there were more officers than usual in the restaurants and hotels. But the war was still young. And dozens of the world's best scientists were still trying to get home safely.

LEAVING AUSTRALIA

The world was about to change. All three fancy ships the BAAS members had boarded for the long trip to the Southern Hemisphere—the *Ascanius*, the *Euripides*, and the *Orvieto*—had been immediately requisitioned by the local government and turned into troop ships for the Australian Expeditionary Force. The members were now obliged to sail on less lavish steamers back to England, and through dangerous waters. Newspaper accounts told of notorious German cruisers running wild in the Indian Ocean, patrolling the same route the members had to take. Passengers were given restrictions. There would be no lights at night, whether the lights were in cabins, up on deck, or on the glowing tips of cigarettes and cigars. That benevolent aura of goodwill that had embraced the scientists back in Adelaide was fading fast among many of the meeting attendees. Rumors of babies bayonetted by "the Hun" had been in circulation even before the outbreak of war. That it was untrue made no difference. Now reports of German soldiers destroying rare works of art in Belgium prompted many of the intelligentsia to join the anti-German mood of the day.[9]

Dyson was eager to get back to his family. He and the other passengers feared that at each port stop along the way, they would hear that Paris had fallen. As they passed the Cocos Islands, the steamship rumor was that things were going better for the Allies. At Sri Lanka, when they learned of victory at the Battle of the Marne, hope swelled on the ship. The signs of war became more evident the closer they got to home. At Bombay, Dyson would watch as General James Willcocks came aboard with a large number of Indian officers, all with a cheerful disposition that the war would soon be over. They crossed the Arabian Sea and sailed up the Red Sea to enter the Suez Canal. After reaching the Mediterranean, they were twice boarded by French sailors assigned to destroyers guarding those waters. The refrain among military and passengers alike was "over by Christmas." But Dyson wasn't so certain. He had been concerned for weeks about the whereabouts and welfare of Davidson and Jones. Had they been able to leave Russia safely? He wouldn't know until early October, when he reached Plymouth—the steamer escorted into the harbor by British destroyers—and was finally home again at Greenwich.

With over half the BAAS members leaving after the last meeting, Arthur Eddington was among those who chose to finish sightseeing to the end. Before Dyson left, the group had toured a sugar factory at Brisbane. "It was amusing to see the whole British Association, 150 of us, sucking sugar cane," Eddington wrote. With Dyson gone, Eddington's ship, the SS *Montoro*, began its scheduled stops around Australia as it headed for England. He and the current BAAS president, William Bateson, went exploring for butterflies in Townsville. Bateson was the main popularizer of Gregor Mendel's ideas, was the originator of the term *genetics*, and was considered by many to be a militant atheist. Yet he excitedly counted butterflies, "thirteen on one twig, many very fine ones." Eddington also wrote home of "lizards and fine centipedes." He was excited to find the skull of a spiny anteater. It wasn't until the ship stopped in Singapore and 140 British soldiers boarded, on their way to the Western Front, that the war was no longer distant words in a telegraph.

Yet, the sightseeing continued for Eddington during stops along the way, with Buddhist temples in Sumatra and the botanical gardens in Sri Lanka. Unlike Dyson, he was a bachelor and seemed in no hurry to get back to England. He did note in a letter home that the ship, with so many renowned scientists aboard, would be a "fine prize for the *Emden*." The SMS *Emden*, a light cruiser with ten guns, had been built for the Imperial German Navy. The vessel was the most famous and feared on the Indian Ocean, having already captured two dozen ships and torpedoed a Russian cruiser and a French destroyer. All the passengers crossing the Indian Ocean—this was true for Dyson's ship as well—dreaded the thought of the *Emden* appearing out of the fog since they were all treading the same waters. Tension was widespread on the decks of civilian ships until the *Emden* was run aground off the Cocos Islands just as Eddington was arriving back in England.

DISAPPOINTMENT THROUGH CLOUDS

As Charles Perrine sailed back into the harbor at Buenos Aires on October 3, the Rio de la Plata flowing into the ocean was a welcoming sight from the deck of the *Arlanza*. Proficient by now in Spanish, he wrote in his notes that the map of Europe would be drastically changed once the war was over. "I will be very optimistic maybe, but I cannot stop thinking that the world is too civilized and the great fundamental laws of nature too powerful for the right of force rather than the force of law to win even in war." As he and James Mulvey rode the thirteen hours by train back to Córdoba, he wouldn't know that on his horizon were concerns that no amount of optimism could change.

No doubt speaking for many astronomers who had gone to Russia in 1914, a frustrated William Campbell would remark in his report to the Astronomical Society of the Pacific about the unreliable weather information they had been given. That the maximum cloudiness had occurred at the same time of day as the eclipse spelled certain failure for many expeditions. Had Campbell known this likelihood in advance, Lick would never have sent the expedition to Russia in the first place. "Observers at Minsk, Theodosia, and other places on the line of totality report precisely the same state of affairs, and examination of weather records for the past ten years shows that this is the rule in the summer weather of western Russia." The conflict between the weather information the Russians provided and actual historical weather data was an astonishing piece of information for any expedition that had spent months in planning and funding and whose participants might have been killed in the outbreak of war.

When Campbell finally realized how the discrepancy had come about, it was a lesson so severe he hoped to apply it to future expeditions. The Russian government had done its observations in the mornings and evenings, when the skies were often beautifully clear. But the total eclipse had arrived in Kiev at 2:50 p.m., in the midst of clouds. It was now obvious that weather observations needed to be done more often during the day, not just morning and evening. And they should be taken three or four years *in advance*, at the exact hour when future eclipses were expected to occur. There had been a glitch in the information, that's all.

Campbell's letter to George Hale, on October 16, 1914, best reflects the heartbreak that can come with courting an eclipse in the early twentieth century: "I never knew before how keenly an eclipse astronomer feels his disappointment through clouds. Eclipse preparations mean hard work and intense application, and I must confess that I never before seriously faced the situation of having everything spoiled by clouds. One wishes that he could come home by the back door and see nobody."

There were more total eclipses coming.

6

A MAGIC CARPET MADE OF SPACE-TIME

The British Take Interest: Science Goes to War

It is not any personal attitude of the German scientists that presents a difficulty, but the feeling that we are involved in a general condemnation of their nation. But the indictment of a nation takes an entirely different aspect when applied to the individuals composing it. Fortunately, most of us know fairly intimately some of the men with whom, it is suggested, we can no longer associate. Think, not of a symbolic German, but of your former friend Prof. X, for instance—call him Hun, pirate, baby-killer, and try to work up a little fury. The attempt breaks down ludicrously . . . [T]he worship of force, love of empire, a narrow patriotism, and the perversion of science have brought the world to disaster.

—*Arthur Eddington, letter to the Observatory, 1916*

THE MODUS VIVENDI that the European scientists had sailed away from in the summer of 1914 was changed by the time they returned. The luxuriant steamers that had carried some of the world's wealthiest people and boasted elegant dining rooms, libraries rich with books, full orchestras, palm courts on deck, swimming pools, and crystal chandeliers disappeared as if overnight. Now, instead of stocking 30,000 pounds of prime beef, 16 tons of potatoes, 15,000 eggs, 10,000 clams and oysters, 3,000 pounds of moist sugar, 500 quarts of ice cream, dozens of cheeses, fresh fish, fruits, and vegetables—and this was just until the ships reached the next port of call—the libraries and ballrooms were packed with ready troops, ammunition, artillery, medicines, and tins of corned beef, sardines, and hardtack.

It was the end, at least for a time, of pomp and circumstance. There would be no more astronomers "rolling down to Rio" on those "great steamers white and gold," as Rudyard Kipling's poem had boasted. In due time, Kipling would be overwhelmed with the loss of his only son, John, a seventeen-year-old for whom he had pulled strings to get commissioned and who went missing in action after being wounded in the Battle of Loos in 1915. The Kiplings, riddled with grief, would begin a pilgrimage from hospital to hospital, asking, as did the poem the author would write, "Have you news of my boy Jack?" No one ever did.

It was during this period of world turmoil and personal sorrow that Albert Einstein was still at the drawing board when it came to general relativity. Now was not a good time to be a German physicist with a provocative idea to present in its final draft to the world of science. Indeed, it may have been *the worst time in history* to write a paper that carried so much consequence in redefining our cosmos. Its acceptance came with a scientific price. What had been for two hundred years an English-defined universe would become a German-defined universe. By the turn of the twentieth century, tension had already been brewing between the two countries. With Germany's growing economy, Great Britain was losing its foothold as Europe's leading power.

Just as he was pulling strings to get son John into the military, Kipling would be one of the literary names recruited to assist in the war effort by signing an "Author's Declaration," published on September 14. Fifty-three British writers, including Thomas Hardy, H. G. Wells, and Sir Arthur Conan Doyle, declared it was a matter of honor that England go to war once Germany had invaded Belgium. Hardy would be accused by his detractors of hoping to revamp his career. But still feeling the loss of his son, Kipling was eager to oblige by writing pamphlets and stories that glorified the heroics of British soldiers standing firmly against the brutality of "the Huns." The hatred between countries was snowballing.

Frank Dyson and Arthur Eddington were still on ships sailing back from Australia when *Harvard College Observatory* bulletin 563 published an announcement dated September 23, 1914. It was received by astronomers in the Western Hemisphere and allied countries in Europe: "Owing to the war in Europe the usual interchange of astronomical announcements between this Observatory and the Astronomische Centralstelle at Kiel, Germany, is necessarily suspended." From then on, the announcement continued, all messages about astronomical discoveries should go to director Edward C. Pickering, who would distribute them in the United States "and to foreign countries as far as practicable." The umbrella of scientific friendship was closing, aided by a groundswell of atrocity propaganda.

The German intelligentsia of the day didn't help matters. On October 4, 1914, as Dyson was on a train back to Greenwich from the dock at Plymouth and as Eddington was admiring the flying foxes in Sri Lanka, an astonishing document was published that would further separate members of the scientific community whose countries were at war. It was written and signed by ninety-three of Germany's leading scientists, artists, and scholars. Titled "Manifesto to the Civilized World," it defended Germany's actions early in the war and denied reports of assault on Belgian citizens. "It is not true that we trespassed in neutral Belgium. It has been proved that France and England had resolved on such a trespass, and it has likewise been proved that Belgium had agreed to their doing so. It would have been suicide on our part not to have preempted this." Among the signers, ten had already won a Nobel Prize, and four more would win one in the years immediately following.

Max Planck, who had been so influential in Einstein's career, was one of the signers, although he would later rescind parts of the document. Fritz Haber, who had taken Mileva Einstein and her sons into his home, had also signed it. "Have faith in us!" the statement ended. "Believe, that we shall carry on this war to the end as a civilized nation, for whom the legacy of a Goethe, a Beethoven, and a Kant is just as sacred as its own hearths and homes." The manifesto would become known as the go-to essay on how German intellectuals felt about the war, and it infuriated their foreign colleagues. Three German scientists, however, did *not* sign the manifesto but drafted their own instead. In "A Manifesto to Europeans," Einstein, Willem Forster, and G. F. Nicolai all protested that "never has any previous war caused so complete an interruption of that cooperation that exists between civilized nations." Their document was sent out privately. It was unknown outside Germany and received very little attention *inside* Germany. But Einstein had announced his pacifistic nature.

GREENWICH AT WAR

Finally back at Flamsteed House, where the previous eight astronomers royal had also lived, Frank Dyson found things greatly changed.[1] At forty-six years old, he would not be called into service by the military, and his sons, at thirteen and nine, were too young. But his employees, fellow astronomers, and friends would walk away to join the fight or would watch their loved ones go. He was relieved to find Charles Davidson safely back home and to learn that he and Harold Spencer Jones had been successful at the eclipse. But he was dismayed to hear that they had returned from Russia without the valued instruments.

Gone, also, were most of the observatory's assistants and computers of military age. They had volunteered to fight, being caught up in the wave of heroic patriotism that was engulfing the country. Dyson was determined to keep their positions open until and *if* they returned and to continue their salaries. This plan put a strain on how to pay new computers, if he could find any not gone to the war. In all, he lost thirty-six members of his staff. A few retirees, women, and local boys too young to fight were hired. And Dyson found more workers among the Belgian refugees who had just begun arriving in England by the thousands. His permanent staff, however, soon dwindled and he lost Jones to the arsenal at Woolwich, in southeast London, where he was attached to the optical department. Jones's job was in cleaning and adjusting the mountains of binoculars sent to the arsenal by the military. Because they had been manufactured hurriedly in the United States when the war broke out, the optical lenses were often unstable. Unable to service the great bulk of these field glasses, the arsenal sent many pairs on to Dyson and his staff at the observatory.[2]

Dyson had hired John Jackson, a Scottish astronomer, as his chief assistant before leaving that summer for Australia. Jackson was still there. Yet without a full workforce and proper equipment, work at the observatory was restricted. Solar photography and observations of the sun, moon, and planets were maintained, and the time ball still went up every day, religiously.[3] But with no astronomer left to operate the twenty-eight-inch refracting telescope, Dyson was forced to discontinue an important study of double star systems—a study that had begun in 1893, when this telescope was installed. Chronometers had always been tested at the observatory, but now this work became an important part of the war effort. Dozens of them were sent to Dyson after sea battles to be repaired and readjusted. Amazingly, arriving at his door one day from Australia was the chronometer from the notorious *Emden* that had been sinking ships in the same patch of the Indian Ocean that Dyson and Eddington had crossed days earlier. All in all, the observatory's work was crippled by the war, as if the study of the universe also had to be rationed.

When Dyson sponsored and hired a French refugee named Robert Jonckerhèere, an astronomer who owned a private observatory back in France, the Englishman finally had an assistant to oversee the twenty-eight-inch telescope. The study of double stars could begin again. But with money short and the need to support his English wife and two children, Jonckerhèere had to find additional work adjusting binoculars at the optical arsenal with Jones. Dyson and his wife also helped in the war effort by hosting garden parties at the observatory to raise money. When he was knighted in 1915 for his important astronomical work, the Dysons' invitations were more eagerly accepted. At first reluctant with their new titles, Sir Frank Dyson became president of a local Red Cross chapter and Lady Dyson, for her part, was a volunteer police officer who walked the heath at nights. One of her duties soon became ordering the giddy girls hanging around the military camp near Greenwich to go home to their mothers.

The Kipling-esque propaganda of glory in combat began to wane as the causality lists of dead soldiers grew longer each day, demanding more and more space in the local newspapers. By 1915, zeppelins loaded with incendiary bombs were soaring in the skies above England. They followed the Thames River upstream, flying over the arsenal at Woolwich and then Greenwich, on their way to London. Once, when Dyson was away and the large dome of the observatory was open, a zeppelin dropped two bombs close by. No one was injured, although Jackson was inside the dome at the time. The only damage was some burned fence rails and the piercing of the papier-mâché domes by splinters of shrapnel.[4] These were the original domes that for over two hundred years British sailors sailing up the Thames had looked for and found comfort in seeing. From then on, when the air raid alarms sounded and searchlights swept the dark skies over the domes, Dyson ushered his family, staff, and any visiting guests into the stone cellars below.[5]

EDDINGTON, AT CAMBRIDGE

In November, Arthur Eddington arrived home from Australia, three months after the outbreak of war. Waiting for him in the east wing at the Cambridge Observatory where the director's residence was located, just off the library, were his mother, his sister, an Aberdeen terrier, and a cat. He had enjoyed the comfort of this new home for just a few weeks before sailing to Australia. Now he was back to settle in, not just in his position as Plumian Professor, but also as the observatory's new director. He soon faced the same situation that Dyson had found at the Royal Observatory, a debilitated staff and limited finances. With no plans to join the war effort, he went to work. A zodiacal catalog had been begun at Cambridge in 1900 but was left unfinished. Eddington took it on as a special project while the war intensified, single-handedly completing the transit observations. With England still relying on volunteers to meet its military needs, he could continue his work at the observatory, despite the restrictions and without reproach.

Eddington was still just thirty-one years old, which was certainly not too old for the military at a time when elderly officers, some of them over the age of sixty, were being called out of retirement to train soldiers. Physical fitness was not a great issue. Many working-class men were not properly nourished to begin with. One medical grading would accept men who were "Unfit, but could be fit within 6 months." Given these lax physical requirements, Eddington would have been a prize. He was not married and had no children. While in Greenwich, he had been a member of the field hockey team, hardly a sport for the timid. He played golf and tennis poorly but, as with hockey, was known more for his vigor than skill. However, he excelled at cycling, an almost daily routine and for many long miles at a stretch.[6] And he was a hiker. He and his close friend C. J. A. Trimble would walk twenty or thirty miles a day on holidays with no effort, sometimes climbing three thousand feet up steep hills and glissading down them, once while being pelted by a snowstorm.

If anyone was physically fit for the military, it would be Eddington. But he was not just a scientist doing important work; he was a Quaker whose religious beliefs prohibited him from the taking of a human life. With England still not conscripting men to fight, he would be left to his work. But his later refusal to serve on religious grounds would cause disapproval among some of his fellow scientists. It didn't help that two of Eddington's own computers at the observatory were finally called off to war. Neither man would return.

The optimism that the war would be "over by Christmas" was over by Christmas. But at least a truce, which appeared to be more spontaneous than official, took place as soldiers from both sides along the Western Front fraternized during Christmas Eve and Christmas Day, 1914. This gesture of hopefulness on the part of already-wearied soldiers, in the form of a holiday truce, would also disappear by the next yuletide.[7]

PERRINE, AT CÓRDOBA

While Charles Perrine might have been a long way from the lines of battle, in Argentina, the economic crunch of war was felt there, too. With German U-boats patrolling the northern waters, all shipping commerce between South America and Europe was now suffering from blockades. European investments in those countries seemed to vanish overnight. What Perrine was experiencing more directly, however, was the money he personally lost during the Russian expedition. Once war had been declared, the rubles that he had brought with him to Crimea were immediately devalued. He was forced to use his own money to finance the weeks he and Mulvey would be in Theodosia, as well as their journey home. The expense would amount to one and a half times his yearly salary as director at Córdoba. Perrine had tried unsuccessfully to turn the rubles into gold. When leaving Russia, he had given them to Backlund, at the Pulkovo Observatory. Backlund promised to safeguard the money until the country's currency would again appreciate. When Backlund died in August 1916, chances became bleak for Perrine to recoup those funds.[8]

Perrine was soon faced with a loss more personal than money. A few months after his and Mulvey's return from Crimea, over the Christmas holidays, Mulvey had fallen seriously ill from an attack of what was most likely typhoid. The illness was so severe that he had to be hospitalized for two months. Hoping to recuperate, Mulvey traveled to the Sierras Chicas, the hills northwest of Córdoba beloved by tourists for their beauty. As he was recovering there, he was suddenly stricken with food poisoning and had to be hospitalized a second time. On the night before he was to be released, he began hemorrhaging and died almost immediately. His body was brought back to Córdoba for burial, whose details Perrine oversaw himself.

Perrine's last actions speak loudly of the affection he felt for his mechanical engineer and friend. Loyalty was not a trait to be taken lightly. As Perrine had once been William Campbell's right-hand man, so had James Mulvey been his. Sending Mulvey's body home to the United States for burial wasn't possible in those days, unless one was cremated. Perrine buried him in the plot he had purchased for himself, in El Cementerio de Disidentes. This was the Cemetery of Dissidents, where Protestants and anyone else who had died outside the Catholic faith, including Muslims, Jews, and atheists, were buried. He paid for a large, pyramid-shaped stone to be placed on the grave. Then he sold what possessions Mulvey had and mailed the money to his aging father, who was by now in a nursing home in Arkansas, where the family had moved twenty years earlier. One of the first men to test for light deflection was now not even a footnote in history.

Perrine wrote a glowing tribute to Mulvey for the Astronomical Society of the Pacific, detailing his genius in the design of astronomical instruments and his strong work ethic. Then he concentrated on what lay ahead: the February 1916 total eclipse. Two annular eclipses had occurred in 1915, mostly over ocean water. With Europe deep into the war and the United States teetering, Perrine intended to take advantage of the fact that this 1916 path would at least clip the upper part of South America. It would cut across northern Colombia, Venezuela, and most of Guadeloupe, with a totality duration of over 2½ minutes. It was not exactly in his backyard, but it was close enough to send a limited expedition to Tucacas, Venezuela. The expedition was so limited, in fact, that it would only consist of Enrique Chaudet, the third astronomer who had gone with him to Brazil in 1912. Perrine would not make the trip. Chaudet would find his volunteers and workers once he arrived. What must have been most disappointing for Perrine was that the forty-foot telescope was still in Russia, his expenses had not been recovered from the Crimean trip, and his indispensable mechanic was dead. Therefore, testing for light deflection regarding Einstein's theory would not be on the program.

PROPAGANDA VERSUS SCIENCE

The spontaneous comradeship that had arisen briefly between opposing soldiers during that first Christmas of the war would soon be impossible to revisit in a theater of aerial bombings, trench warfare, armored tanks, and, the most heinous of all, the introduction of poison gas as a weapon. None other than Albert's and Mileva's old friend Fritz Haber, now a captain in charge of the Chemistry Section in the Ministry of War, was there in person to witness the release of nearly two hundred tons of chlorine gas by the Germans at the Second Battle of Ypres, on April 22, 1915. Haber relied on Mother Nature by waiting for a day of perfect wind that would carry the massive yellow cloud of gas, released from six thousand canisters, over to French troops lying in the trenches. Several thousand would perish from the gas, which destroys tissue, such as eye tissue, and lungs when inhaled.

Haber and his unit of three future Nobel Prize winners continued to perfect the deadly gases used during the war.[9] There was phosgene in 1916, which was said to smell like moldy hay, and sulfur mustard gas in 1917, ultimately earning for Haber the moniker "father of chemical warfare." His wife, also a chemist, committed suicide in their garden eight days after the gas release at Ypres, and on the night before Fritz would leave to oversee the second attack. This act might be considered a symbol of the widening gap not just between men and women of science, but also between science and humanity. Clara Haber used her husband's service revolver to shoot herself in the heart.[10]

If Kipling and others were enraged when the *Lusitania* was sunk by the Germans on May 7, 1915, they were also blessed by the perfect timing of that event when it came to atrocity propaganda. Scheduled for print just a week later, on May 12, 1915, was the *Report of the Committee on Alleged German Outrages*, a sixty-one-page document published in London. The committee that prepared the document included some of England's best-known names in politics, education, and the law. The document was supposedly based on twelve hundred firsthand accounts by unnamed victims of body mutilations, gang rapes, deaths by clubbing, and unimaginable tortures carried out by German soldiers on Belgian civilians, including women, children, and the elderly. Bayonetting babies was a particular horror and gave rise to illustrated poster drawings sometimes showing several at a time impaled on a German soldier's spear. Pillaging, plundering, and destroying fine works of art now appeared to be the nice crimes.

Germany's denials only brought more attention to the awful charges, so the country retaliated with its own booklet of atrocities perpetrated on *them* by Belgians. It attracted little notice. Within two weeks, however, the United States had grabbed up the British report, and the document was reprinted in almost all national newspapers, including the *New York Times*. The British War Propaganda Bureau would immediately ship forty-one thousand copies of the booklet to the United States. When American critics saw the effort as propaganda designed to enlist the United States into the war and asked for hard evidence, it was apparent that the committee had never even *read* those first-person accounts, which had somehow disappeared. This was the state of world affairs as the general theory of relativity was taking final shape.

FREUNDLICH'S DILEMMA

Back at work after his failed expedition to Crimea, Erwin Freundlich still felt like a prisoner when it came to his duties at the observatory. He kept up a steady correspondence with Einstein, in Berlin, both men discussing ways to test the general theory in the future. They were back to courting Jupiter as an option. But Freundlich felt that his steadfast interest in relativity was stymied by the duties that his director, the strict Hermann Struve, was imposing on him. Struve came from a family chock-full of astronomers and was a no-nonsense scientist who had offered no financial or moral support to Freundlich's expedition to Russia.

Freundlich and his former math teacher, Karl Schwarzschild, seemed to be the only Germans who were showing the general theory any support. Added to that, Freundlich was the one with the connections to other astronomers. And Einstein needed astronomers if his light deflection prediction was to be verified by observation in nature. He would try twice again to emancipate Freundlich from his job at the observatory, with salary intact, so that he could concentrate more fully on the general theory. The physicist failed each time. For now, Freundlich would stay put.

Freundlich was apparently casting a wider net than perhaps Einstein and others in Germany realized. On May 31, 1915, Perrine wrote from Argentina to George Hale, at Mount Wilson, informing him about the failed expeditions to Russia and noting Freundlich's good luck in being able to return to Berlin. The German astronomer had recently sent Perrine a letter stating that he wished to continue his work, which was then focused on Jupiter, but outside Germany. Because the large telescope at Córdoba would not be installed for a few more years, Perrine wondered if Hale might have a place for Freundlich at Mount Wilson. He was quick to mention that Freundlich would *prefer* life in Germany, although his mother was English and his wife, Käte, was Jewish. But the astronomer wanted to prioritize his work, which he couldn't do at the Berlin Observatory under Struve's watchful eye. Hale answered briefly on June 20. He couldn't employ Freundlich, because of strict orders by President Woodrow Wilson preventing Americans from hiring Europeans who were directly or indirectly related to the war.[11]

THE EINSTEIN FIELD EQUATIONS

As the war waged on, Einstein again realized that he had made a mistake. In Newton's theory, all three masses (the inertial mass, and the active and passive gravitational masses) are equal. Einstein wanted to know why they couldn't be different. Earlier in his life, before he published the special theory, he had said that he felt inner turmoil as he conceptualized a universe whose laws were not coming together as he knew they should. "At the very beginning, when the special theory of relativity began to germinate in me, I was visited by all sorts of nervous conflicts. . . . When young, I used to go away for weeks in a state of confusion." But he was a decade older now, with different responsibilities in that external world beyond his mind, with two young sons and a bitter marriage that was headed for divorce. He had published papers. His reputation was growing. But geniuses aren't equipped with all-powerful searchlights that can magically find answers in the dark.

He dusted himself off and went back to work. His intuition since 1911 was that light *should be* bent by gravity, as his early papers predicted, but he needed a rigorous mathematical proof, which he still lacked. Finding this proof would mean that his mind had to unlock the problem *with no assistance from prior human knowledge*. The poet William Wordsworth, in writing about Newton's memorial statue that stands at Trinity College Chapel, perhaps best described this kind of intellect as "a mind for ever Voyaging through strange seas of Thought, alone."[12]

In late 1915, Einstein began writing equations that would support his theory. A breakthrough came when he correctly calculated results that would explain the shift in Mercury's orbit. Developing what became known as the *Einstein field equations*, he wrote on one side of the equations the terms that describe how the complex system of space and time can be defined by a mathematical field. In doing this, Einstein remembered that James Clerk Maxwell had shown that the influence of electricity and magnets could be understood in terms of an electric field and a magnetic field. Maxwell demonstrated this relationship between electricity and magnetism in mathematical equations (so-called Maxwell's equations) he developed more than fifty years earlier.[13] Einstein came to the conclusion that the properties of time and space must be described by a gravitational field. This gravitational field would become known as the *metric tensor*, or just metric. On the other side of Einstein's field equations is a description of how the matter and energy of everything else in the universe is related to the *curvature* associated with the metric. Just as a magnet, if surrounded by iron filings, will have an influence on them, so do Einstein's field equations imply that the structure of space and time is influenced by all matter and energy in the cosmos.

Einstein had arrived at what would be his final calculations on the theory of general relativity. His astronomical predictions now revealed an answer to the universe, one far more dramatic than he had anticipated. This discovery, this oracle, this *prophecy*, made him almost giddy with joy, as he would later relate to colleagues. If correct, the pedestal that had been firmly built beneath Newtonian physics, and the pillars that had held it aloft for two centuries, would need remodeling. During his many long hours of work, he had come up with his own third-eye answer. What was woven into those handwritten pages that lay on the table before him, what was embedded in those mystifying equations, was a universe built from a magic carpet known as space-time. Sir Isaac Newton was no longer voyaging on those strange seas alone.

In a series of lectures to the Prussian Academy in the autumn of 1915, he presented these new results. They explained that Mercury's wobble—the planet's stubborn refusal to conform to Newtonian calculations at its *perihelion*, the point at which Mercury was closest to the sun—was due to the warped space-time it was traveling through near the sun. On November 25, he delivered his final lecture, "The Field Equations of Gravitation," which revealed the full theory of general relativity: light would actually bend 1.7 arc seconds, a result *twice* as large as Newton's 0.87 arc seconds. Now, with an even better chance at observing and measuring starlight, given the larger calculations—a single arc second is comparable to the size of a dime at a distance of 1.3 miles—he was ready for astronomers to test again for light bending. That meant waiting for the next total solar eclipse. He wrote to his oldest boy, Hans Albert, then eleven years old. "In the last few days I completed one of the finest papers of my life. When you are older, I will tell you about it."[14]

What may have been more difficult to explain to Hans Albert at the time was that three months later he would entice Mileva to give him a divorce by offering more money than he was already sending her. He also promised to set up funds for the two boys. He was now being pressured not just by Elsa, but also by her parents. He no longer tried to conceal his relationship with his cousin from his wife. Rumors were circulating, he wrote to Mileva. It wouldn't look right to the public since Elsa had two daughters with reputations of their own to consider. He then added a directive on how she might shield their boys from a calcium deficiency. Mileva wavered on the divorce but declined his offer. When he refused to visit her in person, even though Hans Albert begged him, it marked the beginning of Mileva's emotional and physical decline.[15]

THE UNTOUCHED SOUTHERN SKY

In the late summer of 1915, William Campbell had written to encourage Charles Perrine to take advantage of the upcoming eclipse on February 3, 1916. It would be another opportunity to test for light deflection, and Perrine "would probably be the only man of experience on the line of totality." Perrine knew that this observation was correct. But there was no way he could go on this expedition. Not only was Bell expecting a child in January, their third, but he was also facing some disheartening obstacles: pressing finances, the skepticism of many who were now overseeing observatory funding, and the loss of his skilled mechanic, James Mulvey, who was lying beneath a pyramid headstone in the Cemetery of Dissidents. He wrote in his notes, "It was impossible for the observatory to send anything but a small expedition to Venezuela which, along with the unfavorable conditions, prevented including the relativity problem on our program for that eclipse." Among those "unfavorable conditions" was the fact that he was still waiting on the return of his forty-foot telescope.

On February 9, 1916, the *Harvard College Observatory* bulletin 598, from director Pickering, had this to say: "The following cablegram has been received from Professor C. D. Perrine, Director of the Argentine Observatory at Cordoba:—'Argentine Expedition Venezuela announces observation total solar eclipse through thin clouds.'"

But Perrine wasn't ready to give up yet, not with two future total eclipses waiting in the wings. He would unwaveringly petition the government to finance expeditions from Córdoba. The first eclipse, in the summer of 1918, was near his old stomping grounds in California. The second, in 1919, was back in Brazil. Like his friend and mentor William Campbell, Perrine was not a man to quit easily. He had moved to Argentina, a country in which his wife would never feel at home, not just because the salary was enticing. As he would write years later, there was an even greater temptation in that hemisphere, "the untouched southern sky in so many fields."

The Argentinian government would refuse both his petitions. Having chased Vulcan, Charles Dillon Perrine was now out of the race to test for light deflection. The 1914 eclipse expedition to Crimea would be his last.

THE GREAT LOSSES OF THE GREAT WAR

With Europe now caught in the downward spiral of a war so catastrophic it would obliterate a generation of young men, the world of science began a bitter battle within its own ranks. Something else besides financial strain and political bickering would further hamper the relationships both collegiate and friendly among scientists in the various countries now at war. It was more personal and human than any poster created from the inkwells of propaganda or from manifestos quickly written and signed. It was the deaths of their sons, brothers, fathers, colleagues, friends, and employees. No observatory or classroom, no laboratory or office, is so insulated or isolated that it can shut out the carnage of war, especially when there are empty spaces where young chemists, astronomers, physicists, and mathematicians had once pondered the unknown. Unfilled were the university desks where scholars had mused over philosophy, geography, medicine, and history. Many class barriers put firmly in place during the Edwardian period began to dissolve in the wake of this swift mortality. Gone along with those aristocratic sons were millworkers and bank tellers, farmers and shop owners, fishermen and dock workers, tailors and shoemakers. English anger and hatred toward Germany and all things German grew daily.

The older scientists left behind to manage the home front were now met with this loss. By the time conscription was introduced in England in the summer of 1916—Arthur Stanley Eddington would be expected to serve—the roll call of the dead was an endless banner being unfurled with no seeming end in sight. In the beginning, the notion of honor in war was so poignant that young men dashed off willingly to the battlefronts. This patriotism was still evident in the words that J. J. Atkinson had to say about the death of his own son, who was killed in action a few months into the war. At a memorial service held at Atkinson's estate, the heartbroken father offered this sentiment: "My son was killed in battle with a smile on his face, so his brother said. . . . There is not a real man here who will not wish for such an end as his."

Many other tributes to fallen soldiers would mention that farewell "smile," as if it were proof that a young life had not been given in vain. This was not the affable "Atky" who had gone on eclipse expeditions around the world with his famous astronomer friends and kept them entertained with his comical antics. Atkinson and his wife became devoted fund-raisers for the troops. They attended every funeral and memorial service for the war dead in their village. Thus, Brazil in 1912 was amateur astronomer John Jepson Atkinson's last eclipse expedition.

What Sir Oliver Lodge didn't know the evening he gave his brotherhood of science speech in Australia was that his beloved and youngest son, Raymond, would be killed a year later in France, in the autumn of 1915. Or that he would relentlessly seek to contact his son via spiritualists and mediums. These talks with his dead boy would result in a book published soon after his loss: *Raymond or Life and Death*. In sending messages from beyond the veil, Raymond assured the living that the world of the afterlife was much the same as it was on earth—at least in *England*—but without illness and disease. People still lived in houses surrounded by flowers and trees. As for all those soldiers who had died and were still dying on the battlefields of the war? Raymond sent good news. They were greeted on the other side with fine cigars and whiskey. Having long believed in a spiritual afterlife, Sir Oliver and his grieving wife found a measure of peace from these "visits" with Raymond. Since Lodge firmly believed that both the universe and the afterlife were filled with ether—Raymond's otherworldly existence now depended on it—he would perpetually dismiss Einstein's 1905 special theory of relativity, which invalidated its existence.[16]

The loss most talked of and written about among scientists, however, was the death of the brilliant physicist Henry Moseley. He had voyaged to Australia with his mother on the same steamer that Lodge was on, that momentous summer of 1914. Moseley would give lectures at the BAAS meeting and then travel with his friend, British chemist Henry Tizard, to the United States before returning to England and a job offer at Oxford University. When the war broke out, everything changed, including his Oxford plans. The BAAS meeting over, he and Tizard left Australia for San Francisco as they had planned. In what would seem a college road trip before military service, they traveled across the United States by train to New York City and then sailed from Pier 54 on the *Lusitania*.[17]

If Moseley had accepted the offer from Oxford, he would have been allowed to continue his impressive body of work. At the age of twenty-six, he was already known for Moseley's law and the atomic number. Back in England, there was no amount of pleading on behalf of family and friends that could dissuade "Harry," as they called him, from the military. He convinced the Royal Engineers to accept him as a communications officer. On August 10, 1915, as he was telephoning an order during the failed Battle of Gallipoli, in Turkey—this was Churchill's chance to show the world he was a brilliant military strategist—a Turkish sniper fired a bullet into Moseley's brain.

The loss of Moseley was felt deeply by the scientific community, the men and women who could understand the pure genius of his gifts.[18]

THE BROTHERHOOD OF ENEMIES

On the other side of the war were the imprisonments and deaths of soldiers fighting for the Central Powers, losses that were exacting the same kind of pain and anger. Freundlich's former teacher Karl Schwarzschild had died at Potsdam in early 1916 from a painful autoimmune disease he had contracted at the Russian front. The brilliant astronomer and physicist had been instrumental in helping Einstein with his field equations, remarkably developing the so-called Schwarzschild solution while serving in the German army. When news arrived in England that Schwarzschild had passed away, Eddington wrote an obituary for his German colleague.

"The war exacts its heavy toll of human life, and science is not spared. On our side we have not forgotten the loss of the physicist Moseley, at the threshold of a great career; now, from the enemy, comes news of the death of Schwarzschild in the prime of his powers . . . The world loses an astronomer of exceptional genius."

Other notable scientists were feeling these losses. Early in the war, Max Planck's youngest son, Erwin, had been taken prisoner by the French, and his older son, Karl, was killed in action at Verdun, in 1916.[19] Walter Zurhellen, the astronomer who had traveled to Crimea with Erwin Freundlich, had been released a year after the Russians took him prisoner. As a soldier in the German army, he died in the autumn of 1916 at the Battle of the Somme, in France, one of the bloodiest battles in human history. And Albert Einstein's brother-in-law, Milos Marić, who studied medicine in France, had been called by the Austro-Hungarian army to serve as a battalion doctor. When he was taken prisoner by the Russians in 1915, his family had assumed that he was dead or a prisoner of war. His French wife went to live with the Marić family in Novi Sad, Serbia. They wouldn't know for years that he was still alive and that his medical skills were being put to good use in Russian hospitals.[20]

EINSTEIN'S GOOD LUCK: WHAT IF?

On March 20, 1916, the *Annalen der Physik* received Albert Einstein's final submission of "The Foundations of the General Theory of Relativity." It was published in May. So there it was in cement. Light would bend at 1.7 arc seconds. But what if things had gone differently and the astronomers had had better luck on their expeditions? What if the sun had shone down brilliantly over Brazil on October 10, 1912? What if Perrine, with Mulvey's help, had captured magnificent photographs of the sun in the center of the plates, with longer exposures and all stars in the vicinity perfectly visible? In 1912, Einstein was calculating that the deflection of light would be 0.85 arc seconds.

What if clouds hadn't covered the sky over Campbell's rented dacha near Kiev during the eclipse of 1914? What if, after young Kenneth

Campbell shouted, "Go!" the Lick Observatory plates had caught perfect images of the stars and would finally reveal an answer to Isaac Newton's question? *Do not bodies act upon light at a distance, and by their action bend its rays?* Yes, they do.

What if Freundlich had not been abruptly detained in Crimea? What if he and Perrine had risen to bright sun and cloudless skies over the hillside vineyard?

Better yet, what if Frank Dyson had answered Freundlich's letter of February 1913 with a different response? Instead of no, what if the astronomer royal had agreed to have Charles Davidson test for light deflection at Minsk, in 1914? Davidson had clear skies and perfect images. The observation would have been successful.

Because nature never makes errors in its calculations, human beings would have been held accountable. If measured and read correctly, the universe would have gladly given up the right answer: *1.7 arc seconds.* Newton, the genius of classical physics, would have been mistaken. Einstein, the genius of modern physics, would have botched his calculations. But should either man be considered *wrong*? After all, both of them had played a major role in opening up a door to the cosmos. It was as if for two hundred years, physicists had watched a man wearing a hat walk into a room and then saw the hat blown from his head. Over and over again. What was causing that to happen? Newton couldn't tell us; he just knew that it happened. And then Einstein figured out that a *fan* was blowing away the man's hat. That the fan was turned on *high* instead of *low* had been his early mistake.

But Einstein's contribution would have seemed less dramatic had the 1912 or 1914 eclipse expeditions been successful. If Perrine and Campbell had not been undone by inclement weather, Einstein's introduction to the global stage of science would have simply been as an important actor among other actors and not as the superstar of the show. Now he had corrected himself in the last draft of his paper. There may never be another man whose impressive reputation, although deserved, owes its place in history to bad weather, to precipitation and clouds, and to the ironic fact that it's *gravity* that causes water droplets in clouds to fall to earth in the first place. Einstein was as lucky with weather conditions on the days of the eclipses as Perrine and Campbell were unlucky.

Not long after the publication of his paper in 1916, his book, *Foundations of the General Theory of Relativity*, was also printed. In it, Einstein again challenged astronomers to take up the gauntlet: "As seen from the earth, certain fixed stars appear to be in the neighborhood of the sun, and are thus capable of observation during a total eclipse. At such times, these stars ought to appear to be displaced outwards from the sun by 1.7 seconds of arc, as compared with their apparent position in the sky when the sun is situated at another part of the heavens. The examination of the correctness or otherwise of this deduction is a problem of the greatest importance, the early solution of which is to be expected of astronomers."

IN SEARCH OF AN AUDIENCE

Freundlich had studied mathematics with no less than renowned mathematician Felix Klein. But astronomers like Perrine and Campbell didn't need the brilliance of a Klein to correctly test for light deflection. And it wasn't as if an eclipse had to wait for a mathematical mind like Arthur Eddington's. Everything an astronomer really needed to know was contained in one handwritten letter penned on tan stationery, the first letter that Freundlich wrote to Perrine, in the autumn of 1911. "Dear Sir," it began. "Although I am not acquainted with you . . . The known physicist Prof Einstein has derived in a paper he intends to publish in the course of the next month an effect, that any field of gravitation produces upon electro-magnetic phenomena for instance an effect of the gravitation of the sun upon the light of a star passing near to the sun." Reading a few more details in Freundlich's letter, and the astronomers could take it from there.

Perrine, Campbell, Curtis, and others had read the previous versions of Einstein's papers and articles as he moved from 0.83 arc seconds to 0.85. Dyson, Davidson, Newall, and others in England were at least familiar with the theory. Eddington had published an article titled "Gravitation" in the RAS's journal, *Observatory*, in February 1915. In it, he referred to

Einstein and his 1911 paper, which was predicting 0.83 arc seconds. He noted that it would be difficult, even impossible, during a total eclipse to make such a minute measurement, although "the astronomer has a deep interest in the attempts now being made to complete the law of gravitation whereas Newton has left it ambiguous." He was probably remembering his 1912 meeting in Brazil with Perrine, who was there for that purpose. Or he had read Campbell's paper on his attempt near Brovary in 1914. Eddington then laid out the general theory's consequence to science: "A positive result would mean that gravitation has been pulled down from its pedestal, and ceases to stand aloof from the other interrelated forces of nature."

But then war had broken out before Einstein applied his final brush strokes to the canvas. Much of Europe was fighting. Communications had been broken between Germany and scientific types in the Allied countries. Now that the polished paper had finally been published, in Berlin, in March 1916, channels to receive it were mostly blocked. It was as if Einstein had finally obtained his magic carpet, but there were now few places to watch it fly. In Germany, he had far more cynics than disciples. Since the Netherlands was still neutral, he had kept in touch with Dutch colleagues Paul Ehrenfest and Hendrik Lorentz, who shared his papers with Willem de Sitter, director of the Leiden Observatory. De Sitter mailed a copy of the final paper across the North Sea to Eddington, who was still secretary of the Royal Astronomical Society and who had been outspoken in maintaining an academic relationship with German scientists. Now Einstein was predicting 1.7 arc seconds for light deflection. Any deflection would have groundbreaking consequences if proven, but twice Newton's calculation was even more monumental. De Sitter had hoped that Einstein's final paper could be read in England, but he was aware of the wartime restrictions on printing a German's work. He offered to write an article himself for the RAS journal. Eddington eagerly agreed.

Eddington's decision was courageous, given astronomer Herbert Hall Turner's strong criticism of scientific cooperation with the Germans. Turner, who had been Dyson's friend since their early years at the Royal Observatory, had been teaching at Oxford since 1904. He had been known as an internationalist before the war, working mostly with German astronomers. But earlier that May, in his monthly *Observatory* column, "From an Oxford Note-book," he had begun a bitter attack. The journal had been founded in 1877 by the astronomer royal Sir William Christie. It was widely read by many scientific groups, and thus Turner spoke to a large audience of his peers. He denounced the Germans as untrustworthy and reminded his British colleagues that Lodge's "science above politics" toast they had raised their glasses to in Australia, in the summer of 1914, should now be disregarded. Eddington, who was then editor of the *Observatory*, had been one of the few scientists to respond and disagree: "Think, not of a symbolic German, but of your former friend Prof. X, for instance . . . and try to work up a little fury. The attempt breaks down ludicrously."

As Eddington became more familiar with the calculations, he understood the magnitude of what Einstein's brain had conceived and the revolutionary journey it had taken. He saw the theory as a stunning example of the power of mathematical reasoning. For those few who understood it then, and those who understand it now, the general theory of relativity is an elegant work so pure in shape and meaning that adjectives to describe it are often the same used to describe fine works of art. Into a world beleaguered by an unchivalrous war, with its newly introduced atrocities of trench warfare, poisonous gases, air bombings, and large-scale civilian massacres, had been born this graceful new idea.

The next total eclipse was coming in the summer of 1918. This time, the path of totality lay almost in William Campbell's backyard. He would not have to face the limitations of wartime travel that the English would have. With Charles Perrine unable to get funding down at Córdoba, Campbell would likely stand alone in testing for light deflection. Now all he needed was to get his instruments home from Russia.

De Sitter's paper appeared in *Monthly Notices of the Royal Astronomical Society* that October. On November 1, 1916, he wrote to Einstein. "I am sending you today a separate offprint of a little popular exposition of the general theory of relativity which I have published in an English astronomical journal." In his second paper a month later, he reminded the reader that "Newton in his *Principia* gave only formal laws for gravitation and inertia. He made no attempt at explanation: 'Hypotheses non fingo.'"

I feign no hypotheses. De Sitter's report ended with the view that Einstein's theory "represents an enormous progress over the physics of yesterday." And added that "not only has he entirely explained the exceptional and universal nature of gravitation by the principle of the identity of gravitation and inertia, but he has laid bare intimate connections between branches of science which up to now were considered as entirely independent from each other, and has thus made an important step towards the unity of nature." The theory proved to be, de Sitter concluded, "a powerful instrument of discovery."

7

UNRIDDLING THE UNIVERSE

The British Get Ready: The Americans Persevere

In your schooldays most of you who read this book made acquaintance with the noble building of Euclid's geometry, and you remember—perhaps with more respect than love—the magnificent structure, on the lofty staircase of which you were chased about for uncounted hours by conscientious teachers. By reason of your past experience, you would certainly regard every one with disdain who should pronounce even the most out-of-the-way proposition of this science to be untrue. But perhaps this feeling of proud certainty would leave you immediately if some one were to ask you: "What, then, do you mean by the assertion that these propositions are true?"

—Albert Einstein, *Foundations of the General Theory of Relativity*, 1916

IN HER DIARY, Virginia Woolf jotted down a comment she had made one day during a chat with the poet William Butler Yeats. "Unfortunately, one can't unriddle the universe at tea." Another side to that reasoning, amusing and literary though it may be, is that no one had really *wanted* to "unriddle" the universe, not when it came to an orderly cosmos as viewed through the eyes of Sir Isaac Newton. At least, for over two centuries, no one had done it. The one belief that all prominent scientists of the day agreed on was that the universe was static. It was unchanging and eternal. It took many years for the schoolboy in Albert Einstein to reach the top of Euclid's "lofty staircase." He had asked the question, "What, then, do you mean by the assertion that these propositions are true?" With the answer, he had unriddled the universe, at least on paper. And at least for his place in time. But he still needed someone to prove it.

Arthur Eddington's interest in the final paper, which he saw as pure mathematical artistry, would trigger British interest. Eddington was among the small group of astronomers who, like Karl Schwarzschild, had the mathematical skills to understand what Einstein was proposing. That he believed in camaraderie among scientists was even more good fortune. Not all of Eddington's colleagues were so academically gracious. The German-born Einstein might have attained Swiss citizenship, but the postmark on his letters was still Berlin.

If the history of discovery is fickle, it can also be enigmatic, and Karl Schwarzschild is a perfect example. Schwarzschild might have been able to unriddle the theory of general relativity earlier than Einstein, had he tried. The Schwarzschild solution is, in fact, the introduction of the first mathematically rigorous evidence to humanity's consciousness of the concept of the black hole as a phenomenon in physical reality. How had Schwarzschild, under the arduous conditions of serving on a battlefield, and within a year's time of Einstein's 1915 paper, managed to conceive of it? It surprised even Einstein, and over the years it has puzzled many physicists who are unaware that in 1900 Schwarzschild had presented a paper at a meeting in Heidelberg: "Concerning the Possible Measure of Curvature of Space." This paper meant that he had mastered Bernhard Riemann's mathematics more than a decade before Einstein ever began a similar educational effort. With this longer period of familiarity, then, Schwarzschild had probably acquired a deeper understanding of these mathematical ideas and was able to use it to go beyond Einstein.

With the solution of one puzzle, however, a different one arises. In 1908, after first dismissing the mathematical suggestion of his former teacher Hermann Minkowski, Einstein reversed his direction and declared that Minkowski's idea of the "space-time continuum" would be the basis of his further progress. If Schwarzschild had heeded this declaration, and as he already understood the necessary mathematics, he could have beaten Einstein in the race to general relativity by almost a decade. The new puzzle to solve is, why did Schwarzschild not do it? Had he not died in 1916 from the disease he contracted in the wartime trenches, we may have gotten an answer. Given the circumstances, it's likely lost to history.[1]

SKIRMISHES IN THE WORLD OF SCIENCE

Early in 1916, before Eddington received Einstein's paper from Willem de Sitter, he had been faced with a moral dilemma. In the beginning, Great Britain had relied on volunteers to join the military for a war that was escalating by the day. Given the enormous number of causalities, an all-volunteer military was soon no longer possible. Even with Parliament divided on the subject, the Military Service Act was passed and went into effect in March 1916.[2] Religious ministers, widowers with children, and married men were exempt from serving. But single men between the ages of eighteen and forty were liable to be called if needed. By June, married men were no longer exempt, and the service age was raised to fifty-one. Local tribunals were put in place to resolve exemptions being claimed for poor health, domestic hardship, employment of civilian or national importance, and moral or religious reasons. Conscription would pull in 2½ million men while it was in effect and would recognize over 16,000 conscientious objectors, or "conchies" as they were called. Those the tribunals refused to exempt were either put into service anyway or sent to prison. Others who were exempted from fighting took part in noncombatant military jobs.

Eddington, a devout Quaker, was ready to claim exemption on grounds of conscience until Cambridge University—perhaps having a conscientious objector among their ranks during such national patriotism would be an embarrassment—asked the tribunal to let him remain at his post as a matter of national importance. The Ministry of National Service granted Eddington an occupation exemption until April 1918. He was sent a letter asking for his signature. At the bottom, as a postscript, he wrote that if he were not deferred for his occupation and work, he would

claim conscientious objection. This unnecessary admission on Eddington's part put the ministry in a difficult place and caused an angry back-
lash among many academics at Cambridge. But Eddington's faith was his
faith. He believed that if other Quakers were "peeling potatoes" in camps
in northern England, then he should also be doing so.[3]

H. H. Turner, a seismologist interested in earthquakes as well as an
astronomer, was still penning his monthly column for the *Observatory*.
While the column usually related to the life of an astronomer and was
often humorous, topics sometimes went beyond that. One Turner article, for instance, mentioned the atrocities that German soldiers were rumored to have committed. It wasn't the best time for Eddington to resist
fighting on *any* grounds. Many of his colleagues believed that pacifism
meant taking the side of the enemy, or as Turner put it, "shrinking from
the horrors of war." Before the war, Turner had been known for his comic
entertainment at RAS dinners by reciting light verse and doggerel he
had written.

> *Then let us be glad we're astronomers born,*
> *Handling a glass from the dusk to the dawn,*
> *All observations and all computations,*
> *And in-vestigations can wait till the morn.*

But not anymore. The war was taking a social toll at home, as well as
wiping out lives on the battlefronts. But for the time being, Eddington was
free to pursue his interests, which now included general relativity. After
the publication of de Sitter's paper on Einstein's theory that autumn, Eddington knew it was time to act.

As the war blazed on, with casualties mounting on both sides, more
and more scientists from England, Scotland, Ireland, the United States,
and France severed all ties with their German counterparts. Eventually,
the RAS voted to expel any members of German nationality. Despite this
atmosphere, Eddington became the staunchest supporter of general relativity in England. On the heels of de Sitter's paper, he felt competent
enough to present the theory to the members of the BAAS, at a meeting

held in Newcastle upon Tyne in early December. But he needed an ally, and this support would come from his old boss at the Royal Observatory, Sir Frank Watson Dyson, the astronomer royal.

Dyson, an expert on double stars, was not a mathematician and was personally skeptical of the new theory. While he knew little about this theoretical physicist in Germany, he did trust Eddington's talents and instincts, not to mention his enthusiasm. The astronomer royal knew solar eclipses well, having been on four expeditions that had added much to the data of the sun's corona and atmosphere. He quickly became interested in what the general theory proposed, that light would bend *twice* what Newton had predicted when it passed a large gravitational mass. Speculative as it was, the theory was certainly important enough in scope to warrant British attention, even in wartime. Dyson knew that the only way to test it would be the way others had tried in the past: photograph the stars during a total eclipse. It wouldn't be easy. Not only were men scarce during wartime, but money and metal were as well. But if the British were to get in the game, it was time to start planning. Sir Frank Dyson's voice would hold much sway.

AN ASTRONOMER ROYAL IS BORN

Frank Watson Dyson was born on January 8, 1868, in Measham, a village in the English Midlands in the very heart of the country. He was the oldest son in a family of seven children. His father, Watson Dyson, was a Baptist minister descended from a long line of ministers and farmers. To increase his meager salary of a hundred pounds a year and support his growing family, the Reverend Dyson preached a sermon in Measham on Sundays and then a second sermon later in the day in a neighboring village. Unable to afford a horse and carriage, he was obliged to walk the three miles each way. Dyson's mother, Fanny, gave music lessons and opened a small school for local girls in the stone house where the minister's family lodged.

On the night Frank was born, the Reverend Dyson predicted that his first son would grow up to be a senior wrangler at Cambridge and then astronomer royal.[4] It was an extraordinary prophesy for a country minister.

Astronomy was certainly not among the traditional "learned professions" of that period, such as law, religion, education, and medicine. But the reverend, when still a young boy, had met the astronomer-clergyman William Rutter Dawes—craters on the moon and Mars are named for him—whose family lived in the same village. Frank liked to imagine that the astronomer, who also studied double stars, had let his father gaze through the telescopes at his private observatory. One of the best observers of his day, Dawes had died six weeks before Frank's birth.

The family moved several times over the years. Like many other families of any period, the Dysons knew hardship. All clothes, even shirts for the four boys, were handmade and passed down as the children grew. The two daughters helped their mother with the daily household chores, which included making drinks, such as nettle beer and raspberry vinegar, from the berries and plants that grew near the house. At a very young age, Frank was made the family babysitter when his parents were away. Once, left in charge of his baby sister, he accidentally overturned her pram and, in breaking her fall, hurt his hand quite badly. For years, he would use the scar that resulted to determine which was his left hand and which his right.

Christmas trees were only for the wealthy then, but the Dyson children could expect sweets in their stockings, and a shiny half-crown that they were to save for something important. Their parents believed in education, and the reverend spent a third of his salary keeping his seven children in school. Frank was already intelligent and vivacious when he started Bradford Grammar School. The headmaster believed in finding the smartest boys in the village, giving them scholarships funded by wealthy Bradford manufacturers, and then sending them on to Cambridge or Oxford.

SECOND WRANGLER AND A TAVERN OWNER

Frank Dyson excelled in school. At the age of eighteen, he had earned a scholarship to Trinity College, Cambridge. His father's prophesy on that cold January night when Frank was born was on its way to coming true. But in 1889, he ended up as second wrangler instead, in a close competition with a classmate and friend who became senior wrangler that year.[5] Frank had been given a fellowship at Trinity when he met and fell in love with Caroline Bisset Best. Her father was a doctor who had been raised in a saddle shop by a widowed mother. It was a good fit. Carrie, as she was called, was highly intelligent and had studied abroad, learning French and German and earning a degree in modern languages. Frank wrote to his sister, "Undoubtedly, I have found a much better woman than I am a man." He admired Carrie's sincerity and down-to-earth sensibilities. She liked to note that the village she had grown up in was similar to Middlemarch, the one in George Elliot's famous novel of the same title. He preferred this unpretentiousness. Once, when talking to a snob who was boasting about his family lineage, Dyson said that he had traced his own forebears back to a lowly tavern owner and decided to quit there.

On an early summer morning in 1894, the Reverend Dyson married his oldest son to Carrie Best. The new couple moved to Greenwich, where Frank had been appointed chief assistant at the Royal Observatory. The facility sat atop a steep hill on the south bank of the Thames River. But Greenwich didn't share the pastoral tranquility he had known back at Cambridge. The rising factory smoke of industrialism filled the air, the streets noisy and far from clean, given the numerous horses pulling carriages. Yet the history of the area was rich enough to overlook all that in Frank Dyson's eyes. The observatory had been designed by Christopher Wren. Eight astronomers royal had lived there, with names like John Flamsteed, Edmond Halley, and George Airy. Nevil Maskelyne, the fifth astronomer royal, had dressed in a one-piece thermal suit made of wool, silk, and linen to keep him warm at night while he observed the stars. Henry VIII had been born in Greenwich, as had his and Anne Boleyn's daughter, Elizabeth I, the Virgin Queen. This was the kind of British history that could rise above the factory smokestacks.

Dyson didn't know it then, as he searched for a home atop an even steeper hill where he could bring his bride and raise a family, but most of his life would be spent in Greenwich. Besides Sir William Christie, the current astronomer royal, the observatory then employed fifty or so personnel, consisting of astronomers, assistants, computers, clerks, mechanics, and other skilled workers. When his and Carrie's first child was born the following year, Frank named her Stella, a popular girl's name among astronomer fathers. He was observing the Leonids when his first son was born, but Carrie refused to let him name the baby Leo, which had been the name of Frank's beloved childhood dog. In 1905, when he became astronomer royal for Scotland, he packed up his family—he and Carrie had six children by then, one a new baby—and moved them to Edinburgh in the dead of winter.

In 1910, when Christie retired, Dyson returned to Greenwich to become the astronomer royal, the only man to have occupied both posts. He was now living in Flamsteed House, named for the first astronomer royal. Dyson's father, who had died in 1904, did not live to know that although his son may not have fulfilled his prophesy, he had surely satisfied it.

A SON'S EARLY LOSS

Arthur Henry Eddington, father to the astronomer, was born in Somerset County in 1850, to Quaker parents. He was descended from tenant farmers who cultivated the nearby fields that had been grazed by cattle. Growing up with a love for books, Arthur Henry followed a different path than farming. After receiving a solid education, which included study in Heidelberg and Paris, he taught at a Quaker training college. In 1878, he moved to Kendal, in northeast England. That summer, he married Sarah Ann Shout, who had been born and raised in Darlington, a small market town not far from Kendall. Sarah's roots went all the way back to John Camm, an early supporter of Quakerism and a contemporary of George Fox, a founder of the society. When Arthur Henry became headmaster of Stramongate School—the institution was established in Kendal in 1698 by the Society of Friends—the young couple settled down to raise a family. They had two children, first Winifred and then Arthur Stanley, who was born on December 28, 1882.

When Winifred was five years old and Arthur fourteen months, an epidemic of typhoid fever swept through the country. Many people succumbed to the disease, including their father, on February 14, 1884. Later that same year, pathologists would discover that a certain bacterium was the cause of the disease. People afflicted could pollute the surrounding water supply through their stool, which contained a high concentration of the bacteria. The water supply would then contaminate the food supply. And the bacteria could survive for weeks in water or dried sewage. When the *Leeds Mercury* newspaper published a story about the illness and referred to Kendal as a "Fever Stricken Town," the Society of Friends took offense.

In January 1885, its monthly magazine, *The British Friend*, published a letter that referred to the sickness in an attempt to assure readers that the town was a safe place to live. The letter reminded them that in a population of fourteen thousand, only twelve people had died, and the deaths were due to poor sanitary conditions within the home in nearly every outbreak. Of the dozen dead, only one person was mentioned by name, and that was Arthur Henry Eddington, headmaster at the Quaker school. "We regret to say that in the dwelling house belonging to the school serious defects were found, caused, as in so many instances, by carelessness on the part of workmen employed about the drains." Now that the house had undergone "sanitation," it had received a glowing report by the government inspector.[6]

The death of the headmaster at Stramongate had, no doubt, given more credence to the newspaper's attack that Kendal was an unhealthy town. It was a painful death, and people had good reason to be concerned. In 1884, it was nothing but good luck that two-year-old Arthur, his sister, and their mother hadn't also died in that house.[7]

COUNTING THE STARS

While Arthur Stanley Eddington had inherited his father's intelligence, he nonetheless grew up without a father's guidance. Sarah Eddington was now facing the difficulties of raising Arthur and Winifred on meager finances. She moved them to her mother-in-law's house, in Somerset. Arthur was schooled at home for the next three years before attending preparatory school. A curious child, he tried to count the stars when he was four years old. When he was nine, he borrowed a three-inch telescope and discovered a lifelong passion for astronomy. Two months short

of turning sixteen, he was awarded a scholarship by Somerset County at
sixty pounds a year for three years and was accepted at Owens College,
in Manchester. He was an excellent student with an aptitude for mathematics and literature. With many more scholarships and honors behind
him, he was eventually accepted at Trinity College, becoming the first
second-year student in history to be named senior wrangler, an amazing
accomplishment.

In January 1906, he was offered the position of chief assistant at the
Royal Observatory. He would replace Dyson, who had gone to Edinburgh
as astronomer royal for Scotland. "My work will be mainly in the daytime," Eddington wrote to his mother, "but especially during my first year
I shall have to observe at nights a good deal in order to understand the
instruments thoroughly." He did indeed come to know the astronomical
instruments well. At Greenwich, he also met colleagues who would figure
prominently in his future: comet expert Andrew Crommelin and a skilled
computer named Charles Davidson.

More importantly, when Dyson returned from Scotland in 1910 to
become astronomer royal, he would immediately recognize the mathematical brilliance in the young man named Eddington who had replaced him.
These two would spend the next three years working side by side and forg-
ing a solid and mutually respectful friendship.[8] In early 1913, Eddington
was walking the Cornish coast in a drenching rain with his close friend,
C. J. A. Trimble, when good news reached him. He had been promoted
to Plumian Professor of Astronomy at Cambridge, replacing Sir George
Darwin who had died suddenly a few months earlier. A year later, Eddington's first book, *Stellar Movements and the Structure of the Universe*, was
published, cementing his reputation as a leader in the field of astrophysics.

Eddington was an enigma to some colleagues, especially as the years
passed. He loved classical literature. But he was also a fan of *The Wind
in the Willows*, *Winnie the Pooh*, Kipling's *Just So Stories*, and the works
of Lewis Carroll. Like H. H. Turner, he often wrote his own nonsensi-
cal verses. He was a voracious reader. In 1909, on his way to Malta with
the task of redetermining the longitude of a geodetic station, he spent
time walking the deck of the ship while reading *Tristram Shandy* and *Don*

Quixote and at the same time preparing for an important lecture. He was familiar enough with other languages that he read original texts by Dante, Molière, and Goethe. Contrary to his Quaker upbringing, he became a pipe smoker, loved the theater, and appreciated the occasional glass of wine when aboard. Unlike many physicists and mathematicians, he had no interest in music until Trimble introduced him to the opera *Faust* in their post-university years.

Trimble, a mathematician and teacher at Christ's Hospital school, also never married. He and Eddington had met as students at Trinity College, having in common their working-class backgrounds. Trimble graduated fourth in the class; Eddington first. Nature lovers, they cycled and hiked together on many holiday vacations. Heavy rains, sleet, or blinding snow failed to deter them, even when they went coatless. The nature of their true friendship will most likely remain a secret. Being openly homosexual would not have boded well at Cambridge or any other English institution in those days. Less than a decade before the two men met, Oscar Wilde had gone to prison for his love affair with Lord Alfred Douglas. Eddington was a private man, and his privacy remained unreservedly guarded during his lifetime.[9]

As Plumian Professor, he was also made the observatory's director upon the death of Sir Robert Ball. Eddington's mother and sister moved to Cambridge to become his housekeepers and companions. Outside the French windows in his office, he could see the wide green lawns where his sister would keep bees and raise a few chickens. His beloved Aberdeen terrier often sat near his chair, getting its ears scratched as Eddington met with students, a sleeping cat curled on the sofa. Eddington had come a long way from that unsanitary headmaster's house in Kendal.

THE PLANNING BEGINS

Discussions of a British expedition that could take up Einstein's challenge of photographing starlight during an eclipse began in February 1917. First, they needed a total solar eclipse that would have accessibility and a feasible length of totality. The one coming to the United States in the summer of 1918 would not give the British enough time to prepare, even if the war were over and German U-boats weren't dotting the Atlantic Ocean. Adding to the already complicated state of world affairs, the United States would enter the fray in April 1917 by declaring war on Germany. But another eclipse, predicted in May 1919, would provide a brilliant spread of stars near the sun, as well as a viewing time of almost seven minutes.

The usual questions arose. Was the Greenwich instrumentation up to standards? Frank Dyson and Charles Davidson had tested the equipment during the 1905 expedition to Tunisia. Eddington and Dyson reexamined those 1905 plates and concluded that the astrographic telescope could do the job. Yet back in 1913, Dyson had replied to Freundlich's request that the English test for light deflection: "It would be an extremely delicate research to undertake at an eclipse, if not quite beyond present possibilities." And Eddington had agreed. Now, it seems, British attitudes had changed. But then so had Einstein's calculations, from 0.87 arc seconds to 1.7, twice the deflection. Eddington pointed out that if the weather conditions were as good as those in Tunisia in 1905—and they had not been perfect—then twelve stars would be brightly visible in the Hyades star cluster. But where would be the best place for an observation?

The May 1919 eclipse would touch South America first at Peru, cross over Bolivia and Chile to reach northern Brazil at sunrise. From there it would sweep across the vast and empty miles of Atlantic Ocean, skirt the west coast of Africa in southern Liberia, and then arc down over Príncipe, a tiny island 120 miles off the coast of Gambia. Cutting across Africa, it would pass over Lake Tanganyika and end near sunset in the Indian Ocean. Arthur Hinks, an English astronomer and geographer best known for the astronomical unit—the unit is the distance between the earth and the sun, approximately ninety-three million miles—was consulted for weather conditions and possible sites to set up stations.[10]

In a twist that only academia can know well, Hinks had resigned from the Cambridge Observatory in 1914 when young Eddington was offered the directorship after Ball's death. Hinks thought that the position should have gone to him. But he was competing with a senior wrangler with a degree in mathematics from Cambridge, the top rung in the ladder. He resigned rather than be Eddington's chief assistant. "They must have been mad to imagine that a man who had had the ambition to do what I had been able to do would be content with an inferior position and no fun all his life," he wrote to George Hale.[11] Hinks soon abandoned astronomy to become secretary of the Royal Geographical Society, where past and present luminaries who hung in portraits from the walls had names like Livingstone, Burton, Scott, and Shackleton. Now, as a geographer and skilled cartographer, and with an impressive background in astronomy, Hinks was called on to assist in planning the expeditions.

On February 23, 1917, Hinks wrote a letter on behalf of the astronomer royal to the secretary-general of the Geographical Society of Lisbon, inquiring about the tiny island of Príncipe, an outpost in the Portuguese Empire. Having pored over whatever library books he could find that mentioned the island, Hinks now needed information from the source. He wanted to know about clouds, writing that, on tropical islands, they tended to form on a central mountain and that one side of the island is usually the clearer:

> Speaking generally, it is necessarily desirable to get to the highest point. If one could find a place that was high enough to be above low-lying fogs and at the same time offered facilities for living and supplies, it would be very convenient. Supposing for example that some planter were willing to allow the small expedition to camp near one of his factories, to use an existing building as stores, and to obtain the use of his means of transport and the facilities of a plantation. That would be much more desirable than trying to establish a camp out in the wild where so much time would be occupied merely in the investigation and provision of facilities.

Expeditionary astronomers asked, politely, for a lot. And countries around the world were usually accommodating.

Over the next two months, letters went back and forth between London and Lisbon, with meteorological charts being sent, as well as two simple maps of Príncipe. The secretary-general told Hinks, "I have the pleasure to assure you and to the expedition all facilities from the largest society of planters the Sociedade de Agriculture Colonial." Attention should be given to the portion of Príncipe that was "marked down in pink on the map," he added. But no mention was made of those mountains on the island. Hinks wrote a second time, asking the same question: "It happens very frequently in mountainous islands that the central peak condenses moisture which flows away in the direction of the prevailing wind, so that one end of the island will be cloudy and the other clear."

There was obviously a language problem, with Hinks always writing in English, and the secretary-general replying in Portuguese, except for two short letters that arrived from Lisbon in English. "I know personally the islands of S. Thome and Príncipe and I suppose the S. E. part the most convenient for the instalment of the observers," wrote the secretary-general. Determined to get his answer, Hinks remained concerned about mountains. He asked a third time: "When there is a central peak it not infrequently happens that to windward is clear and to leeward is cloudy. Further information at this point would be useful." It would seem likely that in all of London, Hinks could track down a speaker of Portuguese. Or that the secretary-general would know someone who could read and write English. But after a flurry of polite letters, they managed to understand each other. Eddington was fortunate that the man who resigned from Cambridge rather than serve under him cared this much about his expedition. Newall and Perrine could have put Arthur Hinks to good use in Crimea, where they rose each morning to watch clouds gather on the mountaintop at Staryi Krym.[12]

THE BOYS FROM MOUNT HAMILTON

There would be two annulars and four partials until June 1918, when the next total eclipse would occur and thereby provide an opportunity to test for light deflection. With the war in Europe still crippling funding, not to mention the dangers of crossing an ocean on one of the rare civilian steamers that might be found, all European astronomers were out of the running, and so was the unfunded Charles Perrine in Córdoba. William Campbell, however, peering down from the top of Mount Hamilton in California, would have this eclipse all to himself when it came to light deflection. It was an incredible stroke of luck that the path would cross just north of him. If he could only get his instruments back from Russia, there was a spot along the path of totality that was less than seven hundred miles away, in the state of Washington. Maybe Campbell's time had finally come for what he called "the Einstein problem." But now he had another problem, one much closer to his heart, those flesh-and-blood boys who were born and raised on Mount Hamilton.

The war was personal and on everyone's doorstep. Campbell would become a fervent patriot, soon to decry most things German. "The views of the Observatory community as to the unpardonable starting of an inexcusable war by two irresponsible governments," Campbell wrote in a report, "and as to the methods, purposes and consequences of Prussian militarism may safely be inferred from the following statements: Every male graduate of the little grammar school on Mount Hamilton who is old enough for war service volunteered for war work in that capacity which seemed best in his sight. Not one waited to be drafted." The list of volunteers included his three sons as well as the son of the instrument maker, the son of the custodian, sons of astronomers, and astronomers themselves, who, like Curtis, took part as instructors of various departments to train military recruits.

Several of the youngest Mount Hamilton boys had spent the summer of 1917 on farms in Connecticut and California, assisting in the planting and harvest that would add to the food supply, before taking an active part in the war. At Mount Hamilton, all plans for construction—this included a new building to house a central heating plant and shops for instrument-making, carpentry, plumbing, and other needs—were shut down so that the materials and labor could be used toward "the winning of the war." The women on Mount Hamilton did their share, as women were doing in homes all across the country. They saved food, joined the Red Cross, knitted, and sewed. The universe, filled with its mysterious stars and comets, had to take a back seat.[13]

Campbell accused President Wilson of "lacking a backbone" for not jumping into the fight sooner. As for William and Elizabeth Campbell's own three sons, he was proud that they had met the call. Their youngest son, Kenneth, who had hoped to see a German U-boat on the frantic sail across the North Sea in 1914, became an ambulance driver in Italy. He would be awarded for his actions during one battle. The oldest son, Wallace, having graduated from Harvard a year earlier, became a second lieutenant in the Corps of Engineers and went to the fighting lines in France as a noncombatant. Their middle son, Douglas, and his good friend Quentin Roosevelt, the youngest child of former president Teddy, had a year still to go at Harvard when they immediately dropped out to enter aviation school at Massachusetts Institute of Technology. As Campbell was writing to Russia, pleading for the return of his instruments in time for the next eclipse, Douglas was the first all-American trained aviator, now an ace flying over France with the likes of Eddie Rickenbacker. The boy who had once killed rattlesnakes on Mount Hamilton was now shooting German airplanes out of the skies over France. Meanwhile, his father was planning an eclipse expedition.

SOUTH AMERICA AND AFRICA

On March 2, 1917, Dyson submitted a two-page article to the *Monthly Notices of the Royal Astronomical Society*. He referred to the rich star field that would be observable during the future eclipse of May 29, 1919: "This should serve for an ample verification, or the contrary, of Einstein's theory." He noted the need for "observing at as many stations as possible." Astronomers had long realized this logic in playing it safe. "Unfortunately, the track of the eclipse is across the Atlantic and near the Equator," Dyson wrote. "Mr. Hinks has kindly undertaken to obtain for the Society information of the stations which may be occupied." When Dyson received a circular from Henrique Morize that suggested Sobral, in northern Brazil, as a possible viewing station, that town was also considered. Morize was the astronomer-director from the observatory in Rio who had accompanied and assisted Eddington and Davidson for the 1912 eclipse. His letter included meteorological and other pertinent information that the Joint Permanent Eclipse Committee could study.

When the committee met that November 1917, it agreed to send teams to both Brazil and Príncipe. Planning one single expedition during wartime was ambitious enough. Funds would need to be raised, a daunting thought, given that rationing would go into effect in two months and that young men were continuing to die by the thousands. The British either banked on sheer optimism that the war would end before the expeditions set off, or they believed in divine miracles. Application for funding was made to the Government Grant Committee. Would it see fit to allot one hundred pounds for the needed instruments and modifications, as well as a thousand pounds to finance the two expeditions' travel expenses?[14]

As chairman of the Joint Permanent Eclipse Committee, Dyson appointed a subcommittee consisting of himself, Eddington, Alfred Fowler— it was Fowler's 1914 expedition to Kiev that had been scuttled when war broke out—and Edwin Turner Cottingham, an expert clock and instrument maker. The subcommittee met in May and June 1918. The members decided that Eddington and Cottingham would go to a cocoa plantation on the tiny island of Príncipe. Davidson would go to northern Brazil with Father Aloysius Cortie, the Jesuit astronomer who had been barred from Russia in 1914. Davidson and Cortie would set up their viewing station at a jockey club's racetrack. From those two locations on the globe, over three thousand miles apart, the expeditions would photograph the positions of stars in the Hyades as their light waves passed through the sun's gravitational field. They would do this *if* the weather was agreeable.

WAITING ON RUSSIA

Praying that his equipment confiscated in Russia would arrive home to California in time for the June eclipse, Campbell had delayed purchasing replacements. Mount Hamilton was under the financial and physical pressures of wartime as it was. With Americans now also dying, it wasn't the best time for any observatory to buy new equipment. Therefore, he had not yet made the prerequisite comparison plates. But he could take these photographs even after the eclipse, in early winter, when the same stars appeared in the sky away from the sun. Beginning in early 1917, he had written to Russia, asking about the status of shipping his instruments. He was having no luck. Oskar Backlund, his trusted contact in Russia, had died a year earlier. The whole world was a mess. Finally,

Campbell learned that the telescopes and other paraphernalia had left the Pulkovo National Observatory, near Petrograd, on August 15, 1917, almost three years after being held hostage. Perhaps now he could breathe more easily. The expensive instruments so vital to the 1918 eclipse observance were on their way.[15]

Since the shipment was headed to California, it would first need to cross Russia, a distance of some six thousand miles, to Vladivostok, on the Sea of Japan. It's amazing that it left Petrograd at all. Russia was in greater turmoil. Strikes, demonstrations, and uprisings as part of the Bolshevik Revolution were frequent and often violent. Rasputin had been killed and Czar Nicholas II was warned that he would not have the support of his army in a revolution. He and his family, who had so enjoyed their lavish palace on the Russian Riviera, the one Perrine had admired, would have several months left to live before being assassinated. It was in the midst of this chaos that Campbell's astronomical cargo reached Vladivostok in mid-December, where a business boycott then prevented it from leaving Russia. With the Lick's equipment was aviator-scientist David Todd's instruments, which, thankfully, didn't include an airplane, since Todd had planned to rent one in Kiev.

The next word Campbell received was on March 22, 1918. The boycott had been lifted, and his telescopes and other devices would leave within a week, bound for Japan on a steamer belonging to the Russian government. One can only imagine his state of mind. If any small thing went wrong from here on out, the shipment would never arrive in time. He began rounding up instruments that were still at the Lick Observatory. Other apparatuses—a break-circuit chronometer, a chronograph, a sextant, a theodolite, a telescope driving clock, spectrograph slits, and prisms—he borrowed from the observatories of local universities. Unfortunately, the lenses he was able to borrow could not photograph a large field of stars in sharp focus, unlike his own equipment.

On May 15, Campbell learned that the cargo so precious to any eclipse scientist would leave Kobe, Japan, "in a few days." If all went well, it would cross the Pacific Ocean to San Francisco. With the eclipse coming on June 8, he packed up his assembled equipment, mostly substandard for such a precise task at hand. He had studied weather reports and found a viewing spot where he would set up camp. Since the eclipse's path of totality would clip the lower part of Washington State, his team would travel to Goldendale, near the Columbia River Gorge. He had already scouted it out. Even without his instruments, Campbell was a total eclipse ahead of the British.

RATIONING ARRIVES AND CONSCRIPTION RETURNS

Feeding a nation of citizens in wartime was an immense undertaking for most countries in Europe. Many people were hungry, and many were starving to death. The laborers and horses to work the farms had mostly disappeared. Yet, men on the lines of battle had to be fed. So did war horses. With German submarines patrolling the waters around England, shipments of grain could not make it into ports. England began to feel the crunch that other countries in Europe had been feeling for months while the war dragged on. As a way to avoid a crisis, food rationing had gone into effect for London only, in early 1918, and then for all of the country by summertime. Topping the list was sugar, followed by "butcher's meat." Also scarce were military personnel to continue the fight. Desperate for men at the beginning of the year, when it seemed the war might be reaching its culmination, the Appeal Tribunal for Cambridge came back around in June to revisit Eddington. He had been given three extra months exemption, which would end August 1. Now the war effort needed him.

By this time, Eddington had published his second book, *Report on the Relativity Theory of Gravitation*. This accomplishment didn't seem to impress the tribunal chairman, who noted that Arthur's abilities could be better used in the active prosecution of the war, although the official didn't see him engaged as an ordinary soldier. To this comment, the astronomer boldly replied, "I am a conscientious objector." This status didn't count, since his *occupation*, and not his *conscience*, had been the grounds for his exemption in the first place. After months of planning the expeditions, would he now be called into some kind of service for the war effort? The first day of August was still a month and a half away. He and Dyson would have time to call for another hearing, this time on grounds

of Eddington's religious beliefs. Dyson himself served on an Appeal Tribunal for Greenwich. He knew the ropes. Both he and Eddington would prepare a carefully written statement. Dyson was highly patriotic, but having
his brilliant astronomer-mathematician peel potatoes rather than test for
light deflection on a dot off the coast of Africa was not in his plans. He
had a deal to offer the tribunal on Eddington's behalf.

Two weeks later, Eddington was again before the tribunal. Dyson's
written statement expressing his support had been delivered. Eddington
made it clear he was willing to take part in the Friends Ambulance Unit
or the Red Cross or even work as a harvest laborer.[16] But not in combat.
"My objection to war is based on religious grounds. I cannot believe that
God is calling me to go out to slaughter men, many of whom are animated
by the same motives of patriotism and supposed religious duty that have
sent my countrymen into the field." Eddington spoke his heart that day.
"To assert that it is our religious duty to cast off the moral progress of
centuries and take part in the passions and barbarity of war is to contra-
dict my whole conception of what the Christian religion means." And he
spoke his thoughts boldly, given the patriotic temperaments of such colleagues as Sir Oliver Lodge and H. H. Turner.

Dyson's statement focused on the science, on the brilliant strides Eddington had made, which included having penned a first book on astronomy that helped explain the stellar universe to readers in all countries.
He tossed a jab at the Germans by writing that Eddington's exceptional
work contradicted the current notion that the most important scientific researches of the day were being carried out in Germany. Perhaps
the three men on the tribunal were silently reminded of the great loss of
Henry Moseley a year earlier. Then Dyson's statement got down to business: "The Joint Permanent Eclipse Committee, of which I am Chairman,
has received a grant of £1000 for the observation of a total eclipse of the
sun in May of next year, on account of its exceptional importance. Under
present conditions the eclipse will be viewed by very few people. Professor
Eddington is peculiarly qualified to make these observations and I hope
the Tribunal will give him permission to undertake this task."

Eddington then answered questions about the theory from the committee. Light deflection was one test of Einstein's theory. The stars in the Hyades during this eclipse would be brilliant and thus offer a good opportunity for success. He would make his observation in the path of totality. But if they felt he was better suited in a noncombative job in the military, he would gladly do that, too. The astronomer royal's petition for his cause would not be taken lightly by the tribunal. In that capacity, Dyson had close connections to the Admiralty. The tribunal voted to grant Eddington another year's exemption, provided that he maintain his astronomical work. But in particular, he must head an expedition to photograph the stars during the 1919 solar eclipse.

The astronomers went back to work. As the losses of World War I were still mounting, preparations for the British expeditions went full speed ahead.

IN THE MOON'S SHADOW

A solar eclipse expedition was not for slackers. Every member worked hard, regardless of the heat, storms, snakes, ants, mosquitoes, or illness. Elizabeth Campbell once wrote in her diary of her husband's stamina, commenting on how the local volunteers were amazed at his strength as they watched him rope off the instruments and set up the forty-foot telescope. It was 1898, and they had just traveled over fifteen thousand miles from San Francisco to Bombay, India. They would spend Christmas there before returning home four months later to their boys, ages two and three. The last leg of the outbound journey was a seventeen-day sail from Hong Kong on a ship "headed for the scrap pile," with the team members seasick from watching their luggage slide about the cabins. Arriving in India, they were turned away from their chosen viewing site because of an outbreak of bubonic plague. Still recovering from the "rotten chow" that had been served aboard the last ship and had made him ill, Campbell went to work at the new site. "He is working from before dawn until after the sun has left the sky," Elizabeth wrote. "Stones that four men cannot move he lifts with ease. And he is never tired!"

This was the Ohio farm boy who had worked young in the fields to help his mother support a fatherless family. Earlier in 1918, he had quarantined Mount Hamilton to keep his mountaintop community safe from the Spanish flu. Now Campbell was in the state of Washington for the 1918 eclipse. As usual, Elizabeth went with him. He had been only thirty-six years old for the Indian expedition, his and Elizabeth's first. He was now two decades older and saddled with the business of directing an observatory. He had been chasing light deflection and the general theory since Erwin Freundlich had contacted him in early 1912. He needed this eclipse to be brilliant.

Its path was certainly ambitious after it reached the United States. The tail began in the East China Sea, south of Japan, where American whaling ships had hunted right whales for blubber in the last half of the nineteenth century. From there the path curved up and crossed the North Pacific waters to hit the United States, perfectly positioned between Tacoma, Washington, and Portland, Oregon. It then ran diagonally southeast across Idaho, Wyoming, Colorado, Kansas, Oklahoma, Arkansas, Mississippi, Alabama, and Florida. It would die out in the British Bahamas, about a hundred miles north of where Christopher Columbus is believed to have made landfall.

Many other observatories in North America had chosen a station along this track, but none would test for light deflection. The trail was lengthy enough that Campbell could take his pick of good spots. He had chosen Goldendale, Washington, a town of twelve hundred souls just north of the Oregon border. It was snuggled between the Cascades to its west and the more distance Rocky Mountains to the east. Campbell had studied weather reports supplied to him by the US Weather Bureau. The Cascades would prevent any coastal fogs from drifting in. But June was not a good month in the Northwest when it came to forest fires. Even if distant, their smoke could drift for many miles. Still, he believed that the Washington location was a good choice.

He worried, however, that the same problem they had in Russia would appear again, no matter where he set up camp. The eclipse was scheduled to arrive at 4:00 p.m., Pacific Time. Were the reports he'd been given for observations taken in the mornings and evenings or at the actual arrival time of the eclipse? If not at eclipse time, it was anyone's guess what the weather would be during totality. He had considered viewing places in Wyoming and Idaho, both of which would probably be even better atmospherically. But another important factor had held him back, one that was built on sheer optimism. He wanted to stay closer to the Pacific coast because of his en route instruments. What if they arrived at some port close enough that he could still get them to Goldendale on time? Goldendale had railway service, unlike many other towns along the path.

"WHERE THE RAIN AND SUNSHINE MEET"

The team members arrived three weeks before the eclipse to begin preparations. The first person Campbell had met in Goldendale, back when he was scouting out the town, was the wife of a wealthy mill owner. Mr. and Mrs. E. R. Morgan and family had moved out of their well-furnished home, a mile to the west of town, so that the Lick expedition could rent it. The house had a cellar with double doors, thick walls, and plenty of shelves to use. It would work well as a darkroom in which to develop the plates. The best gift was that the residence was wired for electric lights from the plant in Goldendale and also had running water from the village. "Here were all the elements for an eclipse experience *de luxe*," Campbell would write, stoically, not mentioning that his best instruments were likely to be somewhere in the North Pacific, if they had left Japan at all.

The large group of assistants and volunteers, many with impeccable credentials, set up the instruments on the sprawling lawn. Curtis from the Lick Observatory, having had an early interest in Einstein's theory, was indispensable on this expedition. Perrine had sent a young assistant, Estelle Glancy, from the Argentine National Observatory. An American and future visionary in the field of optics, she arrived on June 4. At least Perrine was there in spirit. Three days before the eclipse, the team began the drills. The excitement was building to the countdown. The residents of Goldendale treated the astronomers as the celebrities they were, offering every gesture to help, even if it meant staying out of their way. "Contrary to the opinion of the public in general that astronomers seek mountain tops and high points to make eclipse observations," said the *Goldendale Sentinel*, "the Lick Observatory is located on a level piece of ground sheltered from the wind by bluffs on the Little Klickitat River."[17]

NEWS FROM THE WAR

On the eve of the eclipse, that is, on the day before all eyes would be on the heavens, a car's headlights came up the road to the rented Morgan house. Perhaps someone was coming to tell Campbell that his equipment had just pulled into the Goldendale station on a miracle train from San Francisco. No matter, since there would be no time now to set it up. No doubt, Campbell worried that the driver had come to deliver a different kind of message. He had. Captain Douglas Campbell, the middle son, the flying ace who had shot down his fifth enemy plane shortly after his parents arrived in Goldendale, had just scored his sixth hit. The aerial battle had taken place two days earlier, on June 5. This time, however, in a dogfight three miles high in the skies over France, he had been shot through the back by a machine-gun bullet. He had nevertheless managed to land the plane, but was their son still alive? Had he died from his wounds after the cable was sent? The Campbells didn't know.[18]

It would be a long and restless night. As an astronomer, Campbell believed in good habits that would help a body function during the stress of an eclipse. In the tropics, he advised his team to keep regular hours, to boil all water themselves and not trust the locals to do it, to keep the hot sun off one's neck by wearing a hat, and to buy and wear white linen suits to stay cool. One suggestion he gave would have worked well in Washington that night before the eclipse, when his heart was heavy. "Lastly," he had told his team, "every day, keep your circulation good by drinking some good whiskey—whiskey and soda—with your evening meals, if the nights are cold, take some whiskey when you turn in. It will help you to sleep as the night comes on. Don't neglect it on any account."

On the evening that Campbell received the news about his son, clouds began rolling in early. In other parts of the country and world, a supernova would be discovered that same night: Nova Aquilae. At Harvard, Edward Pickering was getting nonstop phone calls and cables announcing the discovery. By midnight, the clouds over Goldendale were hiding all the stars in the sky.

THE ECLIPSE OF JUNE 8, 1918

As morning broke over the wheat fields, Goldendale began to buzz with activity. Keeping with the usual hoopla that preceded eclipses, the streets were teeming with excited folk carrying pieces of smoked glass. The *Goldendale Sentinel* announced that the largest crowd in the history of the town would be soon arriving and that the village was a "mecca for automobile tourists" driving in from around Washington and Oregon. Cars were now hurrying to eclipses wherever roads made the drive possible. There would be standing room only on the summit of the Twin Buttes outside town. The paper had noted earlier in the week that Campbell was "a courteous man who would gladly give the public any information that would enlighten them on the scientific aims of the expedition." A jeweler had ordered dozens of protective eyeglasses and taken out a newspaper ad informing potential buyers. As a courtesy, he would send boys out onto the streets to sell them for a dime a pair. Hoping to boost sales, no doubt, he reminded the reader that "at every eclipse many persons are blinded or partially blinded."

Any worries that William and Elizabeth Campbell had about their son, far away in France, had to be put aside for a few hours. Campbell and his Lick team got ready, including the sketch artist he had enlisted from Stanford. This was a botany professor named, amazingly, Douglas Campbell. The clouds that had sifted over Goldendale at midnight had been replaced with new ones by the next morning. They, too, eventually dispersed, but more clouds floated in by afternoon. The eclipse was due at 4:00 p.m., and totality would last only 1 minute and 57 seconds. As the anxious team members waited, ready to perform their rehearsed duties, the chance of success appeared bleak. But at 3:59, less than a minute away from the wonder of wonders, a rift opened in the clouds and the sun could be seen. As the seconds ticked down and totality began, William Campbell shouted, "Go!" A volunteer began counting, "One, two, three." The tiny patch of sky where the sun had peeked through stayed clear until the photographs were taken. A few seconds after totality finished, that area of the sky was again obscured by clouds. The total eclipse of 1918 was the darkest Campbell had ever experienced. On June 10, he told the *New York Times* that "the chickens retired as if for the night. They were heard to give the morning cock crows before emerging a few minutes later. It was probably the shortest night in all their lives."[19]

Even after returning to Mount Hamilton, the team members couldn't know what the photographs would reveal regarding Einstein's theory. They would have no comparison plates until the next January, when nighttime photographs of stars could be taken in the same region of sky as during the eclipse. Curtis could do only a cursory examination of the plates back at Lick before he had to return to his war service job at Berkeley. Campbell would write that Curtis "reports that stars fainter than the eighth magnitude are recorded, but that the clouds to the east and west of the clear sky immediately over and surrounding the Sun interfered somewhat with the recording of the fainter stars in the extreme east and west areas. No further statement as to these plates can be made at the present time, but it is hoped that the definitive results may appear in print without undue delay."

Douglas Campbell had performed heroically on June 5, continuing to fight while wounded and bringing the enemy plane down to the ground, where it was destroyed by artillery. He received the Distinguished Service Cross and the Oak Leaf Cluster for his bravery. He came home to the United States to recover and would not get the chance to return to combat as he had hoped. Unfortunately, his good friend Quentin Roosevelt would die a month after the eclipse, also in a dogfight over France, on Bastille Day. Brilliant, popular with his fellow pilots, and known to be daring in the skies, the young Roosevelt had been the well-known little boy who grew up in the White House and had entertained the public with reports of his antics. Once, when a reporter tried to enlist the child to gossip about his father, Quentin had famously said, "I see him occasionally, but I know nothing of his family life."

On June 20, Campbell received a wire announcing that the Lick instruments were leaving Japan for a Canadian port. Campbell wrote,

"Whether, upon arrival there, the instruments will be nearer home than they were in Kobe, is not a subject for expression of definite opinion." The Lick instruments, impounded in Russia, would not reach Mount Hamilton until August 21, 1918, exactly four years to the day of the 1914 eclipse. There was minor damage done to the chronometer, and the packing crates were "on the way to the state of kindling, but fortunately their strong bindings of strap iron and wire held them together."

THE WAR IS OVER: THE BRITISH START PACKING

Negotiations had been under way between the Allies and the Central Powers, and hope was now on the horizon that the war would soon end. Even though most of the Cambridge Observatory staff had gone off to fight—of the 13,878 members of the university who had served in the war, 2,470 were killed—Eddington requested a leave from his post. In October 1918, he sent a letter to the university's general board, asking to be let go during the Lent and Easter terms, from January to June. If he were granted this leave, he could lead a solar eclipse expedition to Africa and test Einstein's general theory of relativity. Permission was granted.

Father Aloysius Cortie, however, found it impossible to leave his work at the Stonyhurst Magnetic Observatory, in Lancashire, which was managed and operated by Jesuit priests who also ran Stonyhurst College. The director of the observatory had fallen ill, and it was apparent that his end was near. Since Cortie would replace him, this was not a good time to leave England on a months-long expedition. Known for his benevolent mannerisms and even his fine singing at RAS dinners, the priest would have made a most suitable traveling companion for Charles Davidson. But Cortie's replacement was quickly decided on. It would be the eclipse-experienced Andrew Crommelin, recognized for his knowledge of comets and another of Dyson's assistants.[20]

During a November 8, 1918, meeting of the Joint Committee, the members decided to collect all the instruments at the Royal Observatory. The sixteen-inch lens from the astrographic telescope at Oxford, loaned to them by Turner, would go to Príncipe. The sixteen-inch lens from the Greenwich telescope would go to Brazil. Huts covered in waterproof canvas and later assembled at the observation camps had to be constructed in England. They were necessary to enclose the valuable telescopes and clocks in Sobral and Príncipe. But the observatory's carpenter had been conscripted to serve in the military. Only an engineer was left behind, and he had no carpentry skills. A request went out to a civil engineer at the nearby Royal Naval College, and *he* undertook the task of building the huts, a woodworking joiner helping with the frames.

Eddington had insisted that each team take along a micrometer, an instrument used to obtain precise measurements. This way, measures of check plates and eclipse plates could be conducted at each camp, rather than much later, after the long journeys back to England. Anything could happen along the way to alter or destroy the plates. It wasn't that the results needed to be rushed for professional purposes, but Campbell and the Lick Observatory had still not released their findings for the June 1918 eclipse.

One would think travel at this time would concern these men, given that a pandemic influenza had begun sweeping the world early in 1918. Nicknamed the Spanish flu, it was killing tens of millions of people even in remote areas of the globe. Perhaps privately, the team members did discuss the possible danger. In letters, they appear to focus on the problems of travel during wartime. On November 11, 1918, three days after the meeting, the long and costly war finally ended. At least, the fighting did. Still under way were negotiations on the release of prisoners, reparations, and all the other untidiness that follows war until the final armistice would be signed and go into effect. But the Allied Powers could claim a victory. Although people would suffer the horrid aftermath for many years afterward, science was given some relief. Travel restrictions were lifted. While the price of war had strained financial resources, the two expeditions could now concentrate fully on their missions.

On Armistice Day, Eddington wrote to Cesar Augusto de Campos Rodrigues, the director of the Lisbon Astronomical Observatory. He referred to Arthur Hinks's letters the year before and asked for assistance in finding a home for him and Cottingham, transportation to the island, and general support for their mission. "We find that all sailings of boats to Lisbon have been cancelled for the present," Eddington wrote. "I suppose owing to the revolution. I trust that you and the observatory are unharmed." Times were dangerous in Portugal. Less than a month after Eddington wrote this letter, Sidónio Pais, the president of the Fourth Portuguese Republic, was assassinated in the Lisbon train station. The country, which had become a Republican state in the 1910 revolution, had been undergoing a political, financial, and economic crisis with the rise of Pais's dictatorial government.

Over the next weeks, letters went back and forth, with the assistant director, Frederico Oom, also replying. He mentioned a plantation owner on Príncipe willing to support the British for their "every need." It was coming together nicely. Eddington and his British colleagues would sail away from England that coming March. Academia may have discovered Albert Einstein. But it would take the moon passing between the earth and the sun, and a handful of determined astronomers, before the world would know his name. Which astronomers would remain to be seen.

8

THE RMS *ANSELM*
SETS SAIL

A Cocoa Plantation and a Horse Jockey Club

There is one other passenger whom I knew through correspondence, Mr. Walkey, an amateur astronomer. He is going out for the Bible Society to live on a house-boat on the Amazon travelling up and down the various tributaries. He expects to be out there most of his life.

—*Arthur Eddington, aboard the* Anselm, *letter home, March 11, 1919*

FOR MANY SCIENTISTS, the art of discovery is a lifelong pursuit. Albert Einstein had compared the state of mind it requires to that of "the religious worshipper or the lover." That strenuous day-to-day commitment, he believed, came "straight from the heart." He would know. At the age of thirty-seven, he had created a work of harmonious art when he conceived of the general theory. People who are not mathematical may wonder why scientists use phrases like *work of art* or *masterpiece* when they are referring to a certain theoretical discovery. A theory has only calculations to gaze at, after all, and not framed images like *The Last Supper*, or *Starry Night*. An artistic masterpiece, a work created by a human being in his or her lifetime, eventually surpasses the artist who envisioned it. All masterpieces have balance and harmony. And what about harmony? The ancients believed in the mathematical principles of sound. In Plato's day, harmony was an essential component in understanding how the universe behaves.

Being a scientist enabled by mathematics is somewhat like being a composer of music. A musician writes a musical score that gives instruction to the musicians who will play it. When conductors see a certain note for an instrument, for example, they can decipher and transpose the music instantly. Eddington was a mathematician who read the theory of general relativity that Einstein had "composed," and the astronomer saw the artistic and universal harmony in it. Imagine a planet without sound, but with musical scores understood only by those residents on the planet who can read music. That was similar to what Einstein was dealing with. A few mathematicians around him could understand his theory. Now, he was waiting for astronomers to turn on the sound.

A GREENWICH FAREWELL: MARCH 1919

The astronomical instruments and the dismantled huts for both expeditions had been packed by Davidson and shipped to Liverpool. The men would follow a week later, carrying with them the glass lenses and the photographic plates, which were packed in hermetically sealed tin boxes. There was no doubt in Eddington's mind that Einstein's prediction of light deflection was correct. The expeditions were not really necessary, not when it came to convincing *him* of the theory's validity. And with the war ending, conscription was no longer a problem. But Eddington still hoped that this alliance with a brilliant German physicist, on the heels of such a significant discovery by British astronomers, would begin to restore relations between scientists of combatant countries. The unbiased universe, however, was waiting to divulge its answer, regardless of nation, war, or weather. Maybe Campbell would be first in line. Where were the Lick results from the 1918 eclipse?

The war was still going on when the 1918 eclipse plates were made. Curtis had accepted a war-related job at Berkeley. He had taken his brief look at the plates and noted that faint stars had indeed been recorded and that there had been some interference by clouds. And Campbell had released this information to the scientific world, hoping that "the definitive results may appear in print without undue delay." That was it. Curtis had then left his work at Berkeley and gone to Washington, DC, a couple months after the eclipse for a job at the Bureau of Standards, also war-related. Intensely patriotic, Campbell agreed to this move, assuming that Curtis would return by January 1919, when they could take the comparison plates and determine the results. When the war ended that November, Campbell felt it was time to get back to the job of astronomy. But Curtis had become involved in more activities that held him in Washington, and the eclipse results went into limbo. "I almost wish you had ordered me back," Curtis wrote to his boss. That Campbell *didn't* may have been a stroke of luck for the British.

On March 5, Eddington and Cottingham took their personal baggage and left Cambridge by train, with Greenwich Station as their destination. They would spend two nights with Dyson at Flamsteed House before leaving for Liverpool. Being employed by the observatory, Davidson and Crommelin were already in Greenwich. It was a good time for the men to be headed to warmer places. Great Britain was experiencing an extreme cold spell. It was the most frigid March in many decades, possibly a hundred years. Snow still lay on the ground and often blew into thick drifts. On the first day of that month, the zodiacal light, or "false dawn," was reported glowing in the skies over Oxford. The ancients would have taken this sighting as a sign from the gods.

Having Eddington down from Cambridge perhaps reminded Dyson of his university years there. He liked to tell his children of those cold winters with ice on the River Cam and how the snow was deep in Trinity Great Court but would be replaced by colorful geraniums and green turf come spring. All of Dyson's eight children were still living at home, ranging from twenty-three to five years old, with a lot of teenagers in the middle. Eddington knew this family well from his years at the observatory as chief assistant, when Dyson would bring him home at the midday meal and say, "Carrie, can you find some dinner for Eddington?"

With the winds of March whistling down the many chimneys of the historic house, Eddington and Cottingham joined Dyson in his study on the night before their leaving. Dinner was over, and a fire blazed in the hearth. They sat in comfortable chairs to enjoy a pipe smoke and discuss the upcoming expeditions. Cottingham, a brilliant clockmaker by profession, understood and designed astronomical mechanisms. But he was new to Einstein's theory. Dyson explained to him that the German physicist predicted that light would bend at 1.7 arc seconds. Cottingham considered this idea and then asked, "But suppose we find *double* this amount?" Puffing on his pipe, Dyson smiled. "If you do," he said, "Eddington will go off his head and commit suicide, and you will have to come home alone."[1] That night, as the men sat in Dyson's study, warm from the fire and safe from the outside cold, they had no way of knowing that this part of their conversation would still be told a hundred years later.

TO WARMER CLIMES

The next day, a Friday, Dyson came down from the hilltop observatory to see the expeditions leave Greenwich for Euston Station, where they would catch a train to Liverpool. They had tickets to sail on the RMS *Anselm* the next afternoon. It was an exciting moment as the four men boarded the train, carrying with them the hampers packed with the lenses and plates. The astronomer royal bade them farewell.

Dyson had been the driving force behind the expeditions once Eddington fully embraced Einstein's theory. Being left behind at Greenwich Station, he must have felt what his wife and older children knew: he missed the early days of his career when he would study the heavens all night long. At dawn, he would leave the observatory and come wearily home for a breakfast in bed before falling asleep. Those were the years when he was the chief assistant. Back then, he could sail the world with the jovial Atkinson and his other astronomer colleagues. He could circumnavigate the globe, as he had in 1901. But this eclipse path was too far away. He could not be absent from the observatory for months, especially at a time when the luckier members of his staff were returning from the war. Morale would be low, and he had administrative duties as well. Still, on those rare occasions when an assistant became ill and a telescope was vacant, Dyson was always eager to fill in.

Eddington, Cottingham, Crommelin, and Davidson arrived at Euston Station an hour and a half before their train left. Managing their personal luggage took some time. Because the hampers carrying the lenses had been labeled GLASS on the outsides, each case cost thirty extra shillings to ship. By the time the train reached Liverpool, at a quarter to four in the afternoon, a freezing rain was falling. There appeared to be no porters available at the train station to help with the bulky baggage, and they were told the local hotels were mostly filled. It seems they had taken considerable measures to reserve places to stay in Africa and South America, but had neglected to book rooms for themselves in England. A porter who represented a baggage agency finally turned up. He agreed to have the crates and luggage delivered to the dock the next morning. Taking just their hand baggage, the men set out in a taxicab to find a hotel for the night. After driving around in the sleeting rain to several places, they managed to find a comfortable establishment with vacancies. They checked in and stayed in their rooms for the evening, hoping for a good night's sleep. The weather was miserable, and the next day would launch them on their arduous journeys.

After breakfast on Saturday morning, March 8, they went down to the dock to wait for their baggage and equipment to arrive. The company had promised to deliver it by 10:30, but by 11 o'clock, it had not yet turned up. Now, with the emigration officer on deck to check passports, the men were obliged to board the *Anselm* with still no sign of the baggage agency. Because they were astronomers from famous observatories and endorsed by the astronomer royal, the customs official looked quickly at their passports and asked nothing about their cargo, which wasn't there anyway. The great worry now was that they would need to come ashore and make plans to leave the next day. Or, if something unpleasant had happened to their equipment, not leave at all. Once Eddington learned that a dozen other passengers were also waiting for their luggage from the same company, he and the other men relaxed. At 12:30, when the suitcases and crates finally turned up at the dock, the relief was enough to warrant a celebratory lunch aboard ship.

The 400-foot *Anselm*, with over sixty first-class cabins, was a steamer owned by the famed Booth Steamship Co. Founded in 1866 by brothers Charles and Alfred Booth, the company was headquartered in Liverpool. Since its inception, it had offered service that included passenger cruises to Brazilian ports and up the mighty Amazon River. In 1903, the company added lines to Lisbon, Portugal, and Funchal, Madeira. The British middle class was now traveling the world, just as young aristocrats had done in the seventeenth and eighteenth centuries when they embarked on "the Grand Tour," a journey around Europe's various cultural centers as part of their education. The Booth lines were more than eager to carry these travelers to exotic places around the globe.[2]

Having booked passage on the *Anselm*, the astronomers could visit the observatory at Lisbon and meet the men who had helped accommodate the Príncipe expedition. But once in Madeira, Eddington and Cottingham would catch a later steamer on to Príncipe, some five thousand miles of sea travel from Liverpool. Crommelin and Davidson would remain on the *Anselm*, crossing the Atlantic Ocean, to arrive at Pará, Brazil, a distance of thirty-five hundred miles from England.[3] The Booth brothers and their company were famous enough that Eddington was impressed when one of the managers came aboard before departure and spent a few minutes speaking to the astronomers.

At 2 p.m., the ship went slowly through the chain of docks—this interconnected port system was then the most advanced in the world—and down the River Mersey, headed for the Irish Sea and the coast of Wales. "We saw the lights of Holyhead at 9 p.m.," Eddington wrote home, "and stopped for a few minutes to drop the pilot." The captain had apparently hired a ship's pilot who knew the local waters well and could navigate the *Anselm* through the dangerous stretch around the Isle of Anglesey, known for its many shipwrecks. Once the steamer reached open water, the captain "dropped the pilot" and the *Anselm* steamed down past St. George's Channel and Land's End, headed for the Bay of Biscay.

After those evening lights seen at Holyhead, and until they made port at Lisbon, Portugal, the passengers would have only a vague idea of their position or course. The end of the war was only four months old. Peace had yet to be negotiated and the Treaty of Versailles was yet to be signed. The treaty would happen three months in the future, on June 28, 1919, exactly five years to the day since Archduke Ferdinand had been assassinated. In the aftermath, the times were hardly peaceful. Revolts and protests were going on all over Europe, including in England. Given the present status of the political environment in Europe, a maritime regulation still in effect forbade a ship's captain to disclose any sensitive information. But after two years of planning, the British expeditions were finally under way.

ALBERT'S PERSONAL LIFE

And what about the man who had caused all this ruckus in the first place? What had Einstein been up to since the publication of his general theory in English in the autumn of 1916? The Kaiser Wilhelm Institute—he had been promised the directorship as part of his 1913 enticement package to come to Berlin—was finally in the building stages. The outbreak of war had put it on a back burner until private funding arrived. At the end of 1917, Einstein had hired the steadfast Erwin Freundlich as the institute's first full-time employee. Freundlich would be freer to pursue light deflection and even made plans to study astrophotography. But he would not be going on an eclipse expedition. His instruments were still in Odessa, Russia, and it would have been impossible to find funding in wartime Germany. Nonetheless Freundlich was still struggling to support the general theory and still under a barrage of criticism, especially from the naysayers who could no longer attack the more prominent Einstein. If they wished to refute the credibility of his theory, it was easier and safer to attack someone they perceived as his underling. At least Freundlich was free of his former boss at the observatory, the formidable Hermann Struve.

Also in 1917, Einstein wrote his first paper on cosmology, as if wanting now to know the *origin* of the universe that he was hoping to better understand. His pace of research publication continued to explode. In addition to work on gravitation, his papers during this time covered the topics of electricity and magnets, heat, quantum radiation, X-rays, and even water waves and flight.[4] This had to have been an anxious period for him as the British expeditions got ready to sail. Most of Einstein's colleagues still embraced the physics of the nineteenth century. Their universe was orderly and explained by the laws of mechanics, as if it had been constructed from gadgets in a workshop. James Clerk Maxwell's discoveries had already begun the dismantling of "the clockwork universe," as it was called. Would light deflection according to Einstein's theory prove that the universe was very different from clockwork and far more exciting? He wanted others to understand it as well as they could. Between September 1918 and February 1919, he even found time to teach a course on relativity at the University of Berlin, as well as one for the university's war veterans.

On a personal level, things were stressful for the physicist. Three weeks before the expeditions left Liverpool, on Valentine's Day, no less, he and Mileva Marić Einstein were officially divorced, their own personal war finally ending, at least on paper. With Mileva and their sons having left for Zurich two days before war was declared in 1914, and with the divorce made final in February 1919, it seemed as if the family's inner turmoil had mimicked the chaos of the outer world. As with all wars, negotiations and settlements would follow. In asking for a divorce this second time, Albert was far more generous. In January 1918, he had written to Mileva with an amazing offer. Along with a larger annual stipend, he would give her any monies coming from a Nobel Prize, should he win it. At first, Mileva was distraught. She was still battling depression. And now their youngest son, Eduard, was also having mental and physical issues. She eventually agreed, and proceedings for the divorce got under way. This meant that Albert must admit he had committed adultery with his cousin, Elsa Einstein Lowenthal. He made this admission at the end of that year.[5]

No one, not even his mother—Pauline had been battling stomach cancer for some months—was happier than Elsa. She had championed her cousin Albert's divorce for several years, since the earliest days of their liaison. Elsa had not been pleased that it wasn't forthcoming soon after Mileva had left Berlin with the boys. In Albert's opinion, his devotion to his cousin and his separation from Mileva should have been enough. He even counseled her not to feel ashamed of their unmarried relationship. He had written just the opposite to Mileva, three years earlier. But at the outset, Elsa had desired more than just the physical man in her life. She wanted him as her husband, and she wanted to share any success in his career. When Albert became ill with severe stomach pains in early 1917, she had the opportunity to prove to him how indispensable she was. Living alone, and with the restrictions of war, he had not been eating well. He had also lost a considerable amount of weight and suspected the pains might be an indication of the cancer that was afflicting his mother. Elsa moved him to an apartment above her own and began nursing him back to health with good food. She even laid in a supply of cigars.

Thus, the following January, Albert had mailed the letter to Mileva, asking again for a divorce. On Valentine's Day 1919, as the expeditionary equipment was being shipped to Greenwich for packing, Albert Einstein was again a single man.

FUNCHAL, MADEIRA

Arthur Eddington was happy to be taking a steamer trip again after years of grisly war and his personal troubles with conscription. He was even pleased with the size of his and Cottingham's cabin, next door to their companions, both rooms high enough above the water that they were quiet. The men read leisurely in their cabins or in rented chairs on deck. The sunshine and warmth as they sailed farther south were a welcomed change from the English winter. Eddington played chess with Crommelin and another passenger. But what was most obvious to him was the end of rationing after more than a year of limitations. "It seems curious to have done with rationing entirely—unlimited sugar, and large slices of meat, puddings with pre-war quantity of raisins & currants in them, new white rolls, and so on."

At meals, Crommelin and Davidson had been given reserved seating at the captain's table, a designated honor among ship's passengers. This social honor was bestowed on Crommelin probably because, among the scientists, he was the best known to the public for his work on Halley's Comet, which had been visible in the nighttime sky over Britain in 1910. Eddington found it amusing, however, that Crommelin and Davidson were seated too far away from the captain to speak to him.

In the Bay of Biscay, they hit wind and scattered showers, enough so that some of the passengers, including Crommelin and Cottingham, succumbed to seasickness and stayed inside their cabins. On the morning of March 12, they reached Lisbon, the capital of Portugal, on the northern banks of the Tagus River. They were met on the quay by Frederico Oom, assistant director of the observatory and son of its first director of the same name. Eddington was uncertain of the local time when he stepped ashore. In 1916, William Willet's concept of daylight saving time had been adopted by England and other countries. "I cannot say what the time was because we had three times—ship's time, Greenwich Time and Summer Time, each differing about an hour; it was most confusing. Although summer-time is legally in force in Lisbon and, I believe, in Madeira most people stick to the old time."[6]

It had been Oom and the current director, Cesar Augusto de Campos Rodrigues, who had responded over the months to Eddington's letters with questions about Príncipe. The Portuguese astronomers had set up introductions to local dignitaries who would act as hosts on the island and help make arrangements there for the British party. Oom took them by motorcar to tour the observatory. The streets of Lisbon were filled with soldiers, given the revolution, and the police force had been disbanded. But the observatory sat in a peaceful park on a sloped hill facing the Tagus. It was surrounded by almond trees in bloom, a sea of pink and white blossoms, a different world from the snowy one they had left behind. They were introduced to the aging Director Rodrigues, a vice admiral, although Eddington thought he looked nothing of the part. Rodrigues would pass away on Christmas Day, later that year. After signing the guest book, they took a trip around Lisbon in the motorcar, enjoying the red-tiled rooftops of the houses. Back at the dock, they again boarded the *Anselm*.[7]

Forty hours later, they spotted the archipelago that made up the Madeira Islands. The ship sailed around Madeira, the largest island, to its capital of Funchal, and all four men went ashore for a stroll around the city. From here, they would part ways for the time being, until they met again in England several months later. To the eyes of the locals, it would be difficult to separate these four Englishmen from the regular tourists and businessmen visiting Madeira. And yet, ahead of them lay what many consider the most important scientific experiment of the twentieth century. It was an experiment that had eluded Charles Perrine and was still causing William Campbell restless nights. Perhaps the British would be more successful.

After a farewell lunch at a local restaurant, they strolled back to the dock where Crommelin and Davidson set sail again on the *Anselm*. Eddington and Cottingham then took a "bullock sleigh" taxi to a hotel, its runners greased with fat so that the bulls could pull it more easily over the cobblestone streets. They would wait in Funchal for nearly a month before boarding the next steamer to Príncipe.

DAVIDSON AND CROMMELIN IN BRAZIL

Andrew Crommelin and Charles Davidson, both of them having been at the Royal Observatory for many years, were two of Frank Dyson's most valued assistants. From Madeira, they would cross two thousand miles of Atlantic Ocean, headed for Pará, in northern Brazil. They arrived on March 23, fifteen days after setting sail from Liverpool. As was usually the case when scientists visited a foreign country to study eclipses, they were given carte blanche. Thanks to the Brazilian government, arrangements had been made with customs officials so that the astronomers would not be put through the usual formalities when entering the country. The British consul in Brazil made the arrangements ahead of time. The two men were granted free railway travel and guaranteed daily telegraphic transmissions. And should it be needed, they would even be given military and police protection during the eclipse. Brazil, like Portugal, was ready to assist the British expeditions.

Crommelin and Davidson were ushered through customs with their packing cases holding the glass lenses left unopened. The two men would not go directly to Sobral. With the eclipse still more than two months away, and with no recent communication with Morize, they decided to take advantage of the Booth Line's promise to carry passengers "1,000 miles up the River Amazon." The company booklet advised that the cruise was not for invalids, the elderly, or pulmonary tuberculosis cases.

But anyone with "jaded and tired nerves" was encouraged to take it for a change of scenery and the relaxation. Back on board the *Anselm*, the Englishmen set off for Manaós, in the heart of the Amazon rain forest at the confluence of the Rios Negro and Selimões. If they knew that their steamer had collided in 1905 with the *Cyril*, another Booth Line vessel, Crommelin never mentioned it in his writing. The *Anselm* sustained bow damage—and was found responsible for the collision during a later enquiry—but the *Cyril* sank in seventy feet of water, with a fortune in rubber. This was not the kind of information the Booth Line would care to advertise in its booklet of what to expect on the cruise, especially not to a passenger with "jaded nerves."

Crommelin was quite well known to the public. Astronomers had expected Halley's Comet to reappear sometime in 1910. Crommelin, along with his colleague Philip Cowell, had traced historic references to the comet all the way back to 240 BCE. Thus, they were able to predict within three days when in 1910 it would be visible in England. Newspapers had made the public well aware that the comet was coming, pulling a 24-million-mile-long tail behind it. The earth's orbit was such that the planet would pass through this streaming tail for six hours. There was even talk of the comet's colliding with earth and that its tail was a river of poisonous gas.

Its appearance was more a social event than one of astronomical discovery. Nonetheless, given Crommelin's and Cowell's impressive work in predicting its return, the RAS Club had honored the two men at its monthly dinner that June. Dyson was still astronomer royal for Scotland until receiving his appointment as astronomer royal a couple months later. But he traveled from Edinburgh for the event, as did Eddington and Davidson from Greenwich. These were occasions of good fellowship and so, with Crommelin and Cowell in attendance as guests of honor, Turner delivered one of his songs, which began with these verses:

> *Of all the meteors in the sky,*
> *There's none like Comet Halley,*
> *We see it with the naked eye*

And periodically.
The first to see it was not he,
But still we call it Halley,
The notion that it would return
Was his originally.[8]

Crommelin had become an expert on comets. Coming from a well-connected Huguenot family, he bore the full name Andrew Claude de la Cherois Crommelin. He was a descendant of Louis Crommelin, who had founded the Irish linen industry in Ulster before the turn of the eighteenth century. After his birth in 1865 in what is now Northern Ireland, Andrew's family moved to England when he was three years old. From Marlborough College, where he excelled on the shooting team, he went on to Trinity College on a scholarship, becoming twenty-seventh wrangler. For a time, he was on the teaching staff at Lancing College and, later, had even considered electricity as a scientific vocation.

But astronomy was where Crommelin's passion lay. When he was accepted as an assistant at the Royal Observatory in 1891—William Christie was then the astronomer royal—he began computing data of cometary orbits. He left the Protestant faith that same year to become a devout Roman Catholic. Often rumpled in appearance, Crommelin was known for his encyclopedic memory, his courtesy, and his scientific fellowship. He often willingly served as a mentor to young astronomers. In 1897, he married Letitia Noble, daughter of a reverend, and they had four children, two boys and two girls.

Crommelin had previously participated in four eclipse expeditions, his earliest being Norway, in 1896. The Norway mission was also the first expedition for the newly formed BAA group, so J. J. Atkinson and fifty-seven other amateurs were also along, as were luminaries like E. Walter Maunder and Arthur Hinks. Crommelin and his wife had gone to Algiers in 1900, to view the eclipse from the rooftop of the Hôtel de la Régence with Maunder and his family. He observed his third eclipse aboard a ship anchored off the coast of Spain, in 1905. And he had viewed the 1912 partial eclipse in Paris, the same one that had brought Dyson and

Atkinson cruising the city streets in Atky's two-seater. His sister, Constance Crommelin, had married the poet John Masefield, who was already famous for such poems as "Sea-Fever" and who later became poet laureate of England. By the time Andrew Crommelin set sail on the *Anselm* with Davidson to steam a thousand miles up the Amazon River, he had just turned fifty-four years old.

Charles Rundle Davidson was born February 28, 1875, in Walton, Suffolk, the youngest of eight children. His father, Alexander, was chief boatman in charge aboard HMS *Penelope*. The children stayed with their mother, Emma Soper Davidson, in various coastguard cottages while their father was at sea. Alexander later became a gunnery instructor in the Royal Navy. Young Charles was educated at the Royal Hospital School, in Greenwich, which was a part of the Royal Naval College. The school provided an education to boys who might enter the Royal Navy, so he was taught nautical skills. As the Royal Observatory recruited young men from the local schools, particularly the Royal Hospital School, he was appointed in March 1890 as a fifteen-year-old computer.

In those days, the staff was divided into junior and senior ranks. The junior-ranked Davidson was quickly promoted to established computer in 1896, recording astronomical data. Christie eventually considered him the expert and supervisor on all matters pertaining to the instruments at Greenwich. When Frank Dyson first joined the observatory in 1894 as the newly married chief assistant, he would forge a lifelong association with both Davidson and Crommelin. While Dyson admired Crommelin's vast knowledge of comets, it was from the technically precise and meticulous Davidson that he learned much about his profession as an astronomer.

When it came to the design, construction, adjustment, and testing of optical apparatuses used during eclipse expeditions, Davidson had a rare degree of skill. Meticulous, he had a reputation for always keeping the photographic plates in his sight when traveling. It was mostly from Davidson that Dyson would ultimately come to see astronomical theory as having its fundamental place. Both men believed that top-notch work should be based on practical observation and then tested by it. Observation should be the first duty of any observatory. Dyson would occasionally remark on a brilliant theorist's inability to make practical observations: "If he'd ever observed through a telescope, he wouldn't say such things." Sir Frank was a good man to have sent these expeditions to test the calculations of a brilliant *theorist* through hands-on *observations*. He trusted Davidson to head the expedition to Brazil.

Dyson's friendship with his fellow astronomers at the observatory grew over the years. He and Davidson, who was known for a dry sense of humor, had a particular relationship. Dyson was a seasoned pipe smoker who always needed a match whenever his pipe went out, as it often did when he was concentrating on some observation or duty. If anyone smoked more than Dyson, it might well have been Albert Einstein. The joke at the observatory was that you could follow Dyson by the trail of burned matches left in his wake. "Got a match, Davidson?" he would ask, after patting all his coat pockets and finding none. Davidson quickly understood that a box of matches was almost as important to Dyson's work as were the astronomical instruments.[9]

In 1907, Davidson converted to Catholicism when he married Eliza Stanford, who had been born in Hong Kong in 1874. The young couple moved to a house in Greenwich, close to the observatory. Over the years, they raised a family of four children, two sons and two daughters. Charles had been on several previous eclipse expeditions already, including his first to Portugal in 1900, with Christie, Dyson, and the amusing Atkinson who had smuggled a barrel of Portuguese wine through customs. He had gone to Tunisia in 1905. In Brazil with Eddington in 1912, he had joined the dinner with Perrine and Mulvey at the Hotel of Foreigners, before being rained out during the eclipse. In 1914, he had escaped with the other teams from Russia and the explosion of war. Now Davidson was back in Brazil, testing for light deflection seven years after Perrine had first tried. He was forty-four years old.

EDDINGTON AND COTTINGHAM IN MADEIRA

As Crommelin and Davidson were getting ready to chug up the Amazon River for a glimpse of crocodiles, blue morpho butterflies, and giant water lilies, Eddington and Cottingham had been confined to Funchal and its surrounding area. The hotel Bella Vista, which had English proprietors, sat ten minutes away from the downtown area. In the years before the war, more English visitors had been checking in, which meant that English would be spoken at most of the shops. The first three days, the two men experienced excessive heat, unusual for the island, so they remained mostly indoors. The problem was the *leste*. This dry wind from the Sahara blew seasonally over Madeira in sporadic gusts filled with fine red sand. When the *leste* was over, the weather was warm with scattered rain showers. Despite not being acclimatized, Eddington, not surprisingly, made himself a tourist while in Madeira. Much as he did on his outings with his friend C. J. A. Trimble, he climbed mountains and walked through steep gorges until he had learned all he could about the island's geography.

Eddington befriended the other English guests at the hotel and put himself to the task of learning some Portuguese words that would facilitate his stay. He enjoyed the "pottering" older man's company, even though Cottingham preferred talks with the less thrill-seeking guests and leisurely visits to the local shops over climbing mountains. "He is just 50," Eddington wrote home, "so, of course, not fond of very much exercise." Some days, the two had tea on the balcony of the Bella Vista, which was well named. It overlooked the town and harbor, as snow-covered mountains loomed in the distance. Below their room, the hotel gardens were filled with perfumed flowers, date palms, and cacti. When they sat out in the evenings, the town of Funchal sparkled with electric street lamps that stretched far up the mountainside. Nights at that time of year were cold by Madeira standards, yet compared with the factory smoke and grime the astronomers had left behind at Greenwich, the place must have seemed like paradise.

Eddington took a liking to the locally grown bananas, eating a dozen a day. Teatime was offered at the local casino, although he found it inferior to English tea. The casino had a lively band of local musicians performing daily, and the roulette tables were busy, even though the game was illegal in Madeira. Despite being a steadfast Quaker, Eddington tried his hand a few times at the game. No one was concerned with being arrested since the police always phoned ahead to inform the casino that there would be a raid. One afternoon, having finished his tea, Eddington found the casino door locked when he tried to return to his hotel. It was explained to him that the chief of police had come for the dancing in one of the other rooms and, therefore, did not want to know about the spinning roulette wheels. The staff had locked the door to the gaming room. Eddington slipped out a back entrance. By the end of his stay in Madeira, he was down only a pound.

Eddington and Cottingham occasionally took the funicular up the mountainside and then rode in the wicker toboggans that carried passengers back down to Funchal.[10] He and Cottingham played games at the casino, including musical chairs, still a novelty then. They visited the local telegraph cable station, which Eddington thought much improved from the days he had spent at a telegraph station in Malta. With Cottingham more the "pottering-about" type, Eddington struck up a friendship with sixteen-year-old George Turner, an English boy from Mumbles who had come to Madeira after an illness and was a guest at the hotel. They shared a love for chess, butterflies, and swimming. They went together one evening to the local "picture-palace." The headlining film was the funeral of King Edward VII, who had died nine years earlier. "It was rather curious seeing it after so many years," Eddington wrote to his sister.

It would be an idyllic month on a peaceful island. And yet, remnants of the war were evident amid the vined pergolas, misty ravines, and snow-capped peaks. Some buildings still bore the marks of shelling. The historic Santa Clara church, dating back to the fifteenth century, had also been damaged. In the harbor waters were the visible masts of two sunken vessels. On December 3, 1916, a German U-boat with the notorious Max Valentiner as its captain had gone into the harbor at Funchal and torpedoed three ships. Each day that Eddington walked down to the dock, he saw the masts of two of those ships still protruding from the water. If he learned their names, he never mentioned them. As a matter of fact, he wrote of the torpedoed ships and bombardments in two short sentences, dropped into the middle of long and witty descriptions of the hotel guests and the local scenery. But it was time to begin healing from the war, and perhaps Eddington thought such talk would disturb his mother back in

Cambridge. Although he never revealed his own feelings, one might wonder if there was any sense of guilt, given the remarks made by colleagues like H. H. Turner, who wrote about his countrymen "shrinking from the horrors of war."[11]

When the Englishmen learned that the steamer they were booked on had service only to São Tomé, the sister island a hundred miles southeast of Príncipe, they booked passage on the *Portugal*. Eddington was having difficulty getting their passports and other papers in order. When an "unkempt" man in the street offered to help, he feared he was only "after earning a tip." The man turned out to be a local editor who introduced the Englishmen to the governor of Príncipe, who was visiting Funchal and was booked to return on the *Portugal*. After being interviewed about the expeditions for the newspaper, and with their passports now in order, Eddington and Cottingham were ready to say goodbye to Madeira.[12]

THE HEART OF THE JUNGLE

Long before foreign interests sought out oil, which was then unmarketable, and minerals like bauxite and diamonds, it was a milky liquid called *latex*, extracted from rubber trees, that brought foreigners into the Amazon basin. From approximately 1850 to 1913, the only place on earth to find this "white gold" was in the forests of Brazil, Peru, Ecuador, and Colombia. The world was experiencing an insatiable demand for rubber, which was used in thousands of machines, gadgets, and other products. The inflatable tire that was needed to satisfy the bicycle craze was followed by the introduction of the automobile and its own need for tires. This boom spurred European colonization and made the "rubber barons" rich. Missionaries arrived and built chapels. Cities grew up where there had been only jungle. Manaós, which was where Crommelin and Davidson were now headed on their *Anselm* cruise, was the grandest city of them all. Its exclusive English Club boasted swimming pools and tennis courts. Known as "the Paris of the Tropics," Manaós had its own opera house that cost ten million dollars and staged some of the world's best talent. When half the members of one visiting troupe died from yellow fever in a single season, there were more actors to fill their places.[13]

To satisfy this prodigious thirst for rubber, however, there arose an impossible demand for local labor and, in many cases, a dehumanizing and immoral enslavement. A great deal of latex had to be gathered, and rubber trees by nature grew some distance apart. Entire populations of indigenous people across Brazil and neighboring countries died under this brutal treatment. Horrid working and living conditions were often compounded with floggings, rape, dismemberment, and murder. Often, the workers were shackled together to prevent escape. Even under the best of circumstances, their cultures of hunting and gathering were not accustomed to the rigorous schedule under which industrialization flourished. This brutality would continue until the seventy thousand rubber tree seeds stolen in 1876 by Henry Wickham began to flourish on plantations in Asia and thus put an end to the rubber boom in South America.[14]

How much of this human history was known to Crommelin and Davidson is hard to tell. The rubber boom in Brazil had waned a few years before they arrived, but all its footprints were still in place. That Crommelin had read the Booth Line brochure informing passengers on what to expect during the cruise is evident. In his own written notes, he often uses the same adjectives to describe the flora and fauna, as well as the great river and its tributaries. The Booth Line catalog itself is very revealing in its colonial attitude toward the jungle and the people who lived in it. While mentioning wildlife such as the jaguar, puma, ocelot, sloth, tapir, howler monkeys, and anteaters, it also advises the passengers to keep a watch for "a caboclo family's primitive shack. . . . On crude platforms outside these shacks, naked children watch the ship pass by." The magazine goes on to inform the reader that seeing a floating brown log in the river suddenly spring to life means that the ship has "rudely interrupted the siesta of a sleepy alligator," when, in fact, it should be *crocodile*. Passengers were encouraged to leave the ship at Manaós and "penetrate" the jungle.

The rubber demand would boom again, thanks to Henry Ford, who wanted to break the British monopoly on rubber. In 1928, nine years after Crommelin and Davidson cruised up the Amazon, Ford would turn a tiny hamlet called Boa Vista, one hundred miles up the Tapajós River, into what the Booth Line brochure would later describe as "a replica of an American town, with macadamized roads, concrete pavement, well-kept lawns and low-ceilinged houses. It is the capital of 5,000 square miles of territory conceded by the Brazilian government to Mr. Henry Ford, where he hopes to grow rubber to make tyres for the cars produced in his factories throughout the world." Some reports compared the size of this territory to the state of Connecticut, others to Tennessee. It is perhaps telling that on the same catalog page describing Ford's clearing of thousands of acres of jungle to create Fordlandia, the writer also warns visitors that the piranha can "strip the flesh from the bones in an incredibly short space of time."[15]

Despite already being world travelers, their venture into the heart of the Amazon jungle was a new experience for the Englishmen. They were fascinated to see the ground "alive with troops of leaf-cutting ants" and sensitive plants that closed to the touch of a hand. But it was the floating wharves at Manaós, where the *Anselm* was loaded with rubber and nuts, that most captured their attention. These ingenious structures were the brainchild of Charles Booth, one of the founding shipping brothers and well known to the astronomers for his work with London's poor. Designed to overcome seasonal fluctuations in water levels, the wharves could adapt to a sixty-foot change in the level of the river during wet and dry seasons. Booth visited Manaós for the last time in 1912, four years before his death. He had written home to his wife about seeing his wharf again: "So there is my monument."

SOBRAL WELCOMES THE ASTRONOMERS

On their return to Pará, Crommelin and Davidson were met by a host of people, including the English and American clubs, the American consul, and the manager of the Tramway Company, which plied them with free passes to explore the city and experience the "primeval forests" around Pará. The Booth Line had thrown a festive lunch on board the ship to celebrate the reopening of its route between Europe and Amazonia now that the war was over. On April 24, the two men left for Camocim on the steamer *Fortaleza*, arriving five days later. Crommelin described it as a "tedious" journey. In coastal Camocim, the red carpet was again rolled out, so much so that they felt like "personally-conducted tourists." From there, they took the train inland to Sobral, some eighty miles southeast, with the tracks skirting around several picturesque mountain ranges.[16]

Sobral, in the state of Ceará, was *not* jungle. Semiarid, it was just the opposite, hot and dry with a constant need for water since for many months there would be no rainfall. When the Englishmen got closer to their destination, the country outside their windows had turned brown and barren, the Acaraú's riverbed riddled with deep fissures because of a severe drought. Thousands of local inhabitants were unable to feed themselves or water their animals. They became refugees in towns like Sobral, relying on government aid and daily handouts of food. Crommelin took comfort in seeing that the mountains still looked moist and green in the distance. They arrived in Sobral to find a delegation waiting in anticipation of Father Cortie, whose letter explaining why he couldn't make the expedition had not yet reached Brazil. Crommelin sent a telegram to Dyson, informing him that they had arrived safely at the viewing site.

Leocádio Araujo, their interpreter who had studied agriculture in the United States, was also there to meet them. He would stay with them for the remainder of their time in Sobral. The district deputy, Colonel Vicente Saboya, who also owned the local cotton factory, made his spacious home available to the scientists. In front of the house was a jockey club with an enclosed racetrack. The grandstand had a canopy that would provide shelter for the instruments, a perfect place to store them. All racing would be shut down while the astronomers were in town. Arriving later would be two scientists from the Carnegie Institution of Washington: Daniel Wise, an American, and Andrew Thomson, a Canadian. Carnegie had asked Morize if he might do magnetic and electric measurements during the eclipse. When Morize said he was unable to do so, he invited the institution to send its own team. Wise and Thomson would share the comfortable Saboya residence with the Englishmen.

The Spanish flu that was still killing millions of people in its second wave had struck down the president of Brazil just two months before Crommelin and Davidson arrived in the country. Rodrigues Alves had been slated to begin his second term as president on November 15, 1918, when he became ill with influenza. He died on January 16, 1919. But the greatest worry to Colonel Saboya, the wealthy host, was yellow fever. This viral disease had come to the Americas on the slave ships from Africa during the 1600s. While it was not always fatal, it had claimed many lives down through the centuries, including that troupe of actors it had wiped out in Manaós.

Caused by the bite of a mosquito—the scientific name is *Stegomyia fasciata*, known today as *Aedes aegypti*—yellow fever was again rampant in the state of Ceará, and particularly in Sobral. While it most often affected children, Saboya was concerned about the Englishmen and Americans. As foreigners, they were much more susceptible to the virus and often succumbed to it. What if these visiting scientists became seriously ill or even died? It would not be good press for Brazil in the eyes of the scientific world. His brother, a doctor, had alerted Saboya to this possibility, and now the colonel wanted preventative measures taken beyond wire screens on all the windows. "Imagine the shame and damage to Brazil if some of these scientists die of yellow fever," Saboya wrote. "It was to avoid this that I sought the sanitary authorities." Safety measures would be taken so that the visiting scientists would not regret their trip to Ceará.[17]

Colonel Saboya contacted the sanitary authorities, who dispatched medical technicians to Sobral to be on hand. The best way to protect the visitors, it was decided, was to keep them as isolated as possible from the general population. For their entertainment, an automobile would be brought to Sobral, the first one seen on the streets there. This way, the two teams could be shown some of the local countryside, despite the devastation of the drought. More importantly, a car could carry them to the nearby mountains. If the excursion left Sobral early in the mornings, or after four in the afternoons, they would avoid the times of day when the sun is hottest and these particular mosquitoes more likely to bite. Plus, the mosquito didn't thrive at high altitudes, making the foreign sightseers safer in the mountains. With this plan in mind, arrangements were made for Morize to bring an automobile from Rio de Janeiro on the steamer with him. Crommelin and Davidson began unpacking their equipment.

LEAVING MADEIRA: ON TO PRÍNCIPE

On April 9, which was Cottingham's fiftieth birthday, he and Eddington left Madeira aboard the *Portugal*. The three-thousand-mile journey on to Príncipe—the word *príncipe* is Portuguese for "prince"—would take two weeks. Schools of porpoises followed the ship, and Cottingham saw his first flying fish. The days of calm blue sky passed with games of chess and reading, the nights with moonlit strolls on deck. Eddington slept well for the first time since leaving Cambridge. There was plenty of good food, even if the tea again fell short of his English approval. It was served with ices or, as he noted, "sorbets like we used to have on the Avon." With no good milk on the *Portugal*, he began taking his tea black. He was by now reacquainted with full sugar bowls, unlimited butter, and more meat at one meal than he would have eaten in a week during rationing.

What Eddington missed most aboard this steamer was his daily workout. The *Portugal* didn't even carry deck chairs to rent, so it was no surprise that it lacked the usual facilities for exercise. Unable to do his vigorous mountain climbing for two weeks and with not even deck tennis available, he and Cottingham took part in improvised games that were played by the passengers. It's comical to imagine the Plumian Professor of Astronomy and the director of the Cambridge Observatory engaged in Boots without Shoes, Rings without Strings, egg-and-spoon races, and Threading the Needle. The ladies had potato races, and there was cock-fighting for the gentlemen, a game in which men wrestled to remove each other from a drawn circle on the deck, not by fighting birds.

Eddington had cabled ahead to Príncipe that they would now arrive on the *Portugal*. But he had not received any news from England, or the world in general, since reading an old copy of the *London Times* on March 31. Not expecting to stay so long in Madeira, they had asked that no letters from England be sent there. Any correspondences would be waiting for them on the island. After the ship made two stops in the Cape Verde Islands, a heavy sea mist prevented the passengers from seeing land again until they reached their destination, even though the *Portugal* was sailing just forty miles off the African coast. On the morning of April 23, they spotted Príncipe in the distance. Perhaps the fact that it was also Shakespeare's birthday was a good omen for the British. After the captain dropped anchor, the two Englishmen enjoyed a last breakfast on the ship.

If Crommelin and Davidson had gone into a world built on rubber, then Eddington and Cottingham would come ashore to one that was built on cocoa.

9

PRÍNCIPE AND SOBRAL

In Colonialism's Shadow: The Teams Prepare

I must confess that I am still a skeptic as to the reality of the Einstein effect in question, but I would not be willing to undertake a technical defense of my skepticism. I am quite ready to welcome positive results, though I am looking for a negative one.

—*William Wallace Campbell to Arthur Hinks, June 2, 1919*

I have always been, and am yet, skeptical of any such effect, although I went into it at the Brazilian (rainy) eclipse of 1912, at the request of Freundlich, with two Vulcan cameras. . . . [T]he whole relativity business has seemed to me unreal and so purely philosophical that to accept it is to upset our previously carefully constructed and very material systems.

—*Charles Dillon Perrine to William Campbell, January 30, 1923*

BOTH WILLIAM CAMPBELL and Charles Perrine had their doubts about the predicted light deflection in Einstein's theory. So did Sir Frank Dyson. It may seem incredulous that the first two astronomers would spend several years of their busy careers in difficult pursuit of a theory they didn't fully endorse, and at such inconvenient times and places for traveling around the globe. Or that the third man, as astronomer royal, would put so much at stake during wartime and under restricted funding to send out two expeditions. But the essential nature of science should not be about establishing truths. Instead, science should have a commitment to removing falsehoods from our system of belief. The true work of researchers is to provide the best humanly possible explanation for what they observe in the world around them. Proving Albert Einstein right, or proving him wrong, was not the main purpose for Campbell, Perrine, or Dyson. Refuting him altogether, as many of his colleagues were doing at the time, would have put an end to those important questions he was asking about the universe. The main purpose, and these astronomers understood it well, was to *advance* science, to open up the cosmos a crack wider if possible.

The very nature of science is an unforgiving battleground as new paradigms based on increasingly accurate observation and measurement reveal the limits and flaws of accepted concepts. These concepts may have existed in the scientific canon for centuries, as did the Newtonian viewpoint, unshakable for two hundred years. In this way, perhaps science is unlike any other system that humanity has constructed for making sense of reality. It calls for practicing scientists to constantly question the entirety of the preexisting canon. This approach reveals the secret of science: it drives change and innovation for our species. Sometimes, however, caught in the machinery of this change and innovation are the scientists themselves, the flesh-and-blood human beings behind the calculations, theories, measurements, and observations.

A good case in point might be the work of Ludwig Boltzmann. At a time when the existence of atoms was not universally accepted by many physicists, Boltzmann had gained fame for having invented statistical mechanics by using probability to describe how the properties of atoms determine the property of matter. But his theories were under a barrage of attacks from his contemporaries. Einstein never met Boltzmann, who took his own life in the autumn of 1906. And yet two of the groundbreaking works during Einstein's 1905 "miracle year" were greatly influenced by the Boltzmann tradition. These were his papers on the photoelectric effect and Brownian motion, the motion evidenced by the random movement of molecules. For those two works, in which Einstein recognized "the principle of Boltzmann," he borrowed concepts from thermodynamics which showed his mastery of it, as well as his deep respect for its intellectual edifice. He would write this about Boltzmann's theory on the dynamic properties of matter and atoms: "It is the only physical theory of universal content which I am convinced, that within the framework of applicability of its basic concepts, will never be overthrown." Not long after he died, Boltzmann's viewpoint would become widely accepted, in part due to Einstein's work on Brownian motion.[1]

That Campbell kept an open mind to the theory of general relativity, as did Perrine and Dyson, was a testament to their occupational ethics. Einstein himself believed that the greatest enemy of truth was a blind respect for authority. But it was now 1919, and results of the eclipse plates taken in Goldendale, Washington, the summer before had still not been measured, let alone released. Curtis had finally arrived back at the Lick Observatory in mid-April. By this time, Crommelin and Davidson were already in Brazil, and Eddington and Cottingham were on a ship sailing to Africa. But the eclipse the English hoped to observe was not coming until the end of May. Despite the unsatisfactory equipment and an interference of clouds that Campbell had had to deal with in 1918, there was still time for the Americans to make their announcement, whether supporting Einstein's light-bending prediction or disproving it. Otherwise, the general relativity ball would be entirely in the hands of the British.

THE "BROWN GOLD" OF THE ISLANDS

The tiny island to which Eddington had traveled to change the foundation of physics was not exactly the Eden he described in letters home. Príncipe and the neighboring island of São Tomé, now one country, were uninhabited when the first Portuguese arrived in the early 1470s. By the turn of the fifteenth century, both were slowly being settled and colonized by Portugal, which sent people to live on the islands. They were usually exiled "undesirables," and mostly Jewish. But those first settlers realized in the rich volcanic earth the perfect soil for growing sugar cane. Since cultivating sugar requires considerable labor, Portugal began shipping enslaved workers from the African mainland to the islands. By the 1550s, Príncipe and São Tomé had become the front-runners in exporting Africa's sugar. Over the next hundred years, with sugar being cultivated elsewhere in the Western Hemisphere, production declined for the islands. The large population of laborers was also difficult to oversee, with Portugal's efforts invested elsewhere. São Tomé then became a prime stop for ships that were trading human beings between the West and continental Africa.

By the early 1800s, two new cash crops were being grown on the islands: coffee and cocoa. Just as Wickham had brought the rubber tree seeds from Brazil to Asia, the Portuguese introduced the cacao tree to West Africa by way of Central America. But it is José Ferreira Gomes, a Brazilian-born Portuguese slaveholder, who is credited with bringing it to the island of Príncipe as an ornamental tree for his garden. That was in 1822, but the plant wouldn't thrive in its new environment until the 1870s, just about the same time that Portugal abolished slavery, at least on paper. But thrive it did. And so did the world's love of cocoa, and then a craving for chocolate.[2]

With a demand now for "brown gold," the Portuguese companies built *roças*—pronounced "rossas"—or plantations, mostly run by absentee landlords. Managers were hired to do the overseeing. These rich soils and hard labor again worked well, and the roças flourished. By 1900, São Tomé was producing more cocoa than any other place in the world. The abolition of slavery meant little to the plantation managers since they worked around it. With government officials in agreement, they devised a state-supported plan for contract labor. Tribal members young enough to work for years were captured in Portuguese colonies in West Africa—they were often from Angola just to the south of the islands—and forced to walk hundreds of miles to the coast. There, a government official would ask each captive through an interpreter if they wished to have a "new master" or to go to São Tomé. Torn from their homeland, often beaten and ill, most answered São Tomé, if they answered at all. Thus, they were contracted to work on the roças for five years.

By the early 1900s, over 4,000 workers were sent to Príncipe and São Tomé annually. None of them returned home. Just a decade before Eddington and Cottingham left England for their expedition to Africa, approximately 320,000 more workers had been laboring on the 230 plantations on São Tomé, the larger of the islands. And 3,000 more were doing hard labor at the 50 plantations on Príncipe. Twenty percent of them would die each year. Workers had no choice but to spend their meager wages at plantation stores. The majority of what they earned was withheld in a repatriation fund that would pay to send them home once their contracts expired. For most plantations, these funds were nothing more than figments of greedy imaginations. In actuality, those contracts ended only with death.[3]

THE CADBURY COMPANY

Back in England, other Quakers had been thinking of the island off the coast of West Africa much longer than Eddington had. Cadbury's had begun in Birmingham, England, in 1824, as a tea and coffee shop that also sold drinking chocolate. By the 1860s, with chocolate becoming its best seller, the family-run company became a chocolate business. Guided by their humanitarian Quaker beliefs, the Cadburys built their own village, Bournville, on the south side of Birmingham. Unlike Ford and his later plans for Fordlandia in the Amazon jungle, Bournville was a superb model town that would greatly benefit its loyal workers and also be efficient for business. The Quaker plan was an idea quite unfamiliar to Victorian times and the novels of Charles Dickens.

With the business growing, Cadbury's began to import cocoa beans from the roças on São Tomé. By 1900, when grandson William and his three brothers were running the business—all three British businesses that dominated the chocolate industry were Quaker-owned—Cadbury's was importing 55 percent of its cocoa from Príncipe and São Tomé. Reports of worker abuse had been filtering into England for years, but nothing had been substantiated, especially if no one went to investigate. Then William Cadbury received an advertisement for the sale of a roça on São Tomé. Among the listed "assets" were two hundred black laborers for 3,555 pounds. The company had no interest in the plantation. But seeing this ad was confirmation that earlier reports of conditions on the roças must be true. Or it certainly should have been a fluttering red flag.

The Cadbury brothers agreed that there should be an enquiry, but they seemed to be in no hurry. They finally decided that William would travel to Lisbon, where most of the plantation owners lived, and meet with them in person. This strategy would be akin to asking the fox about conditions in the henhouse. The year was 1901, and Lisbon was almost four thousand miles from Príncipe and São Tomé. William was assured by the owners that 1903 would bring new regulations that would guarantee better conditions for their workers. William wrote to an abolitionist friend: "I should be sorry needlessly to injure a cultivation that as far as I can judge provides labor of the very best kind to be found in the tropics: at the same time we should all like to clear our hands of any responsibility for slave traffic in any form."

HENRY WOODD NEVINSON

On the question of slave traffic in Príncipe and São Tomé, the media got involved. In 1904 and 1905, an investigative journalist named Henry Woodd Nevinson went to Angola on behalf of *Harper's Monthly Magazine*. Unlike William Cadbury, he wanted to see firsthand what was happening, both in the interior of Angola, where many of the "volunteer laborers" originated, and on the cocoa plantations themselves. In his middle age, Nevinson had become a British war correspondent. He covered several wars, including World War I, during which he was wounded at Gallipoli, in the same battle that took the life of physicist Henry Moseley. As a journalist, he "brought warfare to British breakfast tables."[4]

For his investigation, Nevinson journeyed 450 miles into the dense forests of central Angola, to the origin of the trading route. He then followed the trail to the sea, the same path along which men, women, and children were forced to walk. He would witness floggings, dismemberments, rape, and murder. The trail was already so littered with human skeletons that Nevinson felt outrage. Hanging from the trees were wooden neck and leg shackles, removed from the bodies for the next trader to find and put to use. "It would take an army of sextons to bury all those poor bones that consecrate that path," he wrote.

Henry Nevinson was then forty-nine years old and gravely ill. Believing he had been poisoned by Portuguese dealers to stop his investigation, he swallowed the antidote he carried and made arrangements for his jour-nal to be sent to England should he die. And yet he walked twenty miles a day in the unforgiving heat. The fury he felt as a firsthand witness to this inhumanity drove him onward. The beaten and starving souls who sur-vived were "freed" by the Portuguese officials waiting for them at the end of the march. They were free, that is, to work for five years on São Tomé or Príncipe. These were the places from which no one ever returned. In their language, São Tomé had become synonymous with the word *oka-lunga*, which means "hell." Still suffering from fever, Nevinson followed one group of captives down to the water's edge and boarded the ship wait-ing to transport them to the cocoa plantations on both islands, a voyage that would take eight days. On this particular ship were 273 captives, not counting 50 babies. What Nevinson saw after he reached the plantations was just as deplorable. "As horrible as anything recorded in human his-tory," he wrote. His columns were printed in *Harper's* over the next sev-eral months, causing a public outcry. In 1906, Nevinson's book titled *A Modern Slavery* was published. [5]

William Cadbury had begun an investigation of his own by enlisting a fellow Quaker and former bank employee named Joseph Burtt to inves-tigate. What were the conditions really like on the plantations that were providing him with "brown gold?" To learn the language, Burtt first lived for several months in Portugal. He left England for Africa in 1905 and returned after two years, having spent six months on Príncipe and São Tomé, and then traveling to the interior of Angola. His account, made public to British citizens in 1908, was as shocking as Nevinson's had been. Other Quakers were now calling for a boycott of West African cocoa. Cadbury again traveled to Lisbon and let the plantation owners know that if the company were to continue to purchase cocoa from their roças, "in the future it is to be produced by free labor." It was almost eight years since he had seen the ad that included human beings for sale.

In the autumn of 1908, Cadbury traveled to the islands with Burtt to see for himself. He asked that a missionary named Charles Swan go to An-gola, as Nevinson and Burtt had done, and conduct another investigation there. Swan also interviewed missionaries in the country who knew the situation well. These were not volunteer workers who trudged for miles to the waiting ships at the coast, past the many shackles hanging from trees. Swan counted ninety-five shackles in one day, many with skulls and bones still attached. They were captives tricked into working on the plantations, and the abuses they endured made any Nevinson had written of pale in comparison. Swan published a book that was filled with photos of many of the enslaved workers. They were now actual human beings, with names, and horrific personal stories to tell. The misery endured in Angola was even worse than what was to come on the two islands.[6]

Back in England in 1909, after having seen firsthand the deplorable conditions on the plantations, Cadbury began work on his own book, *Labor in Portuguese West Africa*. He searched for and found an alternative supply of cocoa from the Gold Coast, in what is today's Ghana. Labor conditions were reportedly better there, and the cocoa of a higher quality. He finally announced a boycott on slave-grown cocoa from Príncipe and São Tomé. He encouraged American and British companies to join him, and they did. But the media in England had long been asking what had taken these companies so long.

Adding to an impossible life and staggering death rate for the workers on Príncipe, sleeping sickness hit the island, having already killed thousands back on the mainland. The disease was believed to have been brought aboard ships from Angola as they transported the so-called laborers. Nevinson noted in his 1906 book that scientists "are now inclined to connect it with the tsetse-fly."[7] In a decade, the sickness killed 2,525 people on an island only ten miles long and six miles wide. While São Tomé had received workers from Angola already afflicted, they were somehow spared the fly itself, and thus remained free of the disease. Local authorities initiated measures to eradicate the carrier. Because the tsetse fed on blood, all animals should be killed, except for the needed oxen and mules. And those spared beasts should be protected by mosquito screens. But the money-hungry plantation owners in Lisbon complained that they needed their servants, or *serviçais*, to work harvesting cocoa, not fighting a fly from Angola. Because of the high mortality rates, the landowners were in constant need of more laborers as it was. The authorities nevertheless insisted. By the following year, 1914, the flies had completely vanished.

Less than five years later, on the morning of April 23, 1919, the *Portugal* steamed into the bay at Santo António, the capital of Príncipe, and dropped anchor in the warm waters. On board, enjoying a breakfast of coffee and biscuits, were the Quaker Arthur Stanley Eddington and Edwin Turner Cottingham.

A QUAKER ON PRÍNCIPE

Eddington and Cottingham had seen no mail from England since leaving Liverpool on March 8. They still didn't know if the peace treaty had been signed, making the end of the war official. The governor of Príncipe, whom Eddington had met briefly in Funchal, had gone ashore while the men were having breakfast. Now he came out to the ship with a launch to greet them. He had with him two plantation owners who had offered their roças to the visiting scientists. This warm welcome was recorded by Eddington: "We were met on board by the Governor, Mr. Carneiro, and Mr. Gragera, and we soon found that we were in clover." They were escorted from the dock to Jeronimo Carneiro's private home, which was still under construction. It was the wet season in Príncipe. Until mid-May, rain would fall each day, a heavy and steamy downpour that quickly returned to clear skies. The *gravana*, or dry season, would then begin, with clouds dotting the skies, cool nights, and often a light rain before sunrise. The eclipse would occur two weeks after the *gravana* began.

The Englishmen rested for a day before cabling Dyson of their safe arrival. On Friday, they set out in a tram car pulled over railway tracks by a pair of mules. Not counting the cultivated plantations, the island was covered in steaming lush forest and moist tropical growth. Nevinson had referred to the two islands as gigantic hothouses: "The islands possess exactly the kind of climate that kills men and makes the cocoa-tree flourish."[8] They would visit two of the plantations that the Colonial Agricultural Society had suggested, Roca Esperanca and Roca St. Joaquin. Eddington immediately noticed a striking sugarloaf mountain in the middle of the island. It rose to twenty-five hundred feet and was ringed with a thick mass of clouds. Nestled around it were numerous naked peaks wrapped in heavy mist. This was the range that Arthur Hinks had diligently inquired about in his letters two years earlier. Since Roca Esperanca was hemmed in by this mountain, Eddington passed. The last thing he wanted was permanent clouds. The second plantation they visited, Roca St. Joaquin, was a possibility. But they had yet to check the third, Roca Sundy, six miles from Santo António and owned by their host, Carneiro.

The next morning, wearing pith helmets as protection from the sun and bringing along mackintoshes for the sudden rains, they started out again, this time riding the mules. While Eddington could have easily walked the distance despite its being mostly uphill, it would have been impossible for Cottingham to do so in the sweltering climate. Over an hour later, they arrived at Roca Sundy, which was surrounded by sea and sky and was bordered by a long stretch of yellow sand and swaying palm trees. They had noticed its elegant main house as the ship neared Príncipe three days earlier. Located on the northwest end of the island, the house sat atop a plateau that rose five hundred feet above the sea. There was an enclosed piece of ground just beyond the bedroom windows that was sheltered on the east by a wall. It would provide excellent protection for the instruments. To the northwest, the enclosure opened onto the sea, sloping down in the direction of the sun's path.

Roca Sundy functioned like a self-sufficient town, as did all the plantations on the two islands. They grew their own food and raised livestock to feed the numerous staff and laborers. Spread out beyond the main house, the buildings were laid out in good order. There was an administration office, houses for the foremen, and numerous sleeping quarters for the six-hundred-plus workers, small wooden sheds with galvanized roofs. There were kilns for roasting the cocoa seeds that, once dried, became the valuable beans. And there were warehouses to store them. The main plant, called a *sede*, was where the cocoa beans, coffee beans, or palm oil was processed. A four-wheel locomotive then carried this product to the coast, where it was loaded onto waiting ships. There was the plantation store, where workers spent their meager earnings, and a sawmill that provided lumber for building needs. The plantation had a small, sixteenth-century-style chapel used by the owners and overseers. The stables were designed like a medieval castle, in merlon-and-crenel fashion, and a crenulated wall surrounded the courtyard. These structures must have amused the Englishmen, as if King Arthur and his knights would gallop in from the forest.

Eddington decided on Roca Sundy. He asked that workers build a small pier on which the coelostat would stand. Arrangements were made for the baggage and crates to be brought to the plantation on Monday. He and Cottingham then returned to Santo António. Carneiro was the consummate host. Rather than sailing back to Lisbon as planned, he stayed on in Príncipe for the next month to entertain his English guests. Eddington wrote home glowingly about him: "He is rather a young man, and owns the largest private plantation. He has only been out here two years, but his family have had the plantation a long while. In Lisbon he was a well-known bull-fighter." Explaining that in the Portuguese bullfight, unlike the Spanish tradition, the bulls and horses are not slain, he then added, "The Portuguese here are a very superior type to those we have met before—in particular, they do not spit about all the time, and suck toothpicks at meals. Mr. Carneiro is, I believe, very wealthy." Jeronimo Carneiro certainly was wealthy. There can be no doubt that the Carneiro family's fortune was built on the misery and deaths of thousands of indentured laborers.

The Englishmen spent the next few days at Carneiro's house, relaxing and sightseeing. They would later visit Bom-Bom ("good-good"), the ruins of a palace once owned by Maria Correia, the most famous slave trader in the island's history. Unknown to most tourists by 1919, it was her husband, José Ferreira Gomes, who had brought that first cocoa tree to Príncipe as an ornament for their garden. A member of Príncipe's wealthy elite, Correia traveled throughout the world and was said to have been welcomed by the king of England and given diamonds for her hospitality to British ships passing through Príncipe. "Her palace on the beach is all in ruins but it must have been a huge place," Eddington wrote.

One morning, Eddington and Cottingham had a breakfast picnic in the nearby harbor, courtesy of Carneiro. Afterward, they cruised around the island while a large shark followed their boat. In the afternoons, they enjoyed tennis with "the curador and the judge," the only ones on the island who played the game. In naming the people he met as the days unfolded, often only by their occupations, Eddington might have been a contemporary British playwright describing his characters. Nevinson had described the job of a *curador* in his 1906 book. He was the man who asked the captives on the mainland if they chose São Tomé over a "new master," thus forcing them into five years of servitude on the roças. And later, at the plantations, each curador stood before the lined-up workers and tricked them into another five years and, thus, eventual death on the islands. In Eddington's letter home describing the curador at Roca Sundy, his description was noncommittal. "The Curador, who is responsible for the imported labour—quite a young man."

There seems to be no written evidence that Eddington understood the working conditions on the roças. However, he was twenty-seven years old when Quaker-owned Cadbury finally announced its boycott of slave-labor cocoa. Quaker magazines often carried ads paid for by the big cocoa companies in England. That Eddington did not remember the sensational drama carried out in the British press is questionable. But when he unpacked his baggage at Roca Sundy in the spring of 1919, there were several hundred workers toiling in the hot sun to harvest cocoa. It's difficult to imagine they had volunteered to come to Príncipe from the continent. What were the conditions like then? How much had they improved in a decade, with no real supervision by overseas authorities? Eddington doesn't comment in his letters home except to write, "It is very comfortable here and we have all the assistance and facilities we need. About 600 native labourers are at work on the plantation and they have carpenters and mechanics at work so it is easy to get any small things required."

They dined each night with Carneiro and his friends who came to meet the eminent Englishmen. Since almost no one on Príncipe could speak English except for a few words, these were probably not the liveliest of dinner conversations. They had no steady interpreter, as on other expeditions. Eddington mentioned this predicament in a letter home when recounting the people they met: "Two negro's from Sierra Leone who are the sole staff of the cable-station here. They are British, and interpret for us. But, of course they are only with us now and then." But Carneiro owned a player piano and a gramophone with a large supply of records, some of them grand opera. Music would make up for conversation. Before an early bedtime of nine sharp, the landowner and his guests sat on the balcony wearing cool white suits and listening to the soothing waves of the ocean. It was restful relaxation for the complicated work that lay ahead. It's impossible to know how many bones of enslaved human beings who had worked until they died and were carried, their corpses lashed to poles, into the deep forest by loved ones had been buried on Príncipe by 1919. For an island so small, one can only speculate.

THE CLOCKMAKER OF THRAPSTON

Edwin Turner Cottingham was the last man to believe he would travel halfway around the globe, sent by the astronomer royal and in the company of a prominent astronomer. His main interest lay in clocks and time mechanisms, not eclipse expeditions. He was born in 1869, in Ringstead, Northamptonshire, not far from the banks of the River Nene. His father was a cordwainer, or shoemaker, and his mother had been employed as a household helper. Edwin left school very young to be apprenticed to a tailor in the village. Later, as a teenager, he met Augustus Allen, a clockmaker and watchmaker in nearby Thrapston, and became employed by him.

At the age of twenty-five, he married Elizabeth Smith, whose family owned a local ironworks company. Three years later, their only child, a son, was born. When Allen retired, Cottingham took over the business as its owner. The name COTTINGHAM went up above the wide front window of the shop. He was now in business to build, repair, and sell clocks and watches, a worthy profession for a boy with little education. When Edwin and Elizabeth purchased a house just around the corner, he could stroll through his garden each day to arrive at work. In 1900, he built his first work of art, a Winchester chime clock for the little church in his village. He was not yet interested in the stars. His passion was clocks.

Cottingham had found his calling. He learned the trade well and took pride in his work. The necessity of clocks and watches and time balls in everyday life was valued, even admired. He later built clocks for observatories, including one for the Royal Observatory and another for Edinburgh. The latter had electrical components, demonstrating Cottingham's added passion for the "science of electricity." Some of his clocks saw a good deal of the world, one going all the way to the Royal Alfred Observatory on the island nation of Mauritius, and another to Hong Kong. Like many instrument makers and observatory employees, Cottingham repaired and adjusted chronometers for the Admiralty.

A master craftsman and engineer, he soon discovered astronomy. Since most clockmakers over history were also skilled in the design and construction of scientific instruments, it's not surprising that Cottingham appreciated astronomy. It was an exciting time to be alive in science. The introduction of new and better-designed instruments was opening up the heavens in ways Galileo never dreamed of when he turned his small telescope on the Milky Way and saw a mass of undiscovered stars, not the clouds he believed them to be. The cosmos was filled with uncharted waters. In 1905, the same year that Einstein published his first groundbreaking paper, Cottingham joined the RAS. There he would meet Dyson, Eddington, Crommelin, Davidson, and other prominent astronomers. They would have a great impact on his life.

For a decade before he left England on the 1919 expedition, Cottingham had been taking care of the clocks at Cambridge by cleaning, repairing, and adapting them. Thus, he had earned the friendship and respect of Eddington, the observatory's director. Cottingham was modest and unassuming, not one to show off his diverse skills. He was most at home in his cluttered workshop, which stood behind his store, in the garden at Thrapston. Now Cottingham was on the tropical island of Príncipe, a place Nevinson had called "a magic land, the dream of some wild painter." All around him long, yellow pods dangled from hundreds of cocoa trees, enormous butterflies fluttered like winged rainbows, and bands of mona monkeys—they were brought from the mainland on slave ships—swooped and grunted through the moist canopies overhead. He had come a long way from the boy apprenticed to a village tailor.

SETTING UP CAMP IN SOBRAL

Henrique Morize, the French-born director of the National Observatory at Rio de Janeiro, left that city on April 25 to sail north to Sobral. He was bringing with him his wife and many members of his staff, seventeen people in all, including Theophilus Lee, the English chemist from the Geological and Mineralogical Service of Brazil. Davidson had met Lee at Passa Quatro, when he and Eddington were there for the 1912 eclipse. They had thought the chemist unhelpful and self-serving. Now Lee was back. Also on the steamer from Rio to Camocim was the automobile for the visiting scientists to enjoy. This had been the wish of Colonel Saboya and the Ministry of Agriculture. It's likely that Morize had rented the older-model Studebaker from the Studebaker Company in Rio. With the car came a chauffeur, since it was unlikely anyone in Sobral would know how to drive.[9]

Morize's group from Rio would change steamers at Fortaleza and rendezvous there with the Carnegie Institution team. Daniel Wise and Andrew Thomson had departed from New York City on March 25 aboard the steamer *Hollandia*, out of Amsterdam. The North Americans landed in Recife, Brazil, nineteen days later, on April 15. After sightseeing and doing magnetic observations in and around Fortaleza, Wise and Thomson waited for Morize to arrive. Both groups then sailed on to Camocim, where a special train was waiting to take them the eighty miles inland to Sobral. When the chauffeur drove the Studebaker off the ship, the gathered onlookers cheered wildly. As Morize noted in his diary, "The car came down with a few faults but ran, to the general enthusiasm of the populace." It was then loaded onto a railway car, and the special train pulled out, headed south for Sobral.

Known internationally for his support and hospitality to foreign expeditions, Morize had suggested Sobral to the British as an ideal viewing station two years earlier. He had assisted Davidson and Eddington during their 1912 rained-out expedition farther south at Passa Quatro. He had already traveled the two-thousand-plus miles to Sobral from Rio a month earlier to put in place the arrangements that had welcomed Crommelin and Davidson. Morize reminded all four scientists that expenses were courtesy of the Brazilian government: their room and board and any labor or constructions they might need. While most of Sobral's residents carried their water daily from holes they had dug into the nearby Acaraú's cracked

riverbed, the English astronomers were grateful to find that their comfortable house had water pumped to it from the well of the owner's cotton factory, a mile distant. The daily temperatures averaged a peak of ninety-five
degrees Fahrenheit at three in the afternoon, and a low of seventy-five
degrees at five in the morning. They would need plenty of water cooled in
earthenware pots when it came time to develop the photographic plates.
When Morize and his large entourage arrived in Sobral, they were put up
at a house without running water or mosquito nets.

This great generosity shown to the foreigners, these *estrangeiros*, did
not go unnoticed by at least one member of Brazil's intelligentsia. The
severe drought, the devastation of which Crommelin and Davidson had
seen from their train window, had begun in 1915 and had taken the lives
of almost three hundred thousand people in the state of Ceará alone.[10]
Many survivors who *could* do so had left the state to trek south to the
Amazon area. One of the most admired men of letters during this period was Paulino de Almeida Brito, who had abandoned a law degree
to become a fiction writer, a journalist, and an editor. Well known to
Brazilians, he had in the past taken a strong stand against black slavery in Brazil. Brito had read in the local newspaper a two-page article
Crommelin and Davidson had written while on their Amazon tour. The
piece had been translated into Portuguese by their interpreter, Leocádio
Araujo. They explained as best they could to a lay audience this new
theory "by a physicist named Einstein."[11] They were in Brazil to photograph stars during the eclipse to test for this theory if they were "favoured
with clear skies." Not impressed, Brito published his own thoughts on the
subject:

> However, since all that is human of human weakness resents, Astron-
> omy, in the midst of such majesty, does not cease to have its small and
> even puerile sides . . . [W]hy do not we even say comical? When they
> arrive, for example, in Sobral, Drs. Crommelin and Davidson will be
> surrounded by desolate scenes, in the theater of the ravages produced
> by the drought. And for those poor people who starve to death, it would
> certainly be more important for the two illustrious scientists to discover

the means of dropping an ounce of water on the scorched earth, than to remove it once and for all if the solar attraction curves or do not bend the rays of light from the stars . . . What is the use of knowing the weight of the planet Mars and the distance that separates it from Saturn, if we do not know enough of our terrestrial habitation to remedy or even prevent the bewilderments that endanger our existence?[12]

LIFE AT ROCA SUNDY

On April 28, Eddington and Cottingham left Santo António and again rode on mules until they were met by a carriage that took them the rest of the way to the plantation. They were welcomed by the manager, Mr. Atalia, and made comfortable in the main house, which was quite elaborate in furnishings. Its second floor veranda had windows that opened onto the sea. While mosquitoes were not bothersome, their beds were nonetheless hung with curtains. And the astronomers were advised to take three grains of quinine daily, the local practice for guarding against malaria. Their baggage and equipment was delivered by porters later in the afternoon, again on the mule-drawn tram except for the last half mile, when it was carried through woods. Eddington's plan was to spend a week at the plantation, erecting the huts and getting the apparatus ready. He and Cottingham would then enjoy another week of sightseeing with Carneiro before returning to Roca Sundy, where they would stay for the duration of their time on Príncipe.

Although Santo António was nothing more than a tiny village with muddy streets in the wet season, it was still a world apart from country life at Roca Sundy, where even bedtime was a half hour earlier. Virgin jungle wrapped itself around the plantation, a home to parrots, bats, snakes, feral pigs, and rats. A volcanic island in its infancy, Príncipe had few animals before it was inhabited by humans. Over the centuries, it had to depend on the transport of animals to its shores from the mainland, often aboard slave ships. Birds and butterflies, however, were abundant and brilliantly colored. Along with the numerous cocoa trees, there were also banana, breadfruit, and coffee trees. Eddington noted that the cocoa trees were covered with yellow pods. "It was a very fine sight to see the large golden pods in such numbers—almost as though the forest had been hung with Chinese lanterns." His observation meant the trees were past the flowering season, when tiny orchid-like blossoms, pink and white, grow directly from the trunk amid bright green leaves. Thus, the cocoa harvest wouldn't begin until after the Englishmen had left the island.[13]

There were obviously house servants at Roca Sundy to do the cooking, cleaning, and laying in of supplies. But again, at least in the letters that survive, Eddington never comments on them, although we know the name of the waiter in Funchal who carried his baggage down to the ship. He was pleased, however, that a meat or egg dish was served for breakfast since it was closer to what he was used to back in England. The rest of the time, the foods and rituals were typically "foreign," except that the Englishmen were served tea in the afternoon, a compliment of their host since it was not a practice for the Portuguese. There were plenty of fresh fruits. Pineapples and bananas grew wild on the island, as did African custard apples, the latter of which the Englishmen did not find flavorsome. Eddington got back into a routine of eating a dozen bananas each day, large red ones. He was right. He and Cottingham had found themselves in "clover," every need met quickly and hospitably. The path of totality for this eclipse had been most generous to fall across that small tropical island. Yet England was often on their minds as the days passed. In a letter home, Eddington sent birthday wishes to Punch, his terrier, and wondered how the garden was doing.

Eddington and Cottingham enjoyed chats with the plantation manager, Mr. Atalia. A former soldier born and raised in Portugal, he had been at Roca Sundy for four years. Atalia and Eddington were able to communicate in a very basic French and had long talks over dinner. Eddington sensed that the manager was lonely there on the island and happy to have guests. Darkness fell on Príncipe by six o'clock, so in the cool evenings, the three men sat outside before bedtime. "After dinner we used to sit out in front of the house and there was generally a succession of natives came up to interview him on all sorts of matters. They evidently have great respect and confidence in him." It's possible that Eddington witnessed hard work, but not abuse while at the plantation. In his book, Henry Nevinson wrote of a roça he had visited on São Tomé, one considered a "model," or, as the writer put it, "a show-place for the intelligent foreigner or for the Portuguese shareholder who feels qualms as he banks his dividends." Had Roca Sundy been put in shipshape condition before the Englishmen arrived? Regardless, what could Eddington, a scientist, have done to change things when he was two years invested in an eclipse?

They finished the task at hand by first erecting the two huts. Eddington did not want to unpack the mirror just yet for fear it would tarnish in the damp climate. The stone pier he had asked for, on which he would mount the coelostat, had already been built by the workers at Sundy. On April 30, with the huts just up, a heavy downpour of rain commenced. It was good timing to test how waterproof the huts would be. Eddington was satisfied that they "stood the deluge splendidly." On May 1, he began setting up the apparatus, but was still not ready to unpack the mirror. Then he and Cottingham went back to Carneiro's house for more relaxation and sightseeing. When they returned to Roca Sundy a week later, they would remain there until after the eclipse, and then board their steamer back to England.

THE HORSE JOCKEY CLUB

Meanwhile, in Sobral, Crommelin and Davidson had unpacked their instruments in the welcome shade of the racecourse's covered grandstand. The soil of the track itself was firmer, Crommelin noticed, with coarse shrubs holding it in place and resulting in less dust. It would work much better than the sandier soil around the house and grounds, an area known for unexpected winds. Daniel Wise, from the Carnegie Institution, set up his instruments in the basement of the house, where the day-to-night temperature change was minimal and where he would remain during the eclipse itself. Andrew Thomson would make visual observations on the racetrack, not far from where the English astronomers would be set up. Sharing the same residence as they were, and with Wise's equipment in the basement, the Englishmen took an interest in the observations that the Carnegie team was doing. A scientific camaraderie quickly developed. "Incidentally, Mr. Wise redetermined our latitude," Crommelin wrote, "his value, −3° 41′ .5, being 0′ .2 north of the value given by Dr. Morize. As the region had not been systematically surveyed, we found satisfaction in this close accordance."

Bricklayers were brought in to build piers that would hold the coelostats and carpenters for the sixteen-inch astrographic tube. A square wooden tube, nineteen feet long, had been shipped from England for the smaller telescope. Leocádio Araujo was present to interpret for Crommelin and Davidson as they gave instructions to the workers. The carpenters erected the hut, but did not secure its sections. They planned to finish the job after lunch. While they were eating, a sudden whirlwind blew in across the racetrack. When the astronomers returned, the hut was overturned and many of its beams broken. The workers managed to repair the damage using beams that had been brought from Greenwich for the construction of a darkroom, now no longer needed in Saboya's accommodating house.

Finally, Crommelin lay down their meridian line, using the pointer stars in Ursa Major, which was so low on the horizon that Polaris, the North Star in Ursa Minor, was not visible. When taking some early experimental photographs, he noticed a problem with the coelostat's mirror for the sixteen-inch telescope. The smaller eight-inch mirror had been silvered at Greenwich, but the two larger mirrors had been sent away to be silvered.[14] Now it was apparent that the sixteen-inch had an astigmatism. To avoid distortion on the photographic plates, the men decided on an eight-inch stop. They then found that the drive for this same coelostat didn't run evenly, which meant blurred images during the eclipse. They decided that brief five- and ten-second exposures would be necessary to correct the uneven motion. To obtain a few preliminary check plates, they focused the telescope on the star Arcturus and photographed the field around it. With nothing to do now but wait for the eclipse, the scientists were ready for some sightseeing.

A RIDE IN THE STUDEBAKER

Vicente Saboya and the Ministry of Agriculture, in wanting to protect the foreigners from yellow fever, arranged for the car to fetch them early one morning, when the offending mosquitoes were not likely to bite. Plus, the Brazilians had a new road to show off. A previous drought had put in place a relief work project that employed laborers in building a road that corkscrewed to the top of Serra da Meruoca, six miles to the northwest of Sobral. The road had been completed the year before at a cost of two hundred dollars, but an automobile had yet to climb it. The chauffeur, after cleaning the car and taking it on a trial run, had informed Morize that "pieces were missing." Still, he drove it up in front of Saboya's house to collect the visitors. This initial outing began poorly. First off, they were to leave Sobral at 6:00 a.m. "We were picked up at 7:15," Wise wrote in his diary, "and before we were out of town the chauffer began to show just how smart he was by driving very carelessly and recklessly. We had to warn him to drive with a little more consideration for those in the rear seat." The appearance of a headless horse, especially outside town limits as the Studebaker rumbled toward the mountain, four *estrangeiros* bouncing around inside, sent locals fleeing in alarm.

Wise referred to the car as an "old Studebaker." Even if it were a brand-new 1919 model, the vehicle would still give a bumpy and crowded ride. But the car had other problems. As it climbed the steep and zigzagging road to carry the guests twenty-seven hundred feet to the top, the engine overheated several times. Each time, the men climbed out and "prospected the mountain" as they waited for the engine to cool down. Seeing the mists and clouds, Crommelin knew immediately that the mountaintop would not have been a good eclipse station, as had been considered earlier. The Studebaker finally fought its way almost to the top before it had its first flat tire, the left rear. As they now waited for the driver to change it, the scientists walked on ahead to "prospect" some more. They saw occasional farms that were barely surviving, even in that cooler climate, given the drought below. They ate delicious mangoes and guarana seeds, the latter of which held twice the caffeine of coffee beans. At a farina mill, the men watched as workers pressed the wet manioc roots to prepare them for drying. The astronomers nibbled on cashews, which are native to northeastern Brazil.[15]

When they arrived back at the Studebaker, the driver told them that his air pump wasn't working well enough to pump up the spare tire, which was low. They had to forfeit the summit and head back to Sobral at once. The car had driven a short distance when the tire blew out. With the pump not working, they continued down the mountain on a flat, until the wheel came off. Having consumed those guarana seeds probably didn't help on such a downhill ride. "In one of our repairing spells," Wise wrote, "Crommelin and Thomson gave up and started back on foot." The wheel put back on, and the car still running on the flat tire, they picked up the two hikers farther down the mountain. When the wheel came off again, the four men decided that walking was less stressful. They may not have known that this hottest time of day was biting time for the yellow fever mosquito. Or that this first outing in the car would seem pleasant compared with the hair-raising rides still ahead.

Walking the four miles back to Sobral, the men didn't arrive until late afternoon. "We were all very hungry, and done up from the heat, not to mention being very thirsty," wrote Wise. A fine lunch at their interpreter's house put them back in good spirits. Theophilus Lee nonetheless tattled to Morize that night that the city chauffeur was not just rough on the automobile, he was "driving at unbridled speeds" and "running down animals." The chauffeur denied everything. The troubles with the Studebaker continued. The headlights didn't work, and when the radiator burst, the driver repaired it with a bicycle tube. On another trip to Serra da Meruoca with the visiting scientists, he purposely backed the car into a ravine to keep it from careening out of control down the mountainside. This incident, however, was not his fault, for it was discovered that if the Studebaker went faster than five miles per hour, especially going uphill, it stalled and the brakes wouldn't work.

After that death-defying ride, Davidson and Crommelin sent their interpreter to inform Morize they would no longer ride in the car with that same driver. Morize wrote in his diary, "I think the English are right when they say they do not want to go to the mountains anymore." Thus went Colonel Saboya's and the Ministry of Agriculture's plans to keep their famous guests safe from yellow fever.[16]

PRÍNCIPE: THE NIGHT BEFORE THE ECLIPSE

Arthur Eddington and Edwin Cottingham were done sightseeing on Príncipe. The last heavy rain had fallen on May 9, or so they thought. By the time they returned to Roca Sundy a few days later, the gravana, the dry season, had already set in with its scattered clouds during the daytime. Eddington almost preferred the rainy season for wiping out the clouds. But there were a few clear afternoons as the eclipse day approached, so he was optimistic. In mid-May, with the installations and adjustments complete, he took check plates on three nights. Although they had also been done at Oxford earlier that year, it was best to do a second set, given the different conditions of observation between England and Príncipe. Plus, Eddington worried that a systematic error might occur if any changes had been done to the lens during transportation. With the house's water temperature reaching seventy-eight degrees during the daytime, the water was too warm to develop them then. The plates had to be done after midnight, between 12:30 and 1:00 a.m. Thanks to a supply of ice, courtesy of Mr. Gragera, the manager of the Colonial Agricultural Society, the developing was completed. It was a tedious job that kept both Eddington and Cottingham up until the wee hours. During the day, Eddington measured. The check plates were ready. He and Cottingham unpacked the mirror. Now, it was just a matter of waiting for the eclipse.[17]

How to imagine the tension that any of these men felt the night before an eclipse, especially if his observation might lead to an important announcement or discovery? Resting on the whims of nature were years of planning and funding, separation from family and friends, and the perils and strain of long-distance travel. For some, as with Charles Perrine, success or failure could determine the funding for a future expedition. On that eclipse eve, as Eddington settled into bed behind the mosquito curtains, he could have no way of knowing that William Campbell and H. D. Curtis were frantically working to obtain results from their Goldendale observation. The outer world had almost ceased to exist on the moist tropical island, that "magic land, the dream of some wild painter." His thoughts were on the weather the next day. Would it be friend or enemy? Would the gravana carry him through to a cloudless sky?

Along with the image of a four-year-old boy counting stars in the dark skies of England in 1886, another sadly prophetic sign is the poem that appeared on the same *British Friend* page that mentioned the death of the astronomer's father, Arthur Henry Eddington:

> *When clouds are gray and gather low*
> *T'will never do to sorrow,*
> *Above the chilling doom there's light,*
> *All will be bright tomorrow.*

> *Yet mid the darkness there is hope,*
> *All will be bright tomorrow.*

> *On the morrow—the glad tomorrow*
> *The skies will be forever fair.*
> *Trust in the Love that, changing never,*
> *Lifts the clouds of doubt and spare.*

10

"THROUGH CLOUD, HOPEFUL"

May 29, 1919: Does Light Have Weight?

A null result is not necessarily a failure. The present eclipse expedition may for the first time demonstrate the weight of light; or they may confirm Einstein's weird theory of non-Euclidian space; or they may lead to a result of yet more far-reaching consequences—no deflection.

—*Arthur Eddington, Observatory, March 1919*

THE SECOND-CLOSEST STAR cluster to earth is 153 light-years away and is believed to be about 625 million years old. This rich group of stars is in the constellation Taurus, the bull, whose name dates back to ancient times, when bulls were worshipped in the Middle East. The bull's head is marked by a V-shaped cluster known as the Hyades and is cataloged as Melotte 25. For the ancient Greeks, the Hyades were the daughters of Atlas, who was condemned to hold the sky above his head for eternity. Some legends have them as five sisters, but in reality, 15 stars in the cluster shine brighter than fifth magnitude and can be seen with the naked eye. Hold binoculars on the Hyades, and you will see more than 130 of its stars that are brighter than ninth magnitude. According to one legend, Hyas, a brother to the Hyades, was killed while hunting. His death made the sisters start crying from grief, and their tears turned to rain. Thus, the Hyades are known as "the rainy ones" and are associated with wet and stormy weather. It was this bright cluster of stars that would be closest to the sun during the 1919 eclipse.

The last thing that Eddington and Cottingham hoped for as they waited for the eclipse on Príncipe was rainy weather. On the other hand, in Sobral, rain would have been a gift from the gods for that drought-stricken area, despite certain defeat of Crommelin and Davidson's scientific plans. Although May would ordinarily mark the end of the rainy season in northeastern Brazil, the area had seen scant rain since the astronomers arrived, other than a few brief afternoon showers. And brief showers during any drought meant only temporary relief. The impending natural disaster is most likely what had prompted writer Almeida Brito to criticize the very idea that two Englishmen were visiting in the midst of a devastating drought and yet were wishing for no precipitation on eclipse day. But just as Eddington hadn't selected a path of totality that threw the moon's shadow over six hundred "imported labourers" on Príncipe, neither had Crommelin and Davidson chosen a place of parched desolation that was killing thousands. As always, nature had made this decision all on its own.

MORNING IN SOBRAL

Several newspapers in northern Brazil had published articles about the expeditions to Sobral. One was *Correio da Semana*, a Catholic publication pleased to inform its readers that both Andrew Crommelin and Charles Rundle Davidson were devout followers of the faith. Other papers encouraged locals to be on their best behavior while these esteemed guests were in town. Morize had also given interviews, reminding citizens that what was about to happen was a natural phenomenon. He acknowledged that "ignorant men of the wilderness" would panic and resort to beating on cans and drums—the can beating was to encourage plants to continue growing—all to dissuade "the infernal deities" not to destroy the sun god. "The truly civilized man, however, does not commit any of these absurdities," he said. Morize also cautioned against the use of firecrackers since their bright sparks would interfere with the cameras when the scientists were taking photographs. During his visit in 1912, Eddington had noticed and commented that the setting off of firecrackers was "the regular Brazilian way of demonstrating." But the local authorities had agreed to restrict the general public from entering the racetrack area during the eclipse.[1]

In the predawn hours of May 29, at Saboya's house, Crommelin and Davidson woke to clouds filling the dark sky overhead. The partial eclipse would begin there at 7:46 a.m., so they needed to be up early to prepare. There had been rainfall in Sobral four days earlier, which locals had welcomed for the needed precipitation. Crommelin welcomed it, too, since the rain helped control dust at the racetrack. When they were taking experimental photographs, he and Davidson had noticed that the coelostats and clocks were sensitive to the winds that periodically swept over the grounds of the jockey club. From their experience at previous expeditions, they were well aware that gusts of wind during totality were not unusual, given the sudden temperature drop when the moon blocks out the sun. As the open racetrack was already vulnerable to wind, they had arranged for windscreens to be positioned around the hut in places where the screens wouldn't interfere with the observations. The screens would also protect the parts of the telescope tubes that would be exposed to direct sunlight.

After an early breakfast, the Englishmen made their way out to the racetrack. Carnegie's Daniel Wise, from Williamsport, Pennsylvania, stayed behind. He would spend those exciting minutes of totality in the basement, surrounded by his magnetic instruments. But Andrew Thomson, born in tiny Dobbington, Ontario, near Owen Sound, would make a direct observation on the racetrack, not far from Crommelin and Davidson. Since dawn, the churches in Sobral had begun filling up, jammed with superstitious souls who knelt and prayed for divine protection. The more courageous crowded into the town square to wait. One journalist wondered where the citizens had found so many little telescopes, the smoky shards of glass now well known at eclipse gatherings around the world.

Most of South America would witness this eclipse as a partial. The narrow path of totality, however, would begin before sunrise in the dark waters of the South Pacific. The moon's umbra would move onshore at the southern tip of Peru and the northern tip of Chile. It would then cut across the center of Bolivia, clipping the southern reaches of the Amazon rain forest and traveling at an average speed of twenty-five hundred miles per hour, before it reached Sobral. As Davidson and Crommelin inspected their instruments, they had no way of knowing that across the ocean, Eddington and Cottingham were watching a heavy rainstorm beat down on Roca Sundy. This was a rare thing for that time of year on the island of Príncipe and the worst storm they had seen since arriving there a month earlier. But there was still time for a change in the weather, as the total eclipse would not reach Africa until the early afternoon.

But Sobral was now minutes away from totality, and the sky still heavily clouded. One newspaper, Camocin's *Folha do Littoral*, wrote of watching the tall, slender figure of Henrique Morize, striding back and forth among the huts and tents, nervously waiting. After all the funding, all the months of preparation, and thousands of miles traveled, would he get the photographs he hoped for of the solar corona? At 7:46 a.m., the partial eclipse began. All the astronomers were in their places at the instruments, staring up at the cloudy heavens. Earlier, in darkness, the English team had loaded the plates for the sixteen-inch telescope, which Davidson would oversee. And the smaller plates had been loaded for the four-inch aperture, in Crommelin's care. The smaller glass had been decided on almost as an afterthought when Cortie advised that it be taken along as a backup telescope, with the square brown box that served as its tube. Loaned to the Greenwich team by the Royal Irish Academy, it had a nineteen-foot focal length and had been used successfully by Cortie in Sweden, when the Jesuit was banned from Russia in 1914.

But it seemed at the outset in Sobral that neither telescope would be useful if the sun stayed hidden. At first contact, when the shadow appeared to touch the sun's outer edge, the sky was nine-tenths covered in clouds. This was not an auspicious sign. The total eclipse would begin just an hour and thirteen minutes later. Of all the teams representing multiple nations that were spread out from continent to continent along the path of totality, only two would test for the light deflection predicted by one Professor Einstein. The first team was now clouded over on a dusty racetrack in Brazil. The second team was waiting thirty-three hundred miles away on an island cocoa plantation, watching it rain.

A minute before second contact, with the total eclipse seconds away, luck shifted for the astronomers. According to Morize, a sigh of relief was heard coming from all the spectators. "As totality approached," Crommelin wrote, "the proportion of cloud diminished, and a large clear space reached the sun about one minute before second contact. Warnings were given 58s., 22s. and 12s. before second contact by observing the length of the disappearing crescent on the ground glass." When the crescent vanished, Araujo, their faithful interpreter, shouted, "Go!" and set the metronome in motion. Araujo then called out every tenth beat as totality occurred. The exposure times were recorded in terms of those beats. The gusts of wind they expected did not arrive; nor did the temperature drop noticeably, as often happens when the moon's shadow covers the earth below. The sky at totality was not as dark as Crommelin had witnessed during previous eclipses. It appeared as it might a half hour before sunrise. Everything around them was deadly calm except for the sounds of cameras clicking inside the canvas huts on the racetrack. For a minute during midtotality, a ragged cloud floated over the sun. Other than that, the astronomers were able to use about four of the expected five minutes and thirteen seconds promised them by nature. Davidson managed nineteen photographs with the Greenwich telescope, and Crommelin eight with the smaller four-inch lens.

The total eclipse was over at 9:03 a.m. local time. The shadow had long moved on to the coast of Brazil and was now crossing the Atlantic Ocean. The collective mood of the astronomers on the racetrack was upbeat, but no one would really know until the photographic plates were developed. Morize, so helpful to his fellow astronomers around the world despite speaking what Eddington called "the worse English I've ever heard" and a man said to be "timid and discreet" by his colleagues, nonetheless described this wonder of nature in very poetic terms:

At that moment, everyone, even the simple onlookers who surrounded the camp, felt moved by the magnificence of the spectacle that was manifesting itself. The sky darkened, becoming as before dawn, but without the gilded and reddish clarity of the East, and it was a dark blue that quickly brightened in the vicinity of the now completely blackened solar disk. Around it, one could see the crown of changing color, with shades of mother-of-pearl . . . on which stood out in intense red a beautiful protuberance, which is one of the largest that has been observed.[2]

This was the same magic that had frightened Paleolithic hunters and gatherers as they peered heavenward. It had amazed the early Chinese, Egyptians, and Babylonians as they made calculations and drew up their charts. It had summoned modern men and women of science to travel the globe, no matter the difficulties or the dangers. At 10:28 a.m. local time in Sobral, at fourth contact, the moon's shadow disappeared from the sun and it returned in its full glory. It was finished, as if the magic had never happened. In just over two hours, the full eclipse would first reach the southern tips of Liberia, French West Africa, and the British Gold Coast before touching the tiny island of Príncipe.[3]

From Sobral, Crommelin sent a telegram to Dyson: "Eclipse Splendid."

A THUNDERSTORM AT ROCA SUNDY

When Eddington and Cottingham had finished their breakfast on the morning of May 29, their counterparts were just rising in Sobral, in the dark hours of predawn, to prepare. If someone had given the six hundred "imported" workers a scientific explanation of what would occur that afternoon on Príncipe, if they were told that infernal demons were *not* devouring the sun god, as Morize had clarified in Brazil, would the explanation have brought them comfort? Would they have a better knowledge than did the citizens filling up the Catholic churches in Sobral? In the past forty years, the people of Africa had seen many partial eclipses. The path of totality for at least three total eclipses had swept over Angola, the homeland for most of the Roca Sundy workers. In addition, an annular eclipse in 1900 and a more recent one in 1918 had cut across the heart of that country, the moon blocking the sun's center and leaving what would look like a burning ring of fire in the sky. The laborers on the plantation, whether free or not, had experienced the disruption of the natural world when the universe seemed to misbehave. They would have learned and repeated rituals passed down for centuries from their ancestors. The beating of drums was likely heard that day at Roca Sundy.

Eddington had no way of knowing that a layer of clouds was covering the sky over Crommelin and Davidson when he had welcomed excited guests to Roca Sundy that morning. Coming to his plantation to stay throughout the eclipse was Jeronimo Carneiro, the owner. With him were the curador and the judge, the two men with whom Eddington and Cottingham had enjoyed afternoon games of tennis. The British Mr. Wright from the cable station, who translated for the Englishmen, and three local physicians were also in tow. They arrived in the midst of a thunderstorm, with a tremendous downpour of rain at their heels. In England, the Hyades were once known as the April Rainers. But the deluge was most unusual for the gravana season. The rain didn't stop until one o'clock in the afternoon, when rays of sun shone down. Eddington noted that the rain had probably cleared the sky, but the clouds soon collected again overhead. The total eclipse was due in two hours.[4]

It would have been an anxious morning for Eddington. With an annular due late that same year, and two partials coming in 1920, the next *total* eclipse would not occur until October of 1921. Useless to astronomers, it would reach its maximum in the waters of the Southern Ocean before touching land at Antarctica. The next total eclipse would not occur until September 1922, with five minutes, fifty-one seconds of possible observation at Ethiopia and Somalia, before it crossed the Indian Ocean to traverse all of Australia. If this eclipse of 1919 didn't give the British an answer to Einstein's prediction of light deflection—and if William Campbell was unable to secure positive results from Goldendale, Washington—then it was likely that a multitude of national expeditions would be spread out along the path in 1922 to tackle the relativity question.

Hoping for the best, despite the blanket of clouds overhead, Eddington and Cottingham had the Oxford telescope carried outside to the chosen spot beyond the bedroom windows. As had been done in Sobral, a canvas screen was arranged around the instrument to protect the lens and tube from wind and direct sunlight. The coelostat was readied onto the brick pier that had been built for that purpose by the workers a month earlier. Five hundred feet below, sloping down from the three-million-year-old volcanic plateau where the main house stood, the waves of the South Atlantic washed against the sand. This was the path the sun would take. The astronomers and their guests waited.

The total eclipse would occur at 3:13 p.m. At 1:46 p.m., the partial began, amid clouds. At 2:45, the sun's crescent peeked in and out from behind them. At 2:55, the observers could see the shrinking crescent almost continuously, but still through drifting clouds. Around the sliver of sun, clear patches of blue sky now appeared. As totality began, the astronomers could do nothing but take photographs and pray for the best. Cottingham called out the exposures and managed the driving mechanism of the coelostat. Eddington changed the plate slides. "We had to carry out our programme of photographs in faith," he wrote home. "I did not see the eclipse, being too busy changing plates, except for one glance to make sure it had begun, and another half-way through to see how much cloud there was." They managed to take sixteen photographs in all. In five minutes and nineteen seconds, it was over. A few minutes later, the sun floated free in a perfectly clear blue sky. Soon, the mainland of Africa, where the partial had already begun, would receive the maximum eclipse. The shadow would then move on across several countries on that continent to die near sunset in the waters of the Indian Ocean, before reaching land at Madagascar.

DEVELOPING THE PLATES IN BRAZIL

With the clouds having cleared away before the full eclipse began, Crommelin and Davidson had taken quite a few photographs. Transporting glass photographic plates to the other side of the world was worrisome enough. But the handling of them onsite had to be done carefully to prevent any breaking or cracking of the glass. Because of the warm temperature in Sobral and lacking any supply of ice, development was carried out during the late hours of the night leading up to dawn. The plates had to be loaded in total darkness, and they were now unloaded the same way. A safelight could be used to watch the times during development. With locally made earthenware pots as water coolers to lower the temperature—it was seventy-five degrees just before daybreak—Davidson began the job the first night after the eclipse. For added cooling, the developing dish was placed inside a larger dish filled with melting hyposulphite of soda. Because the films would soften in the warm solutions, formalin was used to harden each one and minimize any possible distortion.[5]

Before the teams left England, Eddington had ordered two brands of plates for the expeditions, Ilford and Imperial. When Davidson had developed the check plates days earlier in Sobral, he had already discovered that only the Ilford plates would respond well enough in the warm temperature. He and Crommelin began with the Greenwich telescope, which had taken nineteen photographs with its wide-ranging field of view.[6] By 3:00 a.m. on May 30, they finished the first four of the larger plates. When they were examined, they showed twelve stars. But the images were out of focus, the stars fuzzy and diffused. Since this was not the case for the check plates that had been done on nights before and just after the eclipse, the astronomers concluded that the sun's heat on eclipse day had affected the curvature of the coelostat mirror. The several star images on the remaining large plates also turned out to be hazy.

The eight-inch mirror feeding the four-inch lens, however, had performed well. This smaller telescope, with its eight-by-ten-inch plates—the exposure times were twenty-eight seconds—succeeded in getting eight photographs. Of those, all but one showed seven well-defined stars. The single flawed photograph had been rendered unusable because of cloud covering. Although the solar corona was not the subject of attention for this expedition, the plates still revealed some detailed, even spectacular images. Crommelin noted that to the naked eye during the eclipse, the corona appeared three times brighter than the moon. The two astronomers worked diligently over the next few days, as did the Carnegie team. When both Davidson and Wise came down with a fever during this time, this symptom was most worrisome, given the many cases of yellow fever in the vicinity. Protecting the visitors from this disease had been Saboya's foremost concern in hosting the foreigners. A doctor was quickly called to the house and pronounced the fevers as symptomatic of the common cold. The men went back to work.

By June 5, Crommelin and Davidson were finished with developing the plates. They were now exhausted from the late-night work and the taxing climate. Before they would take a second set of check plates in mid-July, they were ready for the cool ocean breezes at the very urban Fortaleza, the state capital of Ceará. On the racetrack, workers dismounted the driving clocks after Davidson had marked the mirrors so they could later be returned to their original positions. They were then carried into Saboya's house to keep them safe from the blowing dust. The telescopes and coelostats were covered over and left in place in the hut. On June 7, Crommelin and Davidson bid their colleagues farewell since Wise, Thomson, and Morize and his entourage would be gone by the time the two Englishmen returned to Sobral. The chauffeur had already been careening around town to use up the oil in the Studebaker's engine so that it could be later stored. The two men boarded the train and rode north again to Camocim. There they caught the steamer *Turyassu* on to the coastal city of Fortaleza. They needed to recharge.

Since their housemates Wise and Thomson had spent a month in Fortaleza before the eclipse, the Englishmen knew what to expect. The city had a population of seventy-five thousand, numerous cafés and cinemas, and green parks and gardens that overlooked the sea. The brightly colored mud-brick houses with their red-tiled roofs hugged the streets, the windows lit by electricity. Quality shops offered handmade Brazilian laces made on wooden bobbins and spindles. There were plazas, town squares, and cathedrals. Pagoda-shaped refreshment stands sold mangoes, dried bananas, green coconuts, and palm wines. Even automobiles existed in Fortaleza, and they didn't frighten pedestrians when they rolled past, a wild driver at the wheel. Electric tram tracks were laid out for transportation, and bands played music at night in the open-air plazas. Compared with the five weeks they had just spent in Sobral, a mere 125 miles west as the crow flies, this was heaven. With the hotels packed because of scheduled

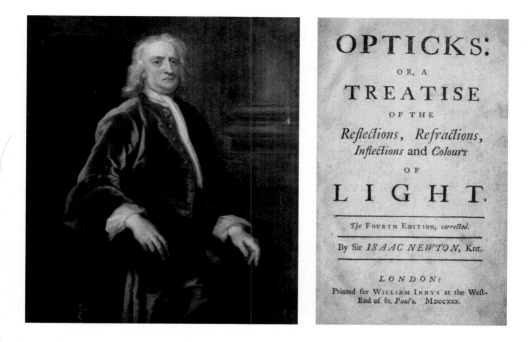

Sir Isaac Newton, in the John Vanderbank painting that hung in the Royal Society's meeting room at Burlington House when the results of the 1919 expeditions were read. Newton's book *Opticks* first asked the question of light bending in 1704. Below, Johann Georg von Soldner, whose 1804 paper calculated the same light deflection as that predicted by Newton a hundred years earlier. Prussian photographer Johann Julius Friedrich Berkowski was the first person to capture the solar corona on a photograph. He made this daguerreotype during the total eclipse of 1851.

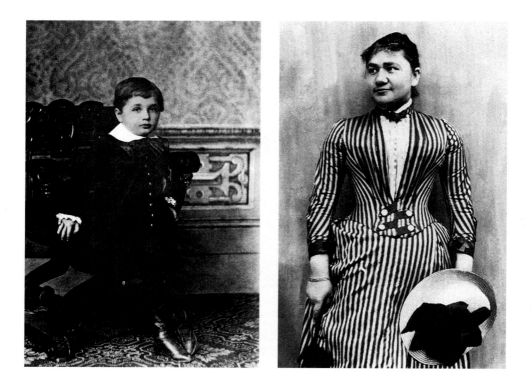

Above, young Albert Einstein and his mother, Pauline Koch Einstein. Below, on January 6, 1903, Albert and Mileva Marić were married at the registrar's office in Bern, with two friends as witnesses. After a celebratory dinner at a local restaurant, the couple returned to their apartment near Bern's famous clock tower. Mileva Marić Einstein with her sons, Eduard and Hans Albert, 1914.

Lick Observatory, on Mount Hamilton, the world's first permanently occupied mountaintop observatory, was funded by the wealthy James Lick. Below left, William Campbell, director of the Lick Observatory from 1901 to 1930. Below right, Charles Perrine, who left Mount Hamilton in 1909 to become director of the National Observatory at Córdoba, Argentina.

Above, Flamsteed House, Royal Observatory, Greenwich. The basement was dug out and bed-rooms added to accommodate Dyson's large family. The wall beneath the top four windows had been vine covered before the arrival of the Dysons in 1911. Below left, Arthur Stanley Eddington, as a student at Trinity College. Below right, Frank Dyson, astronomer royal from 1910 to 1933. In this capacity, Dyson both lived at and ran the Royal Observatory.

Above, the teams at Passa Quatro, Brazil, for the 1912 eclipse. Front row, holding his hat, is Henrique Morize. To his right is James Worthington. Middle row, second from left and holding cigarette is Theophilus Lee. Fourth from left is Davidson, and to his left are Stefanik, Eddington, and Atkinson, looking away from the camera. The rest are volunteers and team members. Below, Erwin Freundlich, the first astronomer to assist Einstein in his quest to photograph starlight. Perrine's camp in Cristina, Brazil, 1912, in the first attempt to test for light deflection.

Above, Henry Moseley, at the Balliol-Trinity Laboratories, Oxford. Moseley died in 1915, at the Battle of Gallipoli. The first photo taken in an attempt to prove light deflection was taken by Charles Perrine, during the 1914 eclipse in Crimea. Minutes before totality, clouds prevented Perrine from becoming the first person to prove that light had weight. Below left, Second Lieutenant Raymond Lodge was killed in action in World War I. Below right, his father, famed physicist and physical researcher Sir Oliver, later sought to contact his son with the help of mediums.

The proportion of the SUN eclipsed as seen in the BRITISH ISLES

From *The Graphic*, August 22, 1914: "By a strange coincidence, at the very moment when all Europe is joined in the clash of battle the world will witness Nature's most awe-inspiring phenomenon, which in olden times—when men were 'dismayed at the signs of heaven'—struck terror into all hearts. Today (Friday) there will be a total eclipse of the sun, visible as a partial eclipse in London, where it begins at 10.59 a.m. and ends at 1.51 p.m., the greatest phase occurring at 11 minutes after noon. The portion of the earth upon which the penumbra, or partial shadow, will fall includes the area involved in the Great War. In Germany and Austria (omen faustum?) the eclipse will be nearly total." *The Graphic* was a weekly British illustrated newspaper published from 1869 to 1932.

The Lick team at Goldendale, Washington, 1918. The night before the eclipse, the Campbells received word that their son, flying ace Captain Douglas Campbell, had been wounded in a dogfight in the skies over France. Only after the eclipse did they learn that their son survived. Below, fighter pilots Eddie Richenbacker, Campbell (center), and Kenneth Marr, of the famous 94th Aero Squadron.

BOOTH LINE. R.M.S. ANSELM. (5500 TONS.)

On March 8, 1919, the second of four steamers named RMS *Anselm* left Liverpool with the British teams on board. Eddington and Cottingham would be left in Madeira for a month, while Davidson and Crommelin would travel on to Brazil. Below, from the Booth Line's passenger catalog, advertising its sightseeing tour a thousand miles up the Amazon, with "unexplored forests" on both sides. With time to spare before the eclipse, Crommelin and Davidson took this journey, from Pará to Manaós, on the *Anselm*. This *Anselm*, launched in 1905, was sold to Argentina in 1922.

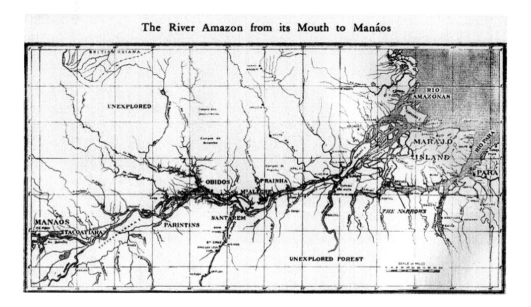

The River Amazon from its Mouth to Manáos

The jockey club in Sobral, Brazil, where several teams were set up. The British and Americans were housed at Vicente Saboya's posh home (the prominent building at left), which faced the racetrack. The Royal Observatory team stored their equipment under the grandstand, center, and then set up the observing hut toward the left of the photo. The Brazilian team was situated in front of the church, in Patrocínio Square, its steeple visible above. Below, spectators and astronomers on the racetrack where the observing tents were set up, on May 29, 1919, the day of the total eclipse, which had occurred that morning.

The Greenwich equipment, in a photo taken by Charles Davidson on June 1, 1919, after the eclipse. In his report, Arthur Eddington described the "square brown wooden tube" brought from Greenwich to hold the four-inch telescope that Cortie, the "Jesuit astronomer," fortuitously advised be included. Below, seated, holding straw hats, are Andrew Crommelin and Charles Davidson. To Davidson's left is Andrew Thomson, of the Carnegie Institution. Back row, second from left is Theophilus Lee, and to his left are Daniel Wise (also of the Carnegie Institution) and Henrique Morize, director of the National Observatory in Rio de Janeiro, and members of Morize's entourage.

Above, a typical cocoa plantation on Príncipe during the time Eddington was there to view the eclipse. Buildings where the many laborers from Angola were housed are shown at right. Below left, discarded wooden shackles in the interior of Angola, 1908, left on the trail for the next slave trader to find and put to use. Attached to the neck, these shackles would prevent any captives from fleeing during the long march to the ships waiting to take them to Príncipe or São Tomé. At right, Edwin Turner Cottingham, the clockmaker who accompanied Eddington on the 1919 expedition.

The Lick team would travel nearly twenty-three-thousand miles to Wallal Downs, Australia, for the 1922 eclipse. William Campbell christened his mess tent "Café Einstein." Below, after the eclipse, workers transport crates of equipment back to the *Gwendolyn*, anchored three miles offshore.

Stars in the vicinity of the sun, during the 1922 eclipse. After almost two decades of involvement, and without a personal belief in the theory, Campbell would travel to a sheep ranch and outpost in Australia to take photographs during the 1922 total solar eclipse. For many scientists around the world, Campbell's results would provide the definitive proof that Einstein's law of gravity was correct. After carefully measuring his plates for several months, he cabled the good news to Berlin. Then he sent a message to Sir Frank Dyson, his friend of many years, at the Royal Observatory: WE NOT REPEAT EINSTEIN TEST NEXT ECLIPSE.

Albert and Elsa Einstein were on a six-month tour to Egypt, Ceylon, Singapore, Hong Kong, China, Japan, Palestine, and Spain when Campbell telegrammed Berlin that his 1922 eclipse photographs verified Einstein's law of gravity. The two men would meet for the first time at Mount Wilson, in 1931. Below left, Campbell, in his role as president of the University of California. Below right, Charles Perrine at his desk as director of the National Observatory Córdoba, Argentina, where he survived an assassination attempt.

Above, Andrew Crommelin (left) and Sir Arthur Eddington. Below left, Charles Davidson in 1963, seven years before his death. Going from horse and wagon to man on the moon, Davidson was the last of the Victorian-born astronomers who tested for light deflection according to Einstein's theory. Below right, Sir Frank Dyson, at the Royal Observatory.

government meetings, the Englishmen were invited to stay as guests at the Catholic Seminário da Prainha, the beach seminary, with its two spires rising above the city. They settled in to rest and recuperate.

THE STARS OVER PRÍNCIPE: WEIGHING LIGHT

Because Eddington had brought along the micrometer, which could make precise measurements, his intention was to evaluate the plates on Príncipe, rather than risk damage or loss during the long steamer journey home. In their makeshift darkroom, on the evening of the eclipse, he started developing, working on two glass plates per night, with Cottingham assisting. Things began well enough, despite the water temperature remaining at a warm seventy-eight degrees. For the comparison plates they had done after arriving at Roca Sundy, they were fortunate to have access to ice, supplied by Gragera. But the supply had trickled out soon after they began work on the actual eclipse photographs. Thus, as had occurred in Brazil, Eddington realized that of the two brands of plates he had brought in sealed cans, only the Ilford Special Rapid could withstand the warm temperature. Again like Crommelin and Davidson, he was left using formalin to harden the film. And alcohol was necessary to speed up the drying of the plates.

Over the next six nights, Eddington developed twelve of the sixteen photographs. Since four were taken with Imperial plates, they would have to go back to England to be developed there. It was a grueling job in the island heat. Accustomed to the mild maritime climate back in England, he wiped sweat from his eyes as he stared down at the images appearing in the dishes before him. But night after night, the results were disappointing. There were some very good exposures of the sun, but not of the all-important other stars in its vicinity. Because of cloud interference, there were almost no star images on the first ten plates. Judging from the last two, Eddington assumed that the clouds had cleared away considerably during those final seconds of totality. The star images on *those* plates appeared to be much better. One was quite good, and the other looked hopeful.

On the seventh day after the eclipse, as Cottingham peered over his shoulder, Eddington continued to measure the plates. "The cloudy weather upset my plans and I had to treat the measures in a different way from what I had intended," he wrote in a letter home. He was already coming down with a fever from the strenuous work. But he needed to know if he had been done in by clouds. Or was there an answer waiting for him on at least one plate? He brought out the pre-eclipse plates they had taken in mid-May, for comparison. As he studied both sets, rough as the measurements were at that point—there would be ample ones done in England under Dyson's supervision—Eddington thought he saw an answer. Like Einstein, he had already seen it in the mathematics. Now he hoped to see it in the stars. "Consequently," he wrote, "I have not been able to make any preliminary announcements of the result. But the one good plate that I measured gave a result agreeing with Einstein and I think I have got a little confirmation from a second plate." It was his first inkling that this eclipse expedition would change the rest of his life.

It had been over two hundred years since Newton asked his question "Do not bodies act upon light at a distance, and by their action bend its rays?" What Eddington saw on the good plate was light from a star *bending* at 1.61 arc seconds. It was not Einstein's full prediction, *not yet*, not until all the plates had been examined carefully back in England. But Eddington had just become the first human being to stare down at the physical proof. Light did indeed have weight. Before him was an answer captured on a glass plate from a distance of ninety-three million miles. It would be an answer formed near those pearly white streamers of the solar corona that were flowing backward into space for millions of miles, an answer born in the gravitational pull of our closest and most brilliant star. Yes, bodies *do* act on light! It might have been Charles Perrine who saw it first, in 1912, but for clouds. Or William Campbell and Perrine, in 1914, if not for clouds. Or Erwin Freundlich, also in 1914, if not for war and clouds. It was Arthur Eddington, in 1919, who saw the *proof*.

The intent of these scientists, including Einstein, had never been to overthrow Newton. But thanks to Einstein, gravity had just become far more exciting than humans had ever imagined. From outside the plantation's open windows came the sounds of a tropical jungle, where monkeys stole the golden cocoa pods and where parrots squawked from the canopies overhead. It was a world most unlike the quiet farmlands and green meadows of England, where a young boy had counted the bright stars over Somerset and first fell in love with astronomy. *Light has weight!* With this glimpse of success, Arthur Stanley Eddington turned to his team member and said, "Cottingham, you won't have to go home alone."

Then he sent a telegram to the anxious astronomer royal, who was waiting for word back in Greenwich: "Through cloud. Hopeful."[7]

LEAVING PRÍNCIPE

When the Portuguese government and the shipping company couldn't agree over passenger rates, boats had been prohibited from leaving Lisbon. A threatened strike would soon cut off steamers from Príncipe. That put an end to Eddington's plans to measure all the developed plates while he was still at Roca Sundy. If he and Cottingham didn't leave the island by June 12 on the SS *Zaire*, they would likely be trapped there until August 1, possibly longer, depending on when the next boat arrived from Lisbon. It was again due to the goodwill of the visiting country that two tickets were "commandeered" by the governor of Príncipe. As Crommelin and Davidson were relaxing in the cool breezes at Fortaleza, the instruments were hurriedly dismantled and baggage packed up at Roca Sundy. The fragile photographic plates were again sealed in tin boxes that would hopefully keep them safe on the journey home.

Meanwhile, back in England, the *Observatory* journal out of Greenwich had stopped the monthly presses to add some new information submitted by the astronomer royal. He had received cablegrams from the two expeditions. Crommelin and Davidson had had perfect weather; Eddington and Cottingham were hopeful. A later message from Sobral informed the astronomer royal that the Cortie lens had captured seven good images.

Eddington and Cottingham had often wondered how their counterparts were getting on over in Brazil. Had the weather been kinder in South America? Although they wouldn't know about the *Observatory* article until after they arrived home, a telegram from Dyson assured them that the Greenwich party had been successful in getting photographs. Eddington was relieved. While he felt strongly that the Einstein prediction was correct, he would only know for certain after he and Cottingham had returned to England. On June 12, five weeks after the *Portugal* had arrived at Santo António, the SS *Zaire* sailed out of that harbor with the two Englishmen on board.

THE GOLDENDALE RESULTS

On June 2, while Eddington and Cottingham were still developing the photographic plates at Roca Sundy, William Campbell had replied to a letter from Arthur Hinks. Hinks complained that the war had thwarted his ability to stay abreast of his own work, let alone take up with a complex new theory like general relativity, which "is beyond the limits of my comprehension." Having already lost two years since Einstein's final publication in 1916, Hinks admitted that he would be "hopelessly outclassed." Hinks was the man who had resigned from Cambridge rather than become an assistant to Eddington but who had nevertheless done his job in securing information about the island of Príncipe for the expedition. Many scientists the world over would understand this loss of time given to the war effort—time taken away from advancements in any discipline.

In his reply to Hinks, Campbell commiserated that "most astronomers" would probably feel the same way. While he was not the strong believer of general relativity that Eddington was, Campbell was nonetheless a committed, curious, and open-minded scientist. "I have not attempted to go through the mathematics," he wrote, "but the applications have interested me very much in a general way." He then acknowledged Eddington's "valuable service in keeping us posted on the applications and implications." While Campbell still preferred that the Goldendale results would disagree with Einstein, he would continue with an open mind. He hoped he could announce the findings soon.[8]

Meantime back at Lick, H. D. Curtis had just begun to measure the Goldendale plates as well as the Vulcan plates that Campbell and Charles Perrine had obtained during the 1900 Georgia eclipse. He believed that the 1918 Lick team *had* captured star images that were measureable. As Eddington and Cottingham were sailing away from Príncipe with their own plates, Curtis believed he had an answer. Time and again, measure after measure, by dividing stars into groups closest and farthest from the sun, he had arrived at the same conclusion. Einstein's final calculation of 1.7 arc seconds was wrong. As for Einstein's earlier, Newtonian value, Curtis could "less definitely" pronounce against it. But with more measuring, he finally did just that by proclaiming, "There is no marked deflection of the light ray when passing through a strong gravitational field," and thus, "the Einstein effect is non-existent."[9]

Few scientists would go into mourning over this news, especially in Germany. The well-respected American astronomer George Ellery Hale, to whom Einstein had written in 1913 asking if photographs of stars near the sun might be obtained in daytime, was, like Campbell, ready to welcome negative results. To be sure, Hale was "really delighted." He had attended a Pasadena meeting of the Astronomical Society of the Pacific, where Curtis's paper was read. As Curtis had not been in attendance, Hale wrote soon afterward to congratulate him. "I confess that I am much pleased to hear that you find no evidence of the existence of the Einstein effect," he wrote. The paper was submitted to the society's journal *Publications*, to appear in its next issue. The Lick Observatory was ready to share its results with the world.

In June, Campbell headed a small delegation of American astronomers that traveled to Washington, DC, for a meeting. This was a preliminary gathering for a later conference in Brussels assembled to form the International Astronomy Union.[10] Before the Brussels event, the delegation sailed to London to attend a special RAS meeting held in their honor. After thanking the Americans for their help in the war effort and praising their contributions to astronomy, Alfred Fowler, the president, asked Campbell to address the members. Campbell began with "the Einstein problem." He detailed Curtis's method of measurement over the past months—Curtis had sent him a telegram while he was still in Washington, DC—and acknowledged that the absence of the forty-foot telescope, which was still in Russia, had hindered the Goldendale observation. "For the one we used," Campbell said, "the stars were too faint and in the long exposure required we suffered from the increased extent of the coronal structure."

Nonetheless, Curtis had not found a displacement that would support Einstein's 1.7 arc seconds prediction. In Campbell's own opinion, he told the listeners, "Dr. Curtis's results preclude the larger Einstein effect, but not the smaller amount expected, according to the original Einstein hypothesis." This observation should be astonishing enough, that Campbell could not rule out that light had weight. All the predictions that had been formulated so far, coming from Newton, Soldner, or Einstein, had been based entirely on mathematical calculations. But it's likely that most of the scientists in the room that day simply rejected Einstein's description of gravity as a property of space-time. They saw no need to improve on a Newtonian universe that was static and unchanging. Campbell, however, remained cautious.

Dyson spoke next. He confessed that the question Einstein had put before them was not an easy one to answer. But it was certainly a *timely* one, given the two British expeditions. Eddington and Cottingham were just changing ships in Portugal, three days from arriving on English soil with the Príncipe plates. And Crommelin and Davidson, back from their ocean-side retreat, had just that morning in Sobral begun taking the comparison plates. Dyson then relayed what he had learned from Eddington in a recently received letter. Eddington was disappointed with the plates he and Cottingham had taken "through cloud." Of sixteen, only the last six had captured any star images, and five of those were flawed. "From his best plate, however, he has some evidence of deflection in the Einstein sense, but the plate errors have yet to be fully determined." What a quandary for two leading astronomers, from famous observatories in two countries, to be in when seeking the answer to such a monumental question on how the universe worked.

One by one, the remaining eight American astronomers were introduced, and each spoke of his current work.[11] The war was still fresh in everyone's minds. Walter Sydney Adams, of Mount Wilson, apologized for the late US entry into the fight. When the Americans had each presented, Turner was asked by the president to give a vote of thanks to the visiting delegation. Turner mentioned the "wonders" shown to British astronomers by the Americans in 1910. He then replied to the remark made by Adams: "Prof. Adams said that they were later than he had hoped in coming to join in the war. That may be, but, at any rate they are not late in the work of reconstruction." He then referenced the passing four months earlier of "a great astronomer," Edward C. Pickering, famous not just for his work and his all-important Harvard bulletins, but also for his female computers.[12]

When Campbell closed the meeting by standing to recognize the vote of thanks, he also turned back to the scars left by the war. Early on, he had criticized President Wilson's hesitation to enter the fight, even though Campbell's pilot son would be shot out of the skies over France. Albert Einstein's German birth was never mentioned, but the fact seemed to hang in the air. With his wife in attendance, Campbell gave a brief parting speech worthy of the stage:

> We have come to assist in reorganizing international astronomical relations, but fundamentally there is something deeper than that in our meeting. We have come somewhat in protest against certain conditions that have existed in international relations, conditions arising from the over-development of militarism. We are not in sympathy with the idea that nations can do no wrong, and that they cannot act towards one another as individuals in a nation are expected to behave to each other. It chances that certain astronomers may be criticized and cut off somewhat from others. We feel, however, that it is clearly our duty to be men first and astronomers later.

As the days passed, Campbell considered the possibility that the Lick Observatory might be jumping the gun on arguing against the Einstein prediction. He had a nagging sense that they might have made errors, despite Curtis's steadfast conviction that Einstein was wrong. From London a few days after Dyson's talk, Campbell cabled Curtis, telling him to use caution. Curtis wasted no time in writing to the editor of *Publications* that same day, asking that his paper be held back from the forthcoming issue. The possibility of miscalculation, as Campbell had reminded him, was too large. They would need more accuracy. Curtis would obtain comparison plates in August that would hopefully lessen the scope of error. A few days later, after arriving in Brussels for the IAU meeting that would install both him and Dyson as vice presidents, Campbell again cabled Lick at Mount Hamilton. This time he took no chances: "Delay publishing Einstein Results."

HOME TO ENGLAND: THE PROFESSOR
AND THE CLOCKMAKER

The cold March night that Eddington and Cottingham had shared with Dyson at Flamsteed House, in front of a roaring fireplace, now seemed a lifetime ago. They were eager to see their colleagues and loved ones again. Most of the English they had heard spoken for the past three months they had spoken to each other. "It seems ages since I started off in a rush in the taxi from the Observatory," Eddington wrote to his mother. He had received no mail from home since a letter written March 28 had arrived in Príncipe on May 11. Coincidentally, they would pass a ship at St. Vincent, in the Cape Verde islands, carrying the next batch of letters from England. He would have plenty of days before the winter snows to go cycling through the countryside or hiking with his friend Trimble. He missed his world at Cambridge, the observatory with its groomed grounds, his sister's garden with its chickens and bees. Edwin Cottingham, too, longed for his house and his beloved shop at Thrapston. He was eager to be back amid the clocks and watches.

On board the SS *Zaire* with them was Jeronimo Carneiro, finished now with hosting the esteemed scientists at his private home and at Roca Sundy. He was headed back to Lisbon for three months, to his exciting world of bullfighting. Four English missionaries serving in Angola, one of them a Quaker, were also aboard the steamer. The ship was not up to par with the *Anselm* or even the *Portugal*. It was overcrowded for one thing, which is not surprising, given the dispute and the impending strike. "Of course, a lot of passengers have been ill," Eddington wrote while at sea. "It is very bad for them being so crowded on the boat. There are lots of children and in some cabins there are as many as seven people. There are three in our cabin—a Portuguese and Cottingham & myself . . . I shall look forward to the strawberries, which are better than anything they have in the tropics." Eddington was still ill with a high fever. One night while walking on deck, he fainted. The fresh sea air soon restored him. Yet it must have been a floating paradise compared with the ship that Henry Woodd Nevinson had taken to São Tomé fourteen years earlier, its belly filled with over three hundred captive human beings, fifty of them crying babies.

Eight days out of Príncipe, there was a stop in Praia, the capital of Cape Verde. After changing ships in Lisbon, they reached Liverpool on July 14, just a month after the bloody race riots had taken place there. The men would soon learn that while they were away, the whole world had erupted in turbulence, with strikes, violence, and revolutions flaring up. In England, fighting men had come home from the war—at least those who were lucky enough to have survived—to find unemployment and even poverty waiting for them. West Indians and Liverpudlians clashed. Liverpool was a port city that had grown rich in the mid-eighteenth century on the flourishing slave trade. Its slave-dealing merchants and ships dominated the transatlantic route that had once included São Tomé as a principal stopping port. Liverpool had helped make Maria Correia, whose castle ruins Eddington and Cottingham had visited on Príncipe, wealthy beyond her wildest dreams.

Despite the upheaval in a world recovering from war, Eddington and his traveling companion, the "pottering" clockmaker from Thrapston, had finished photographing stars off the coast of Africa. They were now safely back on English soil.

SOBRAL: GETTING THE CHECK PLATES

Crommelin's and Davidson's visit to the coast had been courtesy of the Brazilian government. Again, all their needs would be met, gratis, and the same had applied to the Carnegie team. Unlike the isolation of Eddington and Cottingham, marooned as they were on a tiny island and therefore limited to castle ruins and monkey hunts, Davidson and Crommelin socialized with British citizens then living and working in the vibrant city of Fortaleza. After a relaxing month, they returned to Sobral, arriving on July 9. The big house they had shared with the others was quieter now. Wise and Thomson were already doing observations elsewhere in Brazil, and the Rio de Janeiro team was just sailing past the Abrolhos island chain, with its dangerous reefs, and were only two days from home. This eclipse expedition would be Morize's last.[13]

The English team was anxious to obtain the check plates so that they could return to Greenwich. Davidson and Crommelin had been gone from their families for over four months now, and the long journey home would take another five weeks. They wasted no time in uncovering the telescopes and coelostats and having the clocks and mirrors remounted. The instruments were now in the same positions they had been on May 29. On July 11, the men began photographing the eclipse field, now distant enough from the sun that they worked in the premorning hours. Crommelin had estimated that twenty-five minutes before sunrise would imitate the same brightness of the sky during the peak of totality. For a full week, they photographed. The local temperature was unforgiving, compared with the cool sea breezes on the coast. It was demanding work.

By July 18, three days after Eddington arrived back at Cambridge and Cottingham was at work in his beloved clock shop, Crommelin and Davidson felt assured they had enough plates. Again, the dusty racetrack was filled with workers and porters to begin the job of breaking down the equipment and packing it up. The glass plates went back into the tin cans to travel under Davidson's legendary care. The heavier packing cases would be left behind. Brazilian agents who had been assigned to help with baggage would see to their shipment on a later boat. On July 22, the Englishmen thanked their gracious host who had been so vigilant in keeping them safe from yellow fever. And who had made sure they had at their disposal "a motor-car, the first that had ever been seen in Sobral." In the RAS paper they published later, they forgot to thank the chauffeur from Rio.

Crommelin and Davidson said goodbye to the local agents and helpers who had assisted them over the drought-laden months. Then they rode the train north again to Camocim, where they boarded the *Fortaleza* to Maranhão and on to Pará. They were back to where they had taken their thousand-mile journey up the Amazon. There they booked the *Polycarp* headed for Liverpool. It was loaded with the cargo one would expect on a steamship leaving northern Brazil: 900 tons of cork, 500 tons of Brazil nuts, 80 tons of rubber, and 50 tons of animal hides. Also put on board the ship were two marmoset monkeys and two parrots. The animals had been bought by Charles Davidson for his children, monkeys for the boys and parrots for the girls. During his and Eddington's 1912 rained-out expedition to Brazil, Eddington had written home about these monkeys for sale: "Some of the passengers bought little marmoset monkeys at Bahia; they are sweet little things that you could put in your pocket but I was not tempted to go in for one." He had written this at the *beginning* of the expedition. Perhaps this is where Davidson had gotten the idea of taking exotic pets home to England at the end of the 1919 expedition. In late July, he purchased the monkeys and parrots before boarding the *Polycarp*.

On August 25, he and Crommelin were back on English soil. They caught a train from Liverpool to London, and then home to Greenwich. The England they had left behind in white drifts of snow was still green with late summer.[14]

As Crommelin and Davidson were arriving at Liverpool, William Campbell's steamer had almost reached New York, his meetings in Europe finished. After a train ride across the country, he was finally back at his small brick house on Mount Hamilton. By mid-September, the Lick Observatory still had no results they could release with confidence concerning the 1918 eclipse. And now the 1919 measurements were under way in England. Eddington had been evaluating his Príncipe results, using the comparison plates that Frank Bellamy had taken that January and February at Oxford's Radcliffe Observatory. But he believed he still needed the Sobral input before he could more accurately declare for Newton, or for Einstein. Davidson and Herbert Henry Furner, a computer at the Royal Observatory, were now busy measuring the Sobral plates, each man doing it independently to double-check their results. An answer from the astronomer royal would be forthcoming when the time was right. The days dragged on.

ALBERT EINSTEIN'S LONG WAIT

In Germany, meanwhile, Albert Einstein had been publishing a flurry of papers on general relativity. At the end of 1918, a third edition of his book had been printed. In it, he undertook to answer a question about his discovery of $E = mc^2$, the equation that relates energy to mass. In the wake of his now completed general theory, would the 1905 equation remain unchanged? He was gratified to learn that it *would* hold up, and without any modification when considered from the perspective of general relativity. This prolific outpouring continued on throughout the months of 1919. The titles of his papers, aimed solely at the scientific community, show that his major focus was on his finalized theory, now almost three years old. He had learned in mid-June, again thanks to colleagues in Holland, that the British expeditions had obtained usable plates. But what would those plates reveal?

Einstein had focused on his personal life, too, or at least his cousin Elsa had. There had been a monumental event since his divorce became final that spring. On June 2, the day Campbell wrote to Hinks about general relativity and as Eddington was developing the first plates at Roca Sundy, four days after the eclipse, Albert and Elsa were married. Albert had ignored the promise he had made on his divorce papers not to remarry for two years. He moved into the apartment Elsa shared with her two grown daughters. They soon settled into the kind of middle-class life he had never envisioned for himself when he was still in love with his "Dollie." But Elsa was far removed from the more bohemian Mileva Marić, at least the way Mileva had been in her early years. The new Mrs. Einstein was satisfied to be the wife of a well-respected scientist, seeing that he ate well and was cared for domestically so that he could continue his important work. They converted the attic rooms into a study where he would have solitude. It seemed more like a business partnership than a romantic relationship. Albert was forty years old now, and Elsa three years older. The bride's bedroom was at one end of the hallway, the groom's at the other.

Einstein had received at this time a generous invitation to move to Holland and accept a position at the University of Leiden. Colleagues such as Paul Ehrenfest, Hendrik Lorentz, and Willem de Sitter would welcome him there with open arms. They assured him that he would be surrounded by people who loved him for who he was, and not just for his "brain juice." On September 12, Albert wrote an appreciative reply, turning down the "fabulous" offer and stating that he was surrounded by fellowship in Berlin "and this not only from those who lap up the droplets I sweat from my brainy brow." But in a serious tone, he could not turn his back on Max Planck and those who had enticed him to Berlin in 1913. He felt a staunch loyalty and would not forgive himself if he abandoned their trust. "I feel like a relic in an ancient church; the old bones are quite useless, and yet . . ." He ended with an important question. "Have you by any chance heard anything over there about the English solar-eclipse observation?"

EINSTEIN AND FREUNDLICH

That same month, Erwin Freundlich left Potsdam and went to southern Germany on a much-needed vacation with his wife. On September 15, he still took time to write a lengthy letter to his esteemed mentor back in Berlin. He confessed that he was "cut to the quick" after hearing from his editor that Einstein had suggested changes to the third edition of Freundlich's book on general relativity. The foreword for the book had been written by Einstein himself. Freundlich wondered if he hadn't expressed himself properly, for he could not forgive himself if it were "a fundamentally erroneous interpretation of specific things." He promised to remedy the situation when he returned home. The letter then mentioned money, as Freundlich's letters often did. "It is an uncomfortable thought for me that by my request for a raise in salary—although it was determined only by very exceptional financial circumstances—I burden your institute's budget so heavily. With the best of intentions, I cannot give back more than I am dealt."

While Freundlich's appointment was with the Kaiser Wilhelm Institute, he nonetheless did his experimental work related to the general theory at the observatory in Potsdam, whose director was Gustav Müller. "I believe I am not mistaken," Freundlich wrote to Einstein, "when I say that all the gentlemen at Potsdam, even Director Muller, want to cover their rear on the point of the gen. theory of relativity and not advocate its verification any more than by allowing you informally to grant me the opportunity to work independently at their institute." Apparently Freund-lich and "the gentlemen at Potsdam" had read an account of Campbell's remarks at the July RAS meeting in London. Details of the meeting and the speeches by the Americans had been published in journals that August, a brief article in *Nature* and a more detailed account in the *Observatory*. "I noticed this shortly before my departure," Freundlich continued, "just in little things, such as the reception of a short note by *Campbell* at the Lick Observatory in which he reportedly could *not* detect the effect of light deflection at the solar eclipse of 1918 (or '17). . . . The solar eclipse of 1917 or '18 had, to my knowledge, only a very brief totality, and the weather conditions during the same were, as far as I have heard, not favorable either. Thus C.'s result carried no great weight, especially considering that C. is not the right man for doing such a thing. I just saw once again how cowardly the majority really are."

It's *this* part of the letter that is most arresting and singular, the unbe-fitting remark about Campbell's résumé aside. If one needed further proof that communication among scientists during wartime was impossibly curbed, it's this exchange between Freundlich and Einstein. Freundlich didn't know which year the total eclipse had taken place. In fact, there *was no* total eclipse in 1917. And Einstein knew even less. "Did Campbell happen to make any exposures at the time?" he wrote back to inquire. "Were these total eclipses?" He apparently mistook Freundlich's uncer-tainty of the year to indicate two separate eclipses. "I did not know any-thing about it." It's extraordinary that such an expedition, undertaken by a distinguished observatory to test such an important theory, was un-known even to the theory's creator. But the barrier was just coming down between scientists in opposing countries.

From the outside, the relationship between Einstein and his first dev-otee might seem at times like a game of cat and mouse. In a display of the mischievous humor he was known to possess, Einstein often had a bit of fun with his loyal employee: "I just received your letter from your vacation. I am *very* sorry that you cannot relax even during the summertime." Yes, he wrote, there *were* problems that needed attending to in Freundlich's third edition "if gross misunderstandings are to be avoided." He would be happy to discuss his notes with Freundlich when he returned home from his vacation. There would be no harm done if they didn't agree, only that "my brief foreword endorsing it would just have to be omitted." If Freundlich had even considered fighting back over the notes Einstein had sent the editor, this thinly veiled threat would quickly crush his intentions.[15]

Freundlich's letters to Einstein often mentioned reimbursements he hoped to collect for money spent on photographic plates, measuring devices, and other needed materials, not a rare problem between observatories and employees, as Charles Perrine knew so well. Freundlich and his wife's financial difficulties, which were only getting worse, were not the most important issue to Einstein, who *did* offer encouragement and other help from time to time in petitioning on the younger man's behalf. When Freundlich wrote earlier that year about the headway he was making in "determining the redshift of spectrum lines from fixed stars, as I have done for stars in the Orion Nebula," Einstein praised him. "You have *done a great service* to the cause with these two papers."

As for the teaching position Freundlich hoped to land at Potsdam with his mentor's endorsement so that he and Käte would be more secure, Einstein agreed but thought it premature. "The Gen. Th. of Rel. must win acceptance among astronomers beforehand," Einstein wrote to him. He ended his letter with some advice: "Don't you get any gray hairs, now, but enjoy the rest of your vacation. Everything will straighten itself out somehow. You are still a far cry from the necessary nonchalance; your nerves lie much too bare, without any cushion of fat!" That last remark was in reference to Freundlich's tall and very lean build.

It was not a letter Freundlich could take to the bank. But it wasn't as if Einstein was himself living in the lap of luxury. The war had depleted the bank accounts of countries, not just people. The tight blockade that the British Royal Navy had wrapped around the Central Powers greatly crippled the supply of food and coal. Times were hard. "Much shivering lies ahead for the winter," Einstein wrote to his mother in early September. "But at least, if you pay a pretty penny, you can get something to eat!" The elevator was shut down in their building, no minor thing with an apartment up five flights. He and Elsa had been forced to rent out one of their rooms for extra income. The year before, when his mother discovered she had cancer, she was visiting her daughter in Switzerland. Now Pauline Einstein was languishing in pain at a sanatorium in Luzern. Attached as Albert was to his mother—he had visited her that past summer—her illness and situation would have added stress and remorse to his daily life. He asked if Pauline's friend, whom he had encouraged to visit her, had done so. Then he confessed he could not lend this friend money, especially given the exorbitant exchange rate on Swiss francs. "There is still no news about the solar eclipse," he told his mother in the letter. "All the same I am frequently asked, verbally and in writing, about the result."

NEWTON OR EINSTEIN?

Like his still-unmet colleague Arthur Eddington, Einstein had no doubt about light deflection according to his theory. Still, even though he feigned disinterest at times, the answer had to be a burden. Positive eclipse results in his favor would no doubt quickly change his financial situation and status, through lectures and book sales, not to mention future academics prizes. And a fatter bank account would certainly lighten his mother's own dark days. He had been supporting her in Switzerland as best he could and was maintaining his and Elsa's strained finances in Berlin. Part of his salary was also being sent to support his sons.

Then, on September 22, a cable arrived from Lorentz with tantalizing news. Lorentz had received an update from a colleague, the Dutch physicist Balthasar van der Pol, who was in England at the time doing research at Cavendish Laboratory in Cambridge. Van der Pol had attended a BAAS meeting in the coastal town of Bournemouth to hear Eddington speak on the expeditions. While Eddington needed more input from the Sobral plates—they were still being measured at the Royal Observatory by Davidson and Furner—there was nonetheless a light deflection at the sun's rim. "Eddington found stellar shift at solar limb, tentative value between nine-tenths of a second of a degree and twice that," the cable read. Between 0.9 arc seconds and double that, at 1.8? This provocative answer stretched all the way from Sir Isaac Newton, over the distance of two hundred years, to Albert Einstein.

Ever the proud son, Albert wrote to his mother: "Today some happy news!" He had heard from his sister that Pauline was not only in pain, but also in a dark and despondent mood. This announcement may have been the only ammunition the son had to cheer up his adoring mother, given his finances and a separation of nearly five hundred miles between Berlin and Luzerne. It was a jumble of emotions, to be sure. In what might be considered ironic, despite the dire circumstances, was the nonworking elevator in his apartment building. As he waited for a final conclusion from the British, he continued to climb the many stair steps up to his apartment. Albert Einstein, whose famous thought experiments on general relativity included an elevator that accelerated through space, now had an elevator that didn't move at all.[16]

THE RESULTS ARE IN!

Dyson sent Eddington his conclusions after a close study of the Sobral measurements. On October 3, 1919, Eddington replied to the astronomer royal that he had been worrying over the Príncipe plates. He confided to Dyson that he wondered if he had been hasty in basing his early expectations on so few measurements. And was combining the two results the best method? But now, with the Sobral results ready, Eddington, Dyson, and Davidson began writing a paper based on their conclusions. It was titled "A Determination of the Deflection of Light by the Sun's Gravitational Field from Observations Made at the Eclipse of May 29, 1919." It included details of the expeditions themselves and how the astronomers had arrived at their findings. It also mentioned the Lick expedition of 1918 and that its results, yet unpublished, were likely to be inaccurate because the "probable accidental error" was so great. The paper was submitted to the Joint Permanent Eclipse Committee and received on October 30, 1919. A special meeting was called for, to be held November 6, at Burlington House, where the Royal Society and the RAS were housed.

What would the "determination" be? Would it favor Newton, one of the most brilliant scientists the world has ever known, on whose ideas the foundation of physics had rested for two centuries? Or the German newcomer, Albert Einstein?

11

GREEK DRAMA

Searching the Stars: An Answer in the Hyades

Whether the theory ultimately proves to be correct or not, it claims attention as being one of the most beautiful examples of the power of general mathematical reasoning.

—*Arthur Stanley Eddington, Report of the
Relativity Theory of Gravitation, 1918*

No one must think that Newton's great creation can be overthrown in any real sense by this or by any other theory. His clear and wide ideas will forever retain their significance as the foundation on which our modern conceptions of physics have been built.

—*Albert Einstein, Times of London, November 28, 1919*

A QUESTION THAT might be asked is, *What took them so long?* Modern technology had caught up with Newton's question by the turn of the twentieth century. Otherwise, Einstein and Freundlich would not have felt encouraged to send out a circular, in 1911, asking astronomers at large if usable plates from past eclipse expeditions might exist. Nor would Charles Perrine and William Campbell have gone back more than a decade to inspect their Vulcan plates. So light-detecting instruments did exist. There had been several successful eclipse expeditions for Campbell, Perrine, Dyson, Davidson, Crommelin, and numerous other astronomers in spots all over the world. In recounting just the expeditions for the Lick Observatory, one would need to name Chile, India, the United States, Sumatra, Spain, Egypt, and Flint Island. The Lick expeditions to those places (excluding Japan and Labrador, which were clouded over) returned with high-resolution images of the solar corona, the astronomers' interest at the time. Why not light deflection in respect to Newton's long-standing query?

Science is extraordinarily fragile. Until the right question is raised, in the right time and place, it's often impossible to make progress. Scientists are just as human as any other man or woman in a different calling. There is no magical font of wisdom that pours down on the scientific community and guides it in pushing knowledge forward. It's a chaotic playground, a jumbled and messy process with countless false leads and boondoggles, and even a good measure of quackery. But fortunately, the principles of science and the scientific method are powerful enough to overcome all these obstacles. In its wake, science produces the highest quality of accuracy within the human ability to understand nature at its deepest and most hidden levels.

So why did science wait so long to test Isaac Newton's question? Had any astronomer done so successfully during those early eclipses, the glory would have been all theirs in verifying a new law of gravitation. There could be many reasons why no one considered the question urgent before Einstein came along. Being fallible like the rest of humanity, scientists cannot clearly see the future. At any given time, there are many provocative ideas that are worthy of support and that will stimulate research and advance knowledge. But there is rarely enough financial support to allow scientists to pursue more than a few of these projects at one time. Until Einstein raised the issue a hundred years after Johann Soldner did, it is likely that no *astronomer* understood that there was even an issue to be raised. Einstein's persistence, and his pushing Freundlich forward, clearly had its personal incentive. It was his own theory. What greater need could there be for him to stay in the struggle for a verification of ideas that had occupied his brain for years? After all, this was *Einstein's* baby.

THE MEETING AT BURLINGTON HOUSE

The special meeting for members of the Royal Society and the Royal Astronomical Society—the Joint Permanent Eclipse Committee—had been set for Thursday, November 6. Dyson, Crommelin, and Davidson would take the train from Greenwich. Eddington and Cottingham would come in from Cambridge. The day began as the month had, with a chilly wind blowing over London from the north. Rain was mixed with drizzle and occasional sleet as the taxicabs began arriving at Burlington House, on Piccadilly, in the heart of London. Wearing hats and bundled in heavy coats and some carrying umbrellas, the members hurried in from the cold, passing through doors that had welcomed many of the world's greatest thinkers for generations. Isaac Newton was twenty-two years old in 1664, when construction first began on Burlington House, which was then a private mansion. It would be another forty years before *Opticks* was published, his book that had launched the famous question about the fundamental nature of light. Now the members were gathering in the East Wing, home to the Royal Society since 1873, to await the opening of the meeting.[1]

Interest in the 1919 eclipse expeditions had built over the months. Upward of 150 scientists filled the pews and the anteroom. The brilliant Sir Joseph John Thomson, recipient of the 1906 Nobel Prize in physics for his work on the electron, had been president of the Royal Society since 1915. Known affectionately as "J. J." to his colleagues, he opened the meeting by calling on the astronomer royal to give an account of the eclipse results. An expectancy fell over the audience as they waited for Dyson to speak. This was a world recovering from a long war, the ink still dripping like blood from the Treaty of Versailles, signed less than five months earlier. A week before, the German High Seas Fleet that had so plagued the North Sea was scuttled in Scotland, fifty-two of its finest ships going to rest on the ocean floor. If anyone needed proof that the Central Powers had lost the war, the sinking of this fleet would be it. On their way to Burlington House that afternoon, many Royal Society and RAS members, arriving by private autos or taxis, would have driven past the wood-and-plaster cenotaph that had been erected that July in Whitehall for the Peace Day Parade. The monument's presence celebrated the end of the war and commemorated the dead who were buried in foreign soil. There would be a permanent memorial built of stone the following year. But for now, the memory of what had been lost still clung to the base of this wooden one, in the dried flowers and withered wreaths placed there that summer by a bereaved public. It was as if the world needed something remarkable just then to help lighten this burden of grief.

The astronomer royal arranged the pages before him. There was a wave of nervous coughs until the room fell silent. Dyson began his address.

"The purpose of the expedition," he said, "was to determine whether any displacement is caused to a ray of light by the gravitational field of the sun, and, if so, the amount of the displacement." The listeners waited. Even Isaac Newton seemed to be waiting. His three-quarter-length likeness hung from the wall behind Thomson's chair. This was not the younger Newton in his full-bottom wig of cascading curls and ruffles at his throat. At eighty-three, his white hair was thinning. Wearing a long and open brown coat, with a white neck cloth and wide white cuffs, he bore an unsmiling, even grim, expression. Newton had served for over two decades as president of the Royal Society, from 1703 until his death in 1727. Now, from his stern gaze, he seemed tired of all the hoopla surrounding his life. He would be dead a year after the sitting.

After giving the main details of the Greenwich expedition—the planning, the travel, the astrographic instruments taken—Dyson turned to the task at hand. "Einstein's theory predicted a displacement varying inversely as the distance of the ray from the sun's centre, amounting to 1".75 for a star seen just grazing the sun." He went on to recount that Einstein's theory had already explained the perihelion of Mercury, the mystery that Campbell and Perrine had hoped to solve during several total eclipse expeditions in search of Vulcan. This was the ghost planet that had been "discovered" beneath the pencil lead of Urbain Le Verrier. In 1908, and in doing what science does best, Campbell and Perrine had thrown the hypothetical planet into the dustbin. It didn't exist. And then, Einstein's new theory had shown why.

Dyson then told the scientists that previous photographic plates obtained during past eclipses were of too large a scale to show enough stars. Or the scale was too small to test the delicate accuracy necessary for light deflection. He described how the measurements of the Sobral plates had come about, being undertaken by Davidson and Furner, both men independently evaluating each plate twice. The plates taken with the four-inch telescope loaned to the expedition by the Royal Irish Academy—it was plain good luck that Father Cortie suggested taking it to Sobral when the Jesuit was forced to back out—showed good results. Dyson ended his opening remarks with an assessment that would not have pleased many members in the room on that rainy afternoon. "After a careful study of the plates I am prepared to say that there can be no doubt that they confirm Einstein's prediction," said the astronomer royal. "A very definite result has been obtained that light is deflected in accordance with Einstein's law of gravitation."[2]

LAW VERSUS THEORY

One can only imagine the impact those words had that day, especially for the scientists who might not have read the paper submitted a week earlier. Crommelin spoke briefly, mostly to thank Morize and his staff, the interpreter, and the Booth Steamship Company. He was grateful to the party's gracious host, Colonel Saboya, for his house with its constant supply of water, "no small boon in a time of drought, and of great importance in the photographic work." Then it was Eddington's turn. This audience of his peers would have been waiting for him. He was the man who had been so vocal in England, in his papers and lectures, in support of the general theory. Dyson might have been the one to point out that a perfect eclipse was coming on May 29, 1919, and decided to send expeditions. But it was only after Eddington had laid the beauty of Einstein's theory at the astronomer royal's door. After sharing details of his and Cottingham's expedition to Príncipe, Eddington addressed Dyson's earlier announcement.

"The simplest interpretation of the bending of the ray is to consider it as an effect of the weight of light," Eddington said. "We know that momentum is carried along on the path of a beam of light. Gravity in acting creates momentum in a direction different to that of the path of the ray and so causes it to bend." The audience listened carefully. A *half effect* would mean that England's own Sir Isaac Newton would maintain his scientific throne. A *full effect*, and gravity would, as Dyson had just professed, obey the law of Albert Einstein, the native son of Germany, their recent enemy. "This is one of the most crucial tests between Newton's law and the proposed new law," Eddington continued. "Einstein's law had already indicated a perturbation, causing the orbit of Mercury to revolve. That confirms it for relatively small velocities. Going to the limit, where the speed is that of light, the perturbation is increased in such a way as to double the curvature of the path, and this is now confirmed. This effect may be taken as proving Einstein's *law* rather than his *theory*."

So what did Eddington mean? Physicists are rather like storytellers who use mathematics to weave narratives that, when finished, are often called a *theory*. The audience for any theory, the *readership*, is always nature. Because most theories are incorrect from inception, a physicist must be in constant conversation with the fundamental laws of nature. This process, this communication, is called, quite simply, *experimentation*. It becomes nature's job to sort through each mathematical story to decide what's correct and to toss out what isn't. Physics is not a search for the truth, but a search for accuracy. Like all of science, physics can never be 100 percent certain of anything. This uncertainty places a burden on the understanding between scientists and the public. Because the public is accustomed to thinking in certainty, scientists often have difficulty explaining their approach.

Imagine human beings who lived a few thousand years ago. They would have had a sense of what we might call the Law of the Sunrise. They understood that each day began when the sun rose. They had witnessed this event hundreds of times over the years. In science, a *law* is a summary of observations that yield the same result when carried out repeatedly. A law is found to be accurate and valid each time it's tested by observation. What, then, would be the Theory of the Sunrise? Perhaps the sun rises when Helios, the Greek god, drives his golden chariot across the heavens, pulled by fiery steeds. Or the theory is that the Egyptian god Ra floats through the sky each day in his sun boat. A theory is the reason behind why a law is valid. The Theory of the Sunrise has long been proven. We understand now that the sun doesn't rise at all. It maintains its position at the center of our solar system. It only appears to rise because of the earth's rotation every twenty-four hours on its axis. Those two older "theories" have been rejected, just as Vulcan was for causing that glitch in Mercury's orbit.

But Einstein's theory was not home yet. Since his final paper in 1915, he had insisted that the theory must stand up to three tests: (1) the perihelion of Mercury, (2) the deflection of light at 1.7 arc seconds, and (3) the redshift, an example of the Doppler effect that enables astronomers to determine the chemical composition of far-off objects in the universe. This triumvirate was the entire mathematical superstructure and foundation on which the general relativity theory rested. His equations had earlier predicted and explained Mercury's wobble in its orbit around the sun. The expedition results had just proven that not only does light have weight, but it has *double* the weight Newton anticipated. What Einstein needed now to quell his naysayers was the full proof of his theory, the third test. The predicted shift of the spectrum lines would need to be obtained. If so, the theory of general relativity would provide an accurate and new description of how the universe works.

Astronomers had been hard at work on the redshift problem, especially Charles St. John, at Mount Wilson, and John Evershed, director of the Kodaikanal Observatory in India. That St. John, with some of the best equipment in the world, was dubious of the shift had haunted Eddington for many months. As early as January 1918, Eddington had written to a colleague about these doubts: "St. John's latest paper has been giving me sleepless nights—chasing mare's nests to reconcile the relativity theory with the results, or vice-versa. I cannot make any headway."[3] Now, at the Burlington House meeting, Eddington stated that a failure of the redshift to be confirmed would not affect Einstein's *law*, but that his equations that buoyed up the theory, the "views on which the law was arrived at," would be wrong. A blackboard with neatly written equations had been prepared. On it were two sets:

$$ds^2 = -(1 - 2m/r)^{-1}\, dr^2 - r^2 d\theta^2 + (1 - 2m/r)dt^2 \text{ (Einstein's law).}$$
$$ds^2 = \qquad\qquad -dr^2 - r^2 d\theta^2 + (1 - 2m/r)dt^2 \text{ (Newton's law).}$$

"I think the second expression may be accepted as corresponding to Newton's law," Eddington said. "At any rate, it gives no motion of perihelion of Mercury and the half-deflection of light."[4]

General relativity would not be an easy theory to sell. Who but its creator, Einstein, and its foremost spokesman, Eddington, could understand it? The four pioneering mathematicians who had provided Einstein with the rigorous mathematics he needed to encode his theory had died long before Einstein was born: Carl Frederick Gauss in 1855, Nikolai Lobachevsky in 1856, János Bolyai in 1860, and Bernhard Riemann in 1866. His colleague, the brilliant astronomer and physicist Karl Schwarzschild, who had helped him with his field equations from the battlefield, had died in 1916 from the illness he had contracted on the Russian front. Among the living who could comprehend Einstein's theory would be the German mathematician David Hilbert, who was already being falsely credited by Einstein's detractors as having published the theory first. And there would be others, of course, but it would be a select group.[5]

THE AUDIENCE REACTS

Sir J. J. Thomson, from the chair, was ready to open the floor for discussion. But first, obviously impressed with the overwhelming magnitude of this announcement, Thomson, who had once been Dyson's revered professor, shared his thoughts with the members:

> If the results obtained had been only that light was affected by gravitation, it would have been of the greatest importance. But this result is not an isolated one; it is part of a whole continent of scientific ideas affecting the most fundamental concepts of physics. It is difficult for the audience to weigh fully the meaning of the figures that have been put before us. . . . [T]his is the most important result obtained in connection with the theory of gravitation since Newton's day, and it is fitting that it should be announced at a meeting of the Society so closely connected with him. [If the theory passed the third test,] then it is the result of one of the highest achievements of human thought.

Not everyone in the audience agreed. There were probably cynics who wanted no part of a grand idea conceived by a German, but others believed that Einstein was simply wrong. The charismatic Sir Oliver

Lodge, who had just that year retired as the first principal of Birmingham University, was expected to speak. The Royal Society had presented him with its Rumford Medal in 1898 for his studies of the relationship between matter and ether. The members would be curious to hear his comments. It was well known that Lodge had been busy trying to contact his son, Raymond, through the help of mediums after the boy had died in battle in France. According to the grieving father, ether was needed in the afterlife as well as in space. A committed "ether man" of the older school of physicists—Einstein's special theory of relativity had invalidated ether in 1905—Lodge had addressed the Royal Society earlier that year as the expeditions were preparing to leave England. He had conjectured that ether, "responsible for the velocity of light," might be affected by gravity and thus cause a refraction. If so, he would prefer Newton's value over Einstein's. After Thomson was finished heaping praise on these new results, Lodge rose and left the meeting.

One of the next men to speak was H. F. Newall, an astrophysicist and the director of the Solar Physics Institute at Cambridge. "I feel that the Einstein effect holds the day," he said, "but I do not yet feel that I can give up my freedom of mind in favour of another interpretation of the effects obtained." He extended his heartiest congratulations to the astronomer royal and the eclipse observers. Newall had been on at least as many eclipse expeditions as had anyone else in the room that day, beginning in 1898, when his pianist wife had gone with him to India. In 1900, Newall was at the Algiers Observatory a short distance away from where Crommelin stood on the rooftop of the Hôtel de la Régence. In 1914, he waited next to Perrine and Mulvey in the Crimean hillside vineyard, watching clouds float in from the mountain at Staryi Krym. "I prefer to keep an open mind about interpretation," Newall told his colleagues.[6]

It wasn't that the scientists seated at this meeting didn't have a right to be skeptical or at least ambivalent. There were a few weak areas in the evidence presented. There had been problems with the larger telescope in Sobral, some of the plates were troublesome at both places in the warm temperatures, the statistical evidence had been gleaned from a small number of stars, and Eddington had thrown out one of the measurements. But this uncertainty was the nature of observation and experimentation. As a case in point, in the paper Dyson and his coauthors Eddington and Davidson had submitted a week earlier, they conceded that this observation, being of such magnitude in discovery, should be repeated at future eclipses even if they would not be as star-rich as the May 29 eclipse for many years. Moreover, Dyson also offered to send exact copies of the original glass plates to any astronomers who wished to make their own measurements. This was science in the making.

Drama was also in the making. When Ludwik Silberstein took his turn to weigh in, so to speak, he rebuked the high praise that Thomson had leveled on the results. "There is a deflection of the light rays," he said, "but it does not prove Einstein's theory." Silberstein would become a steady nuisance that Einstein would have to endure in the upcoming years. He then brought up the failure of both St. John and Evershed to obtain the predicted shift of the spectrum lines: "The discovery made at the eclipse expedition, beautiful though it is, does not, in these circumstances, prove Einstein's theory." Dyson and especially Eddington had made this quite clear: the *law*, not the *theory*. At this, Silberstein pointed at the portrait of Isaac Newton, who seemed to be eavesdropping from behind Thomson's chair. "We owe it to that great man," he said theatrically, "to proceed very carefully in modifying or retouching his Law of Gravitation."[7]

In the audience that day was Alfred North Whitehead, the esteemed British philosopher and mathematician who had attended the meeting as a member of the Royal Society. Whitehead was then professor of applied mathematics at Imperial College, London. "The whole atmosphere of intense interest was exactly that of the Greek drama," Whitehead wrote. "We were the chorus commenting on the decree of destiny as disclosed in the development of a supreme incident. There was dramatic quality in the very staging:—the traditional ceremonial, and in the background the picture of Newton to remind us that the greatest of scientific generalisations was now, after more than two centuries, to receive its first modification. Nor was the personal interest wanting: a great adventure in thought had at length come safe to shore."[8]

It's the job of Greek drama, after all, to comment on a world-changing event. Einstein was forcing every scientist in the room that afternoon to fundamentally re-conceptualize the universe in which they existed: not only did the earth rotate, mixing time and space, but time and space were also warped. Eddington stood to address Silberstein in what might be taken as a polite dressing-down. "When a result that has been forecasted is obtained," Eddington said, "we naturally ask what part of the theory exactly does it confirm. In this case it is Einstein's *law* of gravitation." The special meeting was over.

Before closing, Thomson thanked his former student, Dyson, and Eddington "for bringing this enormously important discovery before us, and for taking such pains to make clear to us exactly where the problem stands." If learned scholars of mathematics, physics, and astronomy had difficulties grasping the ramifications of what Einstein's theory proposed or what the 1919 eclipse expeditions had actually verified, then how could the public possibly understand even a thread of what it meant? A report would be sent to the *Times of London*, giving an account of events at the meeting.

THE BRITISH RAISE A TOAST

The Royal Astronomical Society began as a dining club when fourteen astronomers met for dinner at the Freemasons' Tavern in London's West End. The year was 1820. During those early decades, when travel to and from monthly meetings could pose an inconvenience for many members, it became customary for them to enjoy a dinner discussion afterward and then to spend the night. Often, guests were invited. The once-yearly "Parish" dinners were more exclusive, with only members in attendance. At these annual gatherings, the astronomers sang songs over a dinner of marrowbones, which were eaten with long-handled scoops designed to extract the marrow. All these social gatherings were beloved by many RAS members, no one more than Dyson. The dinner that night would be memorable, given the special meeting earlier.[9]

The principal players were all there. It was a night for celebration mixed in with the usual songs and toasts. These dinners weren't always devoid of serious conversation or even results. Once, when Dyson mentioned to Eddington that a formula was needed for correcting statistics in errors of observation, Eddington had reached into his pocket for a pen, flipped over his menu card, and soon handed the required formula back to his astonished colleague.[10] As the toasts began, it was time for Dyson to launch the story that would live far longer than would anyone in the room that night. He rose to speak about the importance of the results that the expeditions had revealed. Then he proposed a toast to the four men who had sailed to those far-off continents. He told the members about that March night in his study, at Flamsteed House, the night before both expeditions left for Liverpool and the first leg of their journey. That cold evening by the fire, Cottingham had asked what would happen if the amount of deflection was *twice* what Einstein predicted. Dyson had informed Cottingham that he would have to come home alone because "Eddington will go off his head and commit suicide." When the laughter had died down, he added, "Newton wanted 0.87 seconds of arc. Einstein wanted 1.7, but Cottingham wanted to double this amount!"

The story was given its foothold on that night of rain and drizzle in London. It was the kind of story that would keep the memory of flesh-and-blood men alive long after they had disappeared, more so than formulas on the backs of menu cards. With perfect timing, Cottingham, the clock-maker with a brilliance for astronomical timepieces, sheepishly stood to explain that while he *had* asked the question, he had never doubted the astronomer royal's judgment or Professor Eddington's sanity. And then he added, "All the same, I must confess I was very pleased when Eddington said to me one morning, after making a few plate measurements from those we developed on the island—'Cottingham, you won't have to go home alone.'"

Eddington, being a fan of light verse, had written a parody of the *Rubaiyat of Omar Khayyam*. He delivered several quatrains, such as this one:

> *Oh leave the Wise our measures to collate*
> *One thing at least is certain, LIGHT has WEIGHT*
> *One thing is certain, and the rest debate—*
> *Lights rays, when near the Sun, DO NOT GO STRAIGHT.*

That mention of "the rest debate" was likely the redshift, the final test to prove the theory. The dinner would end with Turner, the club "poet," delivering a new drinking song he had penned to celebrate the expeditions:

> *The idea that light has mass, we got*
> *From Newton, it's bequeather,*
> *They 'waved' aside his news as rot,*
> *And filled all space with ether.*
> *But once more comes a change of scene,*
> *The ether's swept away, Sir.*
> *And space is emptied now as clean*
> *As the bottle of yesterday, Sir.*
>
> *We cheered the Eclipse Observers' start*
> *We welcome their return, Sir.*
> *Right manfully they played their part,*
> *And much from them we've learned, Sir.*
> *No toils or pains they thought too great,*
> *Nor left Einstein unturned, Sir.*
> *Right cordially we asseverate*
> *Their bottle a day they've earned, Sir.*[11]

The lively RAS dinner was over. The members made their way home or to where they were being housed. It had been an exciting day, followed by a dinner of good cheer and camaraderie. Most of England would go to bed that night never having heard the name *Albert Einstein*.

12

THE SEARCH FOR ACCURACY

A New Universe: The Press Juggernaut

Newton, forgive me; you found just about the only way possible in your age for a man of highest reasoning and creative power.

—*Albert Einstein*

THE ANNOUNCEMENT THAT afternoon in the rooms of the Royal Society would come as no surprise to Albert Einstein. He had just returned to Berlin from a relaxing two weeks with his colleagues and friends in Leiden, Holland: Ehrenfest, de Sitter, and Lorentz. While he was in Holland, the English results had been unofficially announced to a standing-room-only crowd of cheering students and professors at a meeting of the Dutch Royal Academy. Einstein was there, but the local newspapers had not been invited. Eddington had apparently sent a letter to astronomer Ejnar Hertzsprung, at the Leiden Observatory, a couple of weeks before Dyson had submitted the results to the Joint Permanent Eclipse Committee. Hertzsprung, known for his work on the life-cycle of stars, had shown the letter to Einstein. From Leiden that same night, October 23, Einstein wrote to Planck about hearing that the results for light deflection supported his

247

prediction: "It was gracious destiny that I was allowed to witness this." But he often told friends that he had no doubt what the deflection would be.

In Leiden, Einstein had been a guest of Ehrenfest and his family. He had almost not gone. That past summer his doctor limited his travel for fear Einstein would not find the proper foods, such as Zwieback or white bread, rice, and macaroni. Thus, his stomach condition would become more aggravated. Ehrenfest had assured him they would stick to the special diet. The visit was a great success. Surrounded by fellowship from his colleagues, Einstein ate the dietary foods his host had promised and enjoyed the well-heated rooms. Ehrenfest had sent him home with a thermos flask to warm the journey. The day after he arrived back in Berlin, Ehrenfest wrote him a letter. "Dear, dear Einstein. You have left waves of sympathy and friendship behind in this land, which will lap back and forth between us all, a long time from now."

The visit had been a reprieve from the constant concern over his mother. It was obvious now that her illness was terminal. He had returned to Berlin to face reality. As the members of the Joint Permanent Eclipse Committee were preparing for their meeting at Burlington House, in London, personal matters were not going well for the man behind the controversial theory. November 6 had dawned cold in Berlin, with gray-clouded skies and snowflakes falling over the city. As the taxis were arriving on Piccadilly, Einstein was penning a letter to the head of Berlin's housing rentals. "A grave injustice is about to happen to me," he wrote. His plan to relocate Pauline Einstein from the sanatorium in Switzerland to her own apartment in his building was now being thwarted by the building's owner. The Einsteins had been promised the next available apartment. Now the owner was letting "a single gentleman" rent it and insisted that Pauline could move in with her son and his wife. "I emphasize, however, that as a university professor and director of the Kaiser Wilhelm Institute of Physics, I have such a tremendous bustle of activity in the flat (secretary, typewriter, incessant phoning, constant visitors) that the flat is absolutely unsuitable for a sick person." He added that his two daughters—Elsa was their mother—were studying music and must practice for hours on end. Considering all this, it's a wonder Albert ever got any work done.

Berlin was now feeling the crunch of new restrictions of a worker strike in the city. Heating with coal was again troublesome. His escalating battle with the landlord was foremost on Einstein's mind on that Thursday in November. As Einstein wrote his letter pleading for the extra apartment, Eddington and Dyson were speaking to the gathered members at Burlington House. "At the current catastrophic rate of exchange," Einstein told the head of housing, "the sick woman's immigration is a vital issue for me. I cannot accommodate her elsewhere than in the building, because she relies entirely on our personal assistance." His landlord, a German architect, could have no way of knowing that he was embroiled in battle over a rental unit with the man who had just rewritten Newton's law of gravity. Despite Einstein's wishes, the "single gentleman" ended up moving into the vacant apartment.

THE FABRIC OF THE UNIVERSE

On November 7, the *Times of London* ran its first story about the meeting that had occurred at Burlington House the day before. The piece appeared in the sixth and last column on page twelve. In the page's first column, which was headlined "The Glorious Dead," King George V invited his subjects to honor the fallen during the upcoming first anniversary of the armistice. All trains in the country would stop running. When village church clocks struck 11:00 a.m., two minutes of silence would ensue in memory of the soldiers lost to war. The other news on page twelve was the usual fare: coal prices, bank rates, labor disputes, and the Bolsheviks acting up. A brief article in column four noted that in Berlin, the general strike was over, except for metal workers, news for which Einstein would be grateful. It would mean more coal for the upcoming winter. At the bottom of column six was the Entertainments Index: *Tristan and Isolde* was being performed at the Royal Opera House, and American war correspondent Lowell Thomas had been booked since that June at Albert Hall. His lecture and moving pictures would help make himself and Lawrence of Arabia famous. Bordering the paper's spine and sitting atop the entertainment news was a three-tiered banner about the special meeting at Burlington House.[1]

REVOLUTION IN SCIENCE.

NEW THEORY OF THE UNIVERSE.

NEWTONIAN IDEAS OVERTHROWN.

The article relayed the pertinent information. A large number of astronomers and physicists had attended a meeting on the previous afternoon to hear the results of the May 29 eclipse expeditions. Dyson's details of the expeditions were recounted. Sir Joseph J. Thomson—well known to the British public, he was referred to only by his position in the society—was given the most eye-catching quotes. "Even the President of the Royal Society, in stating that he had just listened to 'One of the most momentous, if not the most momentous, pronouncements of human thought,' had to confess that no one had yet succeeded in stating in clear language what the theory of Einstein really was." Despite that third test, the redshift still uncertain, "our conceptions of the fabric of the universe must be fundamentally altered." Lodge was then mentioned as having earlier in the year predicted no deflection or, if any, Newton's value. The paper reported that after Thomson's praise, Lodge had gotten up and left the meeting.

It wasn't as if this page twelve news article would catch the British public off guard. Newsboys wouldn't be crying out, "Revolution in Science!" and waving papers at passersby. Readers had been aware of the interest in this eclipse beforehand. Since that January, the *Times* had run a few articles that mentioned the expeditions leaving, then, eclipse day, the cabled telegrams back to Greenwich, and the forthcoming results, quoting Dyson and, once, Crommelin. But the ball had started to roll. The next day's *Times* carried a second article, also on page twelve, column four. The headlines were still somewhat restrained:

The Revolution in Science
Einstein vs. Newton
Views of Eminent Physicists

Einstein, who was then forty years old, was referred to in this second article as a "Swiss Jew, 45 years of age," and a man of "liberal tendencies" who did not sign the German manifesto during the war. He had been a topic of discussion in the House of Commons after the editorial ran the day before. Arriving at an RAS lecture, Sir Joseph Larmor, the Lucasian Professor at Cambridge, had been "besieged by inquiries as to whether Newton had been cast down and Cambridge done in." Davidson was quoted in the interview. While he was certain the results proved Einstein's 1.7 arc over Newton's 0.87, he cautioned that the third test of Einstein's theory, the redshift, was still waiting to be proven. "The latter has been carefully tested by Dr. St. John at Mount Wilson in the United States," Davidson said, "but so far without success."

A letter to the editor from Lodge ran below Davidson's comments, with the heading "The Ether of Space. Sir Oliver Lodge's Caution." He had abruptly left the meeting, as reported by the *Times* the day before, because of a "long-standing engagement and a 6 o'clock train." He also corrected the previous day's assessment of his views. He had not *predicted* that if light had weight, it would be the Newtonian value; he had only *hoped*, and it had been rash of him. Before congratulating Professor Einstein and the eclipse observers, he ended with a word of restraint. "I would issue a caution against a strengthening of great and complicated generalizations concerning space and time on the strength of the splendid result: I trust that it may be accounted for, with reasonable simplicity, in terms of the ether of space."

THE AMERICANS DISCOVER ALBERT EINSTEIN

Getting Einstein's revolutionary ideas past an inflexible old guard would be maddening. And it would seem near impossible to interest the average layperson in them. So how did a juggernaut of acclaim for a German physicist unknown to the public get started in the first place? Page twelve of the *Times* wasn't the level of reportage that would titillate the popular masses and send them out into the streets in celebration. True, the president of the Royal Society said that it was the "most remarkable" discovery since the predicted existence of Neptune. But what Neptune meant to the average person is questionable.

Interestingly, the *New York Times* might have played the biggest role in making Einstein famous.[2] The managing editor of the paper was Ohio-born Carr Van Anda, who had left the *New York Sun* fifteen years earlier for the more successful *Times*. A science enthusiast, Van Anda had studied astronomy, math, and physics at Ohio University before becoming a journalist. A modern editor, he brought coverage of science to a readership he knew was waiting and would be interested. A progressive idea was not written up as entertainment on Van Anda's watch, as was often the case with other newspapers of the day, if they carried science articles at all. Over the past year, he had followed with keen interest the editorials from England about the expeditions. Van Anda was already famous for scooping his competition with headlines declaring that the RMS *Titanic* had sunk off the coast of Newfoundland, while others were reporting only that it had struck an iceberg.[3]

As with most turn-of-the-century newspapers, Van Anda never hired specialists to cover meetings such as the one at Burlington House. The London bureau for the *New York Times* gave the assignment to Henry Charles Crouch, who was mostly their golf reporter. Not a member of the Royal Society or the RAS, Crouch would not have been invited to the meeting. From the London news report on November 7, Crouch quickly pulled parts of the article verbatim, identifying Thomson by name for the American reader and quoting his testimonial that what was confirmed at the meeting was "one of the greatest—perhaps the greatest—of achievements in the history of human thought." Crouch then added new direct comments from Crommelin, the comet expert who had been to Sobral.[4] The entire story was wrapped snuggly around an enormous ad for Oppenheim Collins & Company, the New York clothing store on 34th Street that specialized in women's garments. Next to the four-tier eclipse headlines that included words such as "Epochmaking," and "One of the Greatest of Human Achievements" were fur coats for sale, mink and muskrat, with collars and cuffs made of "Australian opossum or taupe nutria." Van

Anda quickly copyrighted the story and cabled congratulations to Crouch in London.

The next day, on November 10, the managing editor stepped up the pace. The science story on page seventeen might have been far from the front page, but the six-tier headlines towering above it were eye-catching.[5] One of the headlines, A BOOK FOR 12 WISE MEN, was supposedly in reference to what Einstein had said to a publisher when he presented them with his "last important work." He had warned them that only twelve men in the world would understand it, "but they [the publisher] took the risk." This notion of a dozen geniuses would plague him for years to come in questions from the press and the public. It would be the headwaters of a river of invented quotations and fictional "facts."[6]

For this second story in the *New York Times*, Crouch had apparently spoken to, or telephoned, W. J. S. Lockyer, director of the observatory begun by his famous father, Sir Norman Lockyer. The younger Lockyer, known for his interests in aeronautics and photography as well as astronomy, was asked what the 1919 expeditionary results really meant. "The discoveries, while very important, did not, however, affect anything on this Earth," Lockyer was quoted as saying. "They do not personally concern ordinary human beings; only astronomers are affected." Some cynics were by now proposing that starlight bending was just a scientific version of the underwater coin, which was not lying on the spot where it appeared to be. Therefore, wasn't this Einstein stuff just old news?

But the *New York Times* headlines and reports had become more tantalizing. Lights were askew. Scientists were agog. Stars were deceptive. Space was warped. Light had weight. And somewhere wandering the planet, twelve lonely geniuses were reading and understanding the general theory. An important addition was that Americans were now given a snippet biography of the man behind the theory. Albert Einstein was "a Swiss citizen of around fifty years of age." But even more importantly, he had resisted signing the 1914 "Manifesto to the Civilized World," unlike ninety-three of Germany's other leading scientists, artists, and scholars. German born perhaps, but he was not the enemy, at least for readers who could see beyond their still-fresh hatred of the country whose war machine had killed their fathers, brothers, and sons.

As other American newspapers followed suit with the *New York Times*, the headlines eventually calmed down. By November 16, the paper's banners were almost consoling: "Don't Worry Over New Light Theory—Physicists Agree That It Can Be Disregarded for Practical Purposes." Readers were reassured that the "intellectual panic" undergone by British men of science was over. The blame seemed now to rest on the more restrained English. They were "slowly recovering as they realize that the sun still rises—apparently—in the east and will continue to do so for some time to come." This would be comforting news for the public.[7]

EINSTEIN ON HIS THEORY: A HOUSE WITH TWO STORIES

Many astronomers and physicists remained unconvinced of the measurement results from the 1919 eclipse expeditions that supported Einstein's new law of gravity. Some, like Lodge, were too deeply rooted in the physics of the past. But others, such as Campbell, were uncomfortable with the large probable errors that came with the Sobral-Príncipe measurements. Yet Campbell was cautious at the time, saying only that the British results were "especially interesting." Rather than jumping the gun, he would wait "with great interest" for the published details from England. He was also waiting for Curtis's computations for his own 1918 Goldendale results. While pressured by the American astronomer Thomas Jefferson Jackson See, one of Einstein's most ardent detractors and another ether enthusiast, to use the Lick plates to refute the British results, Campbell didn't succumb. See was well known for his arrogance and carelessness in his own work and disliked for his pettiness and collegiate infighting. The 1918 Goldendale eclipse results, it appears, were never published.[8]

Many British scientists were also skeptical of the measurement results on the grounds that Dyson and Eddington had not adhered closely enough to the canonical rules of the scientific method. But Thomson, Dyson, and Eddington were forces to be reckoned with. Their voices that day at Burlington House held much sway in academia and in the press.

There were few astrophysicists and mathematicians as accomplished as Eddington. Other scientists who believed that the measurements were too shaky to fully establish Einstein's law of gravity but whose own skills in astronomy were modest may not have been in the best position to argue with the crème de la crème. And these elite scientists rested easy with the announcements. One thing was definite. Einstein's fame outside scientific circles had begun to explode.

When the *Times of London* asked if Einstein would expound on his theory, he did so on November 28, first by thanking the British astronomers and then commenting on the war that had separated them. "After the lamentable breach in the former international relations existing among men of science," Einstein wrote, "it is with joy and gratefulness that I accept this opportunity of communication with English astronomers and physicists." He commended them on their accomplishment during wartime. In discussing the different theories that exist in physics, he explained that his was a theory of *principle*: "To understand it, the principles on which it rests must be grasped. But before stating these it is necessary to point out that the theory of relativity is like a house with two separate stories, the special relativity theory and the general theory of relativity." This was an assigned task that Einstein obviously enjoyed.

There were *three* principals his theory rested on, he said. Two were already shown "to have received strong experimental confirmation." If the last one, the redshift, failed, then the premise would collapse. Editors would be scratching their heads and sighing for many newspaper articles to come. Then he finished with his Einsteinian sense of humor. "A final comment," he noted. "The description of me and my circumstances in *The Times* shows an amusing feat of imagination on the part of the writer. By an application of the theory of relativity to the taste of readers, today in Germany I am called a German man of science, and in England I am represented as a Swiss Jew. If I come to be regarded as a bête noire, the descriptions will be reversed, and I shall become a Swiss Jew for the Germans and a German man of science for the English!" He apparently did not mind that one paper had aged him five years, the other ten. His droll personality had been shown to the world.

In the same issue, the *Times* replied to his joke, which would be well received by the public: "We concede him this little jest. But we note that, in accordance with the general tenor of his theory, Dr. Einstein does not supply any absolute description of himself." The German physicist and the world press would get along just fine.

Newspaper ink continued to flow. There was a new high priest, a scientific messiah on the stage. A savior had arrived to unriddle the universe, someone smart enough to solve the eternal mystery unfolding overhead. Perhaps it was an amalgam of things that turned Albert Einstein into a celebrity seemingly overnight: the end of a long and painful war, the tantalizing words in the press, the brilliant men of science paying such homage, and the devilish sense of humor his own persona had finally revealed. But it may have been more primitive than all of those combined. Maybe it was the human need to find an explanation of the great unknown, another missing piece of the puzzle, just as the people who lived thousands of years ago longed for an answer behind the Law of the Sunrise.

EINSTEIN'S PERSONAL WAR: BLOOD RELATIONS

The fanfare was speeding up for Einstein. He was on his way to becoming an international star, his name spoken with ease in nonscientific circles, given the public's new fascination with his warped universe. Photographs of him began to appear in the newspaper stories. As the public came to know him more, they would not be seeing the wild white hair and the rumpled clothing of his later years. In the autumn of 1919, Einstein was still dapper, his dark hair barely showing traces of gray and just beginning to suggest that it may later grow into an unruly tumbleweed. After the meeting at Burlington House and as his celebrity grew, his private world was filled with a pressing sadness. In his letters to friends, it's obvious that he was struggling to maintain his rigorous work schedule as he dealt with his mother's arrival in Berlin and her impending death. Because the study where he worked in his and Elsa's apartment had to be made ready for Pauline, her trip from Switzerland was delayed.

Einstein's letters to Ehrenfest reveal a longing for family, for home and hearth—for *peace*—as fame swirled around him. Two days after the

November 6 meeting in London, he wrote of the "untroubled time" he had spent in Leiden with Ehrenfest and his wife, a Russian-born mathematician who collaborated with her husband on his work, and their four children. "Never have I taken part in such a vivacious home setting before; it simply comes from two independent persons not being bound together by compromises!" Perhaps he had once hoped he would find this same partnership with Mileva Marić. "How proud I will be to have a little Ph.D. for a sweetheart," he had once written her. In his letter to Ehrenfest, he described the two Ehrenfest girls, one leaning toward scholastics and the other artistic, the older boy a "blue Tomcat," and the younger, born just a year earlier, as "patient little Crawlikins." He hadn't seen his own two sons in over a year, although he was getting on better with their mother. More months would pass before a visit with Hans Albert and Eduard was possible. He was now famous, but in his letters, he seemed to envy Ehrenfest's perfect life.[9]

Einstein kept up with his manuscripts, papers, and correspondence. Eddington, his second loyal devotee after Freundlich, wrote to him for the first time in early December. He assured the physicist that "all England has been talking of your theory" and that "it was the best thing that could have happened for scientific relations between England and Germany." Eddington was busy writing and lecturing on general relativity. He predicted that "although popular interest will die down there is no mistaking the genuine enthusiasm in scientific circles." He also felt remorse for Freundlich: "Although it seems unfair that Freundlich, who was first in the field, should not have the satisfaction of accomplishing the experimental test of your theory." Freundlich wasn't the only one standing in that line. Perrine and Campbell were at the head of it.

Before Einstein could answer Eddington's letter, he heard from Freundlich, who asked his mentor again for those eternally necessary dollars: "Please forgive me for reminding you to prepare an application that I receive a cost-of-living bonus like the gentlemen at the other Kaiser Wilhelm Institutes. As you know, it is impossible to live today off the amount that I am receiving at the present time, and I am steadily getting in worse straits, despite all my wife's industry in trying to cope with the deficit."

Freundlich, who thought that Einstein could arrange this increase without burdening the institute's science fund, had been carrying out calculations on globular star clusters. He would visit Einstein in a few days and show him the work. He asked to be excused for the printing paper on which his letter was written. It was thin and bleeding ink.

A few days later, Einstein replied to Eddington, congratulating him. He again mentioned the third test. "With regard to science, I would like to make the following comment. I am convinced that the displacement of the spectrum lines towards the red is an absolutely necessary consequence of the theory of relativity. If it were proved that this effect does not exist in nature, then the whole theory would have to be abandoned." Of the loyal Freundlich, he was pleased that Eddington had expressed consideration. "He is very eager," Einstein wrote, "but has been unable so far, owing to objective as well as personal difficulties, to contribute very much to the testing of the theory." This comment was perhaps unkind. Had Freundlich not been done in by war in 1914, he would have later been done in by clouds. Neither were in his control. A lack of funding had prevented him from joining the British on the 1919 expeditions. And he was still petitioning for money, this time so he and his wife might enjoy a better life. His "objective" could hardly be questioned.[10]

The requests from journalists for an interview had become constant, most of them ludicrous attempts to explain to readers what his theory meant. In relying on his sense of humor to sustain him, he referred to this dalliance with the press as the emperor having no clothes. It was during this media pressure that Einstein was faced with perhaps the biggest "test" of his lifetime. It was a far greater challenge than anything his principles or calculations might present. He carried through with preparations for his mother to arrive. Pauline would travel from Lucerne by ambulance, a cold two-day journey. With her would be her doctor; her nurse, who would remain in Berlin; and Maja, her daughter. Three days after Christmas, she was helped up to her famous son's study and made as comfortable as possible. His letters to close friends spoke of her horrendous pain, the morphine, and her gradual acceptance that the end was near. Yet at the same time, she was still fighting to live. Albert did what he knew best to do. He continued working, even keeping up with his lectures at the local college.

Two months later, when Freundlich wrote yet again for his budget, which was now months late, he could have no way of knowing it was probably the same day Pauline Einstein was buried. She died on February 20, 1920, at the age of sixty-two. Albert wrote to his friend Heinrich Zangger a week later. "We are all completely exhausted merely from being with her; you feel in your bones what blood relations mean!" Pauline had been a strong influence in her son's life, strong enough to denounce his love for a woman she called "that Miss Marić." "Mama threw herself on the bed and cried like a baby," he had written to his beloved Dollie, in what must now seem a lifetime ago.

THE QUEST GOES ON: THE NEXT ECLIPSE

Einstein's theory of general relativity did not "triumph" as the first write-up in the *New York Times* proclaimed. It was not proven right by the results of the two expeditions. Nor was Newton proven wrong. It's just that Newton's law of gravity could not describe certain phenomena. Einstein's law, two hundred years later, was more sophisticated. It is not the job of science to prove things. Instead, science is interested in a theory's *probability* in the face of existing evidence. And yet, with the help of the press, this idea that the theory had been "proven" by the 1919 eclipse expeditions would go into cement for years yet to come. The world often desires the story that has a definite ending. But Einstein had put forth such an astonishing concept of the universe that scientists continued to do the *job* of science: they strove to determine its accuracy.[11]

Foremost among those who had worked diligently and under great hardships to verify light deflection were skeptics like Charles Perrine, William Campbell, Sir Frank Dyson, and H. D. Curtis, to name some who left their opinions behind. The discussion was not over in November 1919, not as far as many astronomers and physicists were concerned. As Dyson and his four expeditioners had recommended, observations should be done at the total eclipse coming on September 21, 1922. The path of totality would begin at sunrise in Ethiopia, move over Italian Somaliland,

and cross nineteen hundred miles of Indian Ocean to reach the Maldives. From there it would swoop southeast to plunge Christmas Island into darkness at noon. Cutting a track through the center of Australia, it would pass through New South Wales to end at sundown in the waters of the Pacific. Many expeditions from around the world got ready.

Campbell, whose sleep had been disturbed for years over the "Einstein problem," was at the vanguard. He would not repeat the mistakes of 1914 or 1918. He had the most accurate weather reports and a most impressive collection of instruments, including his forty-foot telescope and two new cameras. He would view the eclipse at a sheep ranch called Wallal Downs, on the west coast of Australia, where totality would be five minutes and nineteen seconds. His expeditionary plans were complicated and yet precisely detailed. Robert Trumpler would travel months ahead to Tahiti, with what equipment he needed to take the comparison plates. Campbell would later ship the remaining instruments from San Francisco. Trumpler landed on the tiny island in French Polynesia on April 11. Nine weeks later, after obtaining the plates, he transported the equipment he had to Sydney, where it joined the remaining half of the Lick paraphernalia, forty tons in all. From there, Trumpler and the expensive telescopes and cameras sailed to Freemantle on the west coast, to wait for Campbell to arrive. Obviously aware of the historic implications behind a successful observation during this eclipse, a photographer would accompany the team. Campbell was now fifty-seven years old.

Of the thirteen expeditionary teams in Australia for the 1922 eclipse, as many as eight would test for light deflection according to Einstein's prediction. Campbell left California on July 18 of that year, with his loyal companion, Elizabeth. They would travel nearly twenty-three thousands miles for this eclipse. From Sydney, they took a train across the country with other teams from Canada, England, and Australia, stopping in several cities for formal engagements before reaching Perth. From there they caught the SS *Charon* on to Broome, an eight-day journey. The eighty-ton schooner *Gwendolyn*, a naval vessel compliments of the government of Western Australia and towed by a smaller steamer, carried them on to the beach near Wallal Downs. It was a cramped journey, with many team members crowded onto the deck, until the *Gwendolyn* finally anchored three miles from Eighty Mile Beach. After breakfast on August 30, Campbell and his team, which now numbered thirty-five and included sailors and aboriginal workers, lowered the baskets with delicate apparatuses from the schooner down into a gasoline launch and a lifeboat. Next came crates of heavy equipment. With the twenty-six-foot tide subsided and the sea calm, the launch towed the lifeboat across open water to land on Australia's western shore. They would unload in this manner for three days. Campbell himself waded in three feet of water holding in a basket above his head the delicate apparatuses. Dozens of donkeys then pulled the wagons through the dust, heat, and millions of sheep blowflies to Campbell's viewing station at Wallal Downs, a telegraph station and a sheep ranch in the middle of nowhere. Once the camp was set up among the wattle trees, he christened the mess tent Café Einstein. On September 1, he cabled home to Mount Hamilton that he and Elizabeth had arrived safely.[12]

Dyson, listening to advice from Arthur Hinks, who had done so well in advising Eddington about the clouds on Príncipe, would send Harold Spencer Jones to Christmas Island, a speck in the Indian Ocean one thousand miles from Australia's west coast. Weather reports predicted habitual clouds, so Campbell had invited the Greenwich team to Wallal Downs, with his own crew. They chose instead the island. Also going to Christmas for his second attempt to photograph the stars for light deflection was Freundlich. The German astronomer was now in league with a Dutch expedition. The Christmas Island teams were met with such severe storms that they were trapped aboard ship for ten days as the sea surged and churned around them. The dark clouds and rain would not abate enough for clear skies on eclipse day. None of the Christmas Island teams would be successful.

Charles Davidson might have gone to Cordilla Downs with an Australian team. The man from whom Dyson had learned so much about astronomical instruments, the man he often asked, "Davidson, got a match?" was invited by the observatory in Adelaide. Had he accepted, Davidson would have crossed parched land where fifty thousand sheep ran free, Australia's economy riding on their backs. The Adelaide team had good chances of success. It had been generously loaned valuable equipment from Campbell and Curtis, who was by then director of the Allegheny Observatory. With no camel wagons available, the team members would ride pack camels, loading and unloading the animals daily for the dusty three-week journey. But Davidson, the man who, in the warm hours of a Brazilian night, had developed what is now the famous photograph Crommelin had captured of the eclipsed sun over Sobral, stayed behind in Greenwich. The Christmas Island plans had already been cemented. Davidson would wait for a future eclipse.

In this attempt to learn more about the universe that perpetually challenged them, these astronomers, like others before them and others yet to come, sought out the path of totality. Beckoned by mystery, beckoned by the astounding idea of a brilliant man whom most would never meet, beckoned by starlight, they went. They crossed oceans. They traveled the planet, not knowing in 1922 that the earth was not one of eight, as was then believed, but one of several thousand confirmed planets orbiting their own suns. And there may be a million more planets waiting to be discovered. This is the nature of science.

EINSTEIN'S BABY: THE ANSWER

Eclipse day was scorching hot. The aboriginal workers placed branches around the instruments to cool them, poured sand to hold down the fine dust, and sprinkled water on the ground around the site. Campbell, thanks to his superlative instrumentation and skilled team, would be gloriously successful. Compared with the few stars photographed at Príncipe and Sobral, he would capture hundreds. The pressure was now intense since many astronomers around the world, as well as the press, were waiting for the results. A month before he returned from Australia, Campbell was sent a letter from astronomer Samuel Mitchell, a respected expert on eclipses—Mitchell's first expedition had been to the US state of Georgia for the 1900 eclipse—with interesting news. "When I saw Dyson last, he said that he would not be in the least surprised if the 1922 photographs did not confirm the Einstein effect. He thought that possibly they in England had stressed Einstein a little too much." And Dyson was far from alone. Ever cautious, Campbell announced the preliminary results several months later. He first sent a cable to Berlin, to Einstein, who had already embarked on his visit to Japan and Israel: "EINSTEIN DEFLECTION . . . BE-TWEEN ONE POINT FIFTY NINE AND ONE POINT EIGHTY SIX SECONDS ARC." Next, he cabled Dyson at the Royal Observatory. After the same brief details, he ended with "WE NOT REPEAT EINSTEIN TEST NEXT ECLIPSE."

Eddington, ever the lover of light verse, remarked, "I think it was the Bellman in 'The Hunting of the Snark' who laid down the rule 'when I say it three times, it is right.' The stars have now said it three times to three separate expeditions, and I am convinced their answer is right."

Einstein had never doubted any of these findings. From these exciting years in his life and onward, he would be sought after for an interview or a photograph, a smile or a funny face, a paper or a lecture, to the very end of his days. Many men admired him, and many women daydreamed about him. He had calculated this, too.

He once wrote to a friend that "the world is a strange madhouse."

MEN MADE ON DREAMS

Our revels now are ended. These our actors,
As I foretold you, were all spirits, and
Are melted into air, into thin air:
And like the baseless fabric of this vision,
The cloud-capp'd tow'rs, the gorgeous palaces,
The solemn temples, the great globe itself,
Yea, all which it inherit, shall dissolve,
And, like this insubstantial pageant faded,
Leave not a rack behind. We are such stuff
As dreams are made on; and our little life
Is rounded with a sleep.

—*William Shakespeare, The Tempest*

IN THE TWENTIETH century, the earth was witness to sixty-two total solar eclipses, one of which made Albert Einstein famous for his law of gravity, and another that proved it. Eclipses will continue to fascinate humans for as long as this planet is inhabited. The longest one that will ever occur in earth's history is scheduled for July 16, 2186, with a maximum time over northern Guyana of seven minutes and four seconds. There will never be one longer, because there is bad news for future eclipse chasers: The orbit of the moon around the earth will become more distant each year. The span between the two bodies will eventually increase by fifteen thousand miles, decreasing the apparent diameter of the moon. Thus, as viewed from earth, the moon will no longer be able to completely cover the face of the sun during a total eclipse. There is no need for immediate concern, but in 620 million years, the last solar eclipse will go the way of the dinosaur.

In the hundred years that have passed since the 1919 eclipse, astounding progress has been made in astronomy and physics. Human beings have gone into space and have walked on the moon. Spacecraft have touched down on Mars. Over a thousand active artificial satellites circle the planet, feeding back information on stars and asteroids, comets, black holes, dark matter, gamma ray bursts, supernovae explosions, and far-off galaxies. Scientists have never stopped mapping the cosmos or seeking to understand the vastness of the universe. The thousands of committed thinkers spread out along the road that begins in folk astronomy and runs through antiquity up to Isaac Newton would not recognize the modern scientific world. The dedicated astronomers who were born in the Victorian era and whose lives are the core of this book would be lost in today's astronomy. It's a very different world now, and yet they *all* helped shape it.

CAMPBELL: "I BELONG ON A MOUNTAIN TOP"

When William Wallace Campbell arrived back in San Francisco from the 1922 eclipse expedition, it was the month of December. His steamer sailed into the Port of San Francisco, where an unlikely delegation was waiting on the dock. The welcomers were the University of California regents, who insisted that Campbell accept the presidency of the university. They needed a man with vision, with executive and administrative skills, and, as one of the deputation put it, "a man with backbone." Campbell had been offered the prestigious position before and had turned it down. He liked his life on Mount Hamilton and his work studying the stars. He was now a world-renowned astronomer with all the impressive medals that could possibly be awarded, along with many honorary degrees.

Campbell had already served as president of the American Association for the Advancement of Science, as well as the Astronomical Society of the Pacific. For three more years, he would be president of the International Astronomical Union *and* the American Astronomical Society. As director of Lick, he couldn't have been any busier. But the regents reminded him that they had always met his past requests for the mountaintop community and observatory. Perhaps now it was time for *him* to help the university. They persuaded him by consenting to his requests, the primary one being that he would retain the directorship at the observatory. And second, the regents would not interfere in the university's internal matters. The Campbells could retain tenancy of the house on Mount Hamilton, where they had raised their sons—it would now be expanded and remodeled—and where they might occasionally find solitude or entertain friends. Planning on a short term of service, President Campbell began his new role the next July, at the age of sixty.

He took leave of his duties in September so that he and Elizabeth could travel south of the border to Ensenada, Mexico, to successfully observe the 1923 total eclipse. For the next eight years, the university prospered and grew under his guidance. Generous private donations arrived from the likes of the Hearsts, the Rockefellers, and others. The campuses expanded. Even his detractors, who believed he ran too rigid a ship, would concede years later that his had been the best presidency during their time. With nature blessing him yet again with another total eclipse in California, he and his wife accompanied the Lick team to Comptonville in late April 1930. The phenomenon would never cease to amaze him. Nor would the cosmos itself. He once said that college graduates know all about the electrical lights in their houses, but nothing of the lights in the sky.

A few days after returning to the university, at the commencement service on the Berkeley campus, he gave his farewell address: "I belong on a mountain top, away from the world—and yet near enough to enjoy occasionally its centers of population." Feeling ill and fearing he would collapse, Campbell was assisted from the platform and first aid was administered backstage. His retirement would not be what he had anticipated. The rigors of running such a large enterprise, combined with the stress of academic infighting, had taken its toll. Two years later, he would lose vision in one eye. His youngest son, Kenneth, the boy who had wanted to see a German U-boat on the family's mad dash across the North Sea in 1914, would later comment on his father's health: "The presidency, so late in life, was very hard on him, and finally killed him."

Recovering his strength, Campbell might have retreated to the dome on Mount Hamilton. He and Elizabeth were still allowed to occupy the house there, even though he was no longer director. Prodigious developments in astronomy and spectroscopy had occurred in his absence, those eight years that he was involved in administrative duties. This was the same feeling that Hinks had experienced when he wrote to Campbell in 1919. The field of astronomy had moved too far ahead while Hinks was concerned with wartime tasks. Campbell felt it now more than ever. One day at the small post office on the mountaintop, he opened a letter from the National Academy of Sciences, asking that he become its president. After some deliberation, he accepted. In the summer of 1931, he and the loyal Elizabeth left the mountaintop once again to settle into a new home in Washington, DC. While Campbell would steer the august body of the academy through four financially critical years, his health was now rapidly failing.

In 1935, his term with the academy finished, he and Elizabeth returned to California. Campbell was now seventy-three years old. He had likely suffered a stroke for he was diagnosed with aphasia, a condition caused by damage to those parts of the brain that deal with language and communication. After a short visit to Mount Hamilton, perhaps to say goodbye, the Campbells moved into a charming apartment in San Francisco. Its windows looked out on the blue waters of the Golden Gate strait, to the Marin Headlands and Mount Tamalpais beyond. The famous bridge had been under construction for two years before the Campbells returned to California. It would be finished and opened for traffic two years later, in May 1937. While visitors to their home found the couple in good spirits, the famous astronomer's health was now declining rapidly as aphasia took its toll. Once so eloquent, Campbell struggled to find the proper words. Even reading and writing had become laborious. The eyesight in his remaining eye was also failing. The astronomical study that had been his passion for so many years, in that heavenly classroom filled with stars, was lost to him.

On June 14, 1938, while Elizabeth slept, he rose before dawn to carry out the plan he had conceived. Apologizing for his poor handwriting—his scribbled words were incoherent in places—the man known as "Wallace" to family and friends wrote five heartfelt notes, the last at four in the morning. He praised his wife and sons—his greatest pride—and thanked his friends. His foremost worry was not wanting to be a burden to his wife. "Goodbye, dearest Elizabeth. Be of great courage," was perhaps his most difficult note to write. "It is better that I go away now with my powers nearly all gone, than I stay and become an incompetent person." He worried that he might not be able to walk unaided by the next day. He put the notes on a table in the hall of their apartment and placed his wristwatch next to them. Pulling on trousers over his pajamas, and then his coat, he leaped, barefoot, from the fourth floor of their apartment building. He was seventy-six years old.

A boy delivering newspapers early the next morning discovered the crumpled body lying on the stones of the courtyard. By the time an ambulance arrived, it was apparent that Campbell had been dead for some time. His middle son, Wallace, who told the press that his father had been in a cheerful mood the evening before, identified the body. The day after his death, the *San Francisco Chronicle* mentioned the "pitiful notes" as it covered his tragic suicide. "All of his life William Wallace Campbell did difficult things with high courage. There can be no doubt that the most difficult thing he ever did, and the one which required the highest courage, was to end his own life." Pallbearers included the governor of California, the president and officials from the University of California, and directors of the Mount Hamilton and Mount Wilson Observatories.

For the RAS obituary, Sir Frank Dyson, who had watched Campbell kill a rattlesnake on Mount Hamilton nearly four decades earlier, mentioned the late astronomer's illustrious career and numerous eclipse experiences. "At the eclipse of 1918, he attempted unsuccessfully to verify Einstein's prediction of the deflection of light by the Sun's gravitational field." Dyson then noted the success of the 1922 expedition to Australia. "From very accordant results they found a value of 1.72 for the displacement near the sun's limb as against the 1.75 predicted by Einstein, confirming the results and conclusions of British observers in 1919."

Elizabeth, who knew him best, was more poetic. "He loved flowers if someone else would raise them," she once wrote of him. She always suspected that the hard job he had as a child hoeing his mother's vegetable garden had ruined the hobby for him. But this work on the farm—he was barely four when his father died, leaving behind six children—shaped young Campbell for the tough work that lay ahead during his expeditions. "Stones that four men cannot move he lifts with ease," Elizabeth had written of their 1898 first expedition to India. William Campbell made numerous decisions throughout his life: where to best observe an eclipse, which weather records were accurate; to remember white linen suits for hot climates; and to boil all water. And to take a shot of whiskey at bedtime to encourage good circulation. But perhaps the best decision he ever made was asking a college girl named Elizabeth Ballard Thompson to be his wife. She died in 1961, at the age of ninety-three.

Years later, in his seventies himself, Kenneth Campbell had a special memory of his father. The family had taken a trip to Europe in 1911, one not attached to an expedition. Vacation over, they boarded a smaller steamer than had been planned, to return to the United States. Kenny Campbell counted the number of lifeboats during a particularly bad storm while at sea. With no way of knowing that the *Titanic* would sink just months later, its lack of lifeboats costing hundreds of lives, he asked his father if their steamer had enough of them. "No, Kenneth," Campbell answered truthfully, "I think probably there are not enough." The boy then said, "I suppose if we founder in this storm, you'd grab Mother and me and we'd all three climb into a lifeboat." To this Campbell replied, "Well, Kenneth, I'd put Mother and you in a lifeboat, but you know, I don't think I would join you. I think I'd stay around."

Just before dawn on that last day, and again thinking of his family first, William Wallace Campbell decided it was time to leave.

DYSON: ASTRONOMY WITHOUT TELESCOPES

In the summer of 1925, Sir Frank Dyson would see his good friend William Campbell at the meeting of the International Astronomical Union, of which Campbell was then president. It was held at Cambridge and coincided with the 250th anniversary of the Royal Observatory. Astronomers from around the world were welcomed with concerts and garden parties once the work was done. Arthur Eddington and his sister, Winifred, hosted a reception at the observatory. Dyson was proud to show Cambridge to the visitors in midsummer. It was the place where Sir Isaac Newton had studied as a young man and had later taught. "There is no place in England which we are more pleased to show them," he wrote, "and no place in which we take a greater pride." Addressing the audience of over two hundred, Campbell stressed that the union's objective should be the care of international astronomy. Dyson also hinted that it was time to welcome the Germans back into the family of scientists.

In June 1927, Dyson traveled to Giggleswick, in Yorkshire, for what he called his "first eclipse." He would observe "unofficially" while Charles Rundle Davidson did the work. The British newspapers were on fire with news of this total eclipse. It would be the first to cross English mainland soil in 203 years. Reporters were ringing the entry gate bell at the observatory and his telephone rang constantly with requests for interviews. Passing through England's backyard as it would, this eclipse was garnering far more attention than the one in 1919 did. When the *Daily Mail* offered a twin-engine, fourteen-seater airplane to carry astronomers above the clouds for a more successful chance at observation, Dyson accepted. As astronomer royal, he would send others to fly above the ground. What was also rare about this eclipse is that Lady Carrie Dyson would accompany her husband for the first time on an expedition. Fourteen-year-old Ruth would also go. She was the baby whose photograph Dyson had carried in his pocket in Australia, after World War I broke out. It would have been a fine time for Dyson's good friend J. J. Atkinson to get out his two-seater and take another car trip, as they had done in 1912. But "Atky" had passed away two years earlier.

The following year, on his sixtieth birthday and according to the terms of his appointment as astronomer royal, Dyson was to retire. Now, he felt he still had much to offer the observatory. The Lords of the Admiralty, who hired him in 1910, agreed. Dyson would continue in his job for another five years. The family could remain at Flamsteed House, which had been their home for nearly two decades. All the Dyson children had grown up in the house, and the scribbled graffiti on the walls was a testament. They knew their father as a disciplinarian. Yet he would sometimes turn up in the kitchen late at night if he heard them whispering and giggling, to join them in his pajamas. Once, he warned his young son about the dangers of climbing up buildings and into trees. "If you must climb, climb that!" the astronomer joked, pointing at the observatory dome. One of his favorite stories was of the day his telephone rang and a voice asked, "Do you know there's a boy on the great dome?"

Carrie Dyson began to travel extensively with her husband, including a trip across Canada by train with him giving lectures along the way so that he could pay for her passage. This was 1932, and another total eclipse was coming on August 24. The Greenwich team would again be headed by Davidson, along with the Scottish astronomer John Jackson. The Dysons and the two astronomers left England in July. They sailed past whales spouting in the Saint Lawrence Seaway, and the steamer continued on through the gray fog until it reached Quebec City. Davidson and Jackson said their goodbyes and rode a train 250 miles north, to tiny Parent, Quebec. The Dysons crossed the Canadian prairies by train and then continued over the Rocky Mountains to Victoria. They would return to New England in time for the eclipse, which they would view in southern Maine.

By August, the couple was touring New England in a motorcar with American friends. Carrie Dyson, who had stayed behind for so many years to raise their large family, was now seeing the world. They checked into a hotel in South Portland, Maine, which sat in the path of totality. They would drive down to Harvard the next day for the first IAU meeting held in America, where they would meet up with Arthur Eddington and other colleagues. On the day of the eclipse, Sir Frank Dyson sat in the sun on the lawn of his "luxury hotel," waiting. Again the Dyson luck was with him, and the afternoon was cloud-free. Sitting less than a hundred miles from where Samuel Williams first saw Baily's beads in 1780, this would be Dyson's last eclipse.

In 1933, Frank and Carrie Dyson said goodbye to Flamsteed House and moved back to the top of the same steep hill where they had started their married life. While his days would be filled with meetings and writing, he felt the loss of the stars that had marked the years of his life, all those times he had trudged home at dawn to fall into bed, exhausted. When a friend consoled him by saying he was only giving up the observatory, not astronomy, Dyson replied curtly, "You can't have astronomy without telescopes." Like Campbell, he knew that his profession was changing, expanding beyond the knowledge of astronomers who had learned their craft during the Victorian and Edwardian periods. The observatory itself was feeling the weight of urban pollution, the optics of its telescopes affected by grime and smog. Street lights were growing more numerous, and the rattling of trains added to the disturbances.

He kept busy with more world travel. He had already written a book on astronomy in which he encouraged readers to learn about the stars and the heaven filled with wonders. He hoped that people would one day know the brightest stars in the sky as they might the most common flowers. He began a second book. In 1935, when Carrie returned from Malta, where she had traveled for the birth of Ruth's first child, the couple escaped to the seaside town of St. Ives, in Cornwall, for a month of quiet vacation. As Dyson was chatting with a friend at his garden gate, an automobile careened around the corner at a great speed. The car collided with both men, who had no time to move to safety. They weren't fatally injured, but both were badly bruised and suffering from shock. Dyson was now sixty-seven years old. He would never entirely recover from the accident and would walk with a cane for months.

Carrie's health began failing that same summer, because of a heart condition. Their children encouraged the older couple to take life easy in those days of retirement. They should hire taxis instead of catching trains or walking. They had noticed that their father had difficulty climbing the hill to home. Once, he had even taken the wrong train, much to the amusement of the observatory staff. He boarded at London Bridge, was carried past Greenwich, and ended up in Gravesend. When the astronomer royal explained to the ticket collector what had happened, the man replied, "The fact is you oughten be allowed out by yourself." No one would have known the story if Dyson hadn't told it on himself. But the truth was that he and his wife preferred the simpler life to speeding taxicabs.

Carrie had a heart attack and died in March 1937, just past her seventieth birthday. Sir Frank was devastated. "I have been a very happy and fortunate man," he wrote to a friend, "to be blessed with her companionship for 43 years." Slowly, he returned to the social life he had always enjoyed, but he saw signs of another war encroaching when the digging of trenches began in London parks. "Hitler should be in seventh hell," he wrote to his children. "One hates to see the dictators get their way, but the sacrifice of millions of lives is too big a price to stop them." He remembered the emptied observatories during the past war and the endless lists of the dead in the daily papers. A heavy smoker for years, he now had a perpetual cough, and his own heart was not strong. Yet he made plans to visit his son in South Africa and his daughter in Australia, as he had done shortly after Carrie's death. His family would not learn until later that he'd had a slight stroke or that his doctor had advised him not to undertake a long voyage.

Dyson sailed from Liverpool in late January 1939, his sister Aggie accompanying him. He had just published the obituary he had written for his friend William Campbell. It would be the last document Dyson would write. While he was at sea, he received word that his colleague Andrew Crommelin had been struck by a motorcyclist while walking to church and was hospitalized with severe injuries. He would pay his colleague a visit on his return to England. Yet, by the time the ship reached South Africa, Dyson was described as "an invalid." The ship's doctor advised that he check into a nursing home for rehabilitation once he reached Australia. And he must return immediately to England when his visit was over. His daughter met him at the port in Sydney and was shocked at his appearance. Having lost the use of his left hand, her father had not been able to shave and now sported a white beard.

In early May, he was taken aboard the *Ascanius* for his trip back to England, the same steamer he and Eddington had taken to Australia in 1914. A steward named Wheeler, assigned to look after him, moved into the cabin next door. Instead of his usual cigars, the ship's doctor recommended no more than five cigarettes a day. Dyson remained cheerful even though he needed to be dressed and bathed. He could walk a little, and so made friends with his fellow passengers. Wheeler was an amateur astronomer who owned a small telescope. At night, as the ship sailed over dark waters, the steward pointed out Southern Hemisphere stars to the famous astronomer. When the ship hit swells and large waves in the Indian Ocean, Dyson's health became worse and he was confined to his bed.

On May 25, he appeared to know the end was near. He told his sister how proud he was of his children. Then he called for the chef and thanked him for preparing special meals. When he asked to hear Big Ben again, Wheeler struggled in vain to get the famous clock's bells on the wireless. Eventually, Dyson told him not to bother; it didn't matter. He seemed in no pain, and Wheeler thought at first he was asleep. When the doctor was summoned and arrived, he announced that Dyson had passed away. The astronomer had held a lifelong dread of London's graveyards and often spoke of a wish to be buried at sea. The astronomer's body was wrapped in the Union Jack. On deck, his grieving sister, the ship's company, and nearly all the passengers lined up for the service. The ship's captain later scribbled in his log:

> 4:30 p.m. 25/5/1939. 29 deg., 55 S, 67 deg. O5 E. To certify that the body
> of Sir F. W. Dyson was buried at sea.

The coordinates put the burial spot about a thousand miles southeast of Madagascar.

When the sad news reached England, memorial services, accolades, and tributes were planned. Eddington wrote his friend's obituary. He remembered the IAU meeting at Harvard in 1932, when the secretary of the US Navy had described astronomy in his welcome address as though its job was solely to assist in naval warfare. "As you have said," Dyson had replied, speaking as president of the union, "some of us are from observatories which supply data for navigation. But in addition, Sir, may I ask you, if astronomers had not discovered that the earth was round, where on Earth would the United States of America be?"

The RAS Club members debated whether to cancel their monthly dinner. Dyson had become president of the club three years earlier. Knowing how much he loved these events and rarely missed one, the members went ahead with the plans. Stories were told, such as the time the astronomer royal rang up a Greenwich clergyman and told him there was a crack in his church's tower. The clergyman looked carefully, then returned to the phone. "I can't see anything wrong," he said. "No, I daresay not," Dyson had answered, "but my big telescope can." The RAS Club members hung a plaque in his memory. *Praeses et Amicus*. President and friend.

CROMMELIN: "A BORN ASTRONOMER"

By the time autumn came to Greenwich Park in 1939, the man known as an expert on comets, Andrew Claude de la Cherois Crommelin, succumbed to the injuries he had suffered in the accident several months earlier. For over forty years, he had lived a short stroll from the park, on the same road where his parents and many assistants at the famous observatory had once resided. Like his colleagues, Crommelin had been a member, or secretary, or president of a good many societies and groups, as well as director of the comet section of the British Astronomical Association, where he had also been president. He had been awarded the Lindemann Prize and an honorary doctorate from Oxford University for the work he and Philip Cowell had done in predicting the return of Halley's Comet.

When Letitia Noble Crommelin passed away in 1921, Andrew never remarried. In 1927, he retired from the observatory where he had spent his entire career. The astronomy books he would write had titles befitting poems, *The Story of the Stars*, *Diamonds in the Sky*, and *The Star World*. For years, the *Observatory* magazine and the BAA's journal had published his well-researched reports on comets and minor planets.

Charles Davidson mentioned the 1919 eclipse in the obituary he wrote for his friend, without referring to himself as having been Crommelin's companion to Sobral. "He observed with a 4-inch lens with a 19 feet focus and obtained seven excellent photographs, the results from which proved the predicted deflection beyond question." Crommelin's knowledge was encyclopedic. Davidson took his description a step further when he referred to the man who had traveled a thousand miles up the Amazon with him as "careless of appearance but always courteous." The obituary that appeared in the *Observatory* referred to him in terms that Crommelin would have liked: "a born astronomer; a student immersed in his work."

Crommelin's last expedition was in 1927, when the eclipse visited the mainland of England. With Dyson and Davidson in Yorkshire, he had gone privately to Colwyn Bay, on the north coast of Wales. While Dyson and Davidson enjoyed sunshine on that day in June, Crommelin would comment that his rainy experience in Wales was "the most hopeless eclipse I have ever taken part in." Nature's unbiased indifference continued. He should have settled into a happily retired life filled with RAS Club dinners, lectures, and spending time with his four children. But in 1933, an incident would change those plans forever.

Crommelin's younger son, Peter, had become a Catholic priest, and the older son, Claude, an electrical engineer. His older daughter, Andrina, was working as a computer at the Royal Observatory, which must have pleased her father. And his younger daughter, Philomena, was a teacher at a Catholic college in Edinburgh, Scotland. Philomena, an avid rock climber, convinced Claude to join her for a climb in Ennerdale Valley, in northwest England. They would ascend Pillar Rock, on the rugged north face of Pillar Mountain. To reach the summit had long been the goal of the earliest hikers, mostly young middle-class and upper-middle-class professionals for whom climbing was an exciting hobby. On July 1, a Saturday, brother and sister left their automobile at Wasdale Head Inn, famous among climbers. They set up camp near Burnthwaite Farm, where they came for their evening meal. They visited the farm again for the Sunday and Monday suppers.

When the Crommelins didn't show up Tuesday evening as expected, the innkeeper called the local police. Search parties were organized and sent out on Wednesday morning. A constable and a farmer found the two bodies, roped together, lying in the gully where they had fallen six hundred feet to their deaths. It took almost a dozen local shepherds three hours to carry the stretchers down the hazardous path, reaching the farm at midnight. A driver's license identified Claude Crommelin. A phone call was quickly placed to London police so that the famous astronomer could be notified. "I knew nothing of their death until a policeman came and told me in the middle of the night," Crommelin was quoted as saying to the press. "They had set out to ascend the Pillar Rock in the Ennerdale Valley, a well-known and difficult climb. My daughter was a greatly experienced rock climber. My son was also a climber, but not as experienced as his sister. I did not want them to go. But she had a strong influence on him and persuaded him to go to Lakeland."

The sadness of this loss would engulf Andrew Crommelin, the man who took the famous photograph showing light deflection in the skies over Brazil, for his remaining years.

COTTINGHAM: A PASSION FOR CLOCKS

Through his natural intelligence and a philosophy of hard work, Edwin Cottingham had earned the respect of men like Dyson and Eddington. After his well-known expedition to Príncipe, he continued working on clocks and repairing instruments at his shop. For thirty years, he had cleaned, repaired, and adapted the clocks at Cambridge University. He took great pride in cleaning the clock of Trinity College Chapel.[1] Clocks were his passion, but he had also nurtured a keen interest in machines, radios, and other electrical inventions.

In May 1910, the Eiffel Tower had begun broadcasting wireless signals to military installations and ships at sea, and the transmissions were being picked up as far away as North America. Cottingham became the second man in his county to be granted a wireless license. Now he could receive the daily time signals from Paris. When all domestic radio sets in the country were confiscated in World War I for security purposes, he built himself a crystal set, what Allied soldiers called a "foxhole radio." Not depending on an outside power source, these simple radios were not subject to enemy detection. The radio kept him connected to the world he was no longer exploring, unlike his famous astronomer friends, who seemed to be in a different country each time a conference was announced.

In 1928, Cottingham closed his shop and retired from official business. He could now tinker about his workshop as he pleased. When the clockmaker passed away on March 20, 1940, the world was again becoming a battleground. Hilter's army had invaded Poland six months earlier, and the Battle of Britain lay just ahead. Cottingham would have turned seventy-one the next month. It's unlikely a birthday ever arrived without his remembering the day he turned fifty, in sunny Madeira, as he and Eddington were boarding their steamer to a tiny island off the coast of Africa. His passing was noted in the *Times of London*. "He had a wide reputation for scientifically built instruments with special pendulum compensations and airtight temperature-controlled cases. One such clock, which was used as standard timekeeper at the Royal Observatory, had a guaranteed mean daily variation of not more than one-hundredth of a second."

Eddington penned an obituary for the journal of the RAS. He noted Cottingham's long history with various important clocks, and that the two had gone to the Gulf of Guinea together to view the 1919 total eclipse. He mentioned the clock that had ignited Cottingham's passion as a young man, the one he had built for the village church. As St. James Church was hidden from the main street, an external clock face would not be seen by the public. Thus, a wall-mounted clock was built for inside. "The first clock which he made was the Thrapston church clock, including the whole chiming mechanism. It was his favourite 'child,' and his last exertion, though ill at the time, was to put it forward for Summer Time in 1940."

In the days before he died, Cottingham had, indeed, been climbing the circular stone staircase that wound up to the bell tower of the Thrapston church, where he could work on the clock's chiming mechanism. It would be a difficult climb for a younger man, and Edwin was already in poor health. Perhaps, because the end was near, it was his way of saying good-bye to the clock he was building in 1900, when the first eclipse of that new century had swept across the globe, catching some of the world's finest astronomers in its shadow. It was a clock that marked the beginning of Cottingham's career, and now, still keeping excellent time, it would also mark the end. The schoolboy who had once been apprenticed to a village tailor had done well for himself.[2]

EDDINGTON: THE THEORY OF EVERYTHING

Sir Arthur Eddington soon realized a new career in lecturing to packed audiences about the general theory of relativity. He wrote more books. Since his first eclipse expedition to Brazil in 1912, with Davidson, he would visit all the continents but Antarctica. He was now famous as the driving force behind Albert Einstein's general relativity theory. In 1923, he collected many of these lectures and published them in a book titled *Mathematical Theory of Relativity*. Einstein noted that it was "the finest presentation of the subject in any language." Eddington's mother, who died in 1924, did not live to see her son knighted several years later. He was more proud of the Order of Merit, which was presented to him in 1930. As the founder of astrophysics, he was the most famous scientist in that field the world over. He may not have been as famous as Einstein, but he was held in awe by other scientists and laypeople alike.

Sister Winifred remained with him at the observatory. Students and colleagues often found him standing at the French windows in his office, gazing into the garden, deep in thought. It was his custom to eat more than one apple during these sessions, consuming even the seeds and stems. His biographer and former student, Allie Vibert Douglas, writes of the time that Edward A. Milne, the astrophysicist and mathematician and then a professor at Manchester, asked Eddington to look at a piece of theoretical work he had just carried out. While a colleague waited in the library, Milne continued on to Eddington's office. When Milne returned, quite amazed, he muttered to his friend that he wondered if the man he had just spoken to were human. Once he had told Eddington his problem, the astronomer rose from his desk and went to the window. Hands clasped behind his back, he gazed out into the garden where Winifred kept her bees and chickens. Several silent minutes passed as Milne waited. Finally, Eddington turned around and told Milne what he expected the answer should be. It was the same result Milne had arrived at earlier.

Students often commented on the untidiness of Eddington's office, the table and desk smothered in documents, books and reprints towering on the sofa and chairs, the floor carpeted with sheets of paper. Of the bookshelves that lined the walls, one was devoted entirely to P. D. Wodehouse and detective stories. Douglas remembered waiting there for her professor to arrive. "In a few seconds there would come the sound of scuffing and of footsteps, and then the door would open and in came the terrier, followed by his master—all three of us would then sit down on the settee and Eddington would proceed to look over the results. I can see him now, intently studying the pages which he supported against one knee with his left hand, while his right hand fondled the ears of the dog between us." Sometimes, on chilly autumn days, he would come in from a brisk cycling ride of twenty or thirty miles, rubbing his hands at the fireplace to warm them before settling down to work.

World War II brought food rationing back to England. At nearly fifty-seven years old, Eddington now had no worries of being conscripted. But he felt out of tune with a Cambridge at wartime. In 1940, the Luftwaffe began its savage nonstop bombing of English cities. A rumor circulated later that the Germans and British had an agreement not to bomb cities like Oxford, Heidelberg, and Cambridge. This wasn't true. The first bombs that fell over Cambridge hit the sugar beet fields on the outskirts of the city, causing only crop damage. But in 1941, several other bombs did inflict damage on a few of the buildings. A loner to begin with, Eddington grew more isolated. He would write to colleagues how much Cambridge had changed in the years of the war. The Observatory Club was no longer meeting, and the campus itself seemed void of graduate students. "We get on pretty well here," he wrote to Douglas. "What I miss most is the stimulus of research students." He would have liked to attend a baseball or football game now and then and enjoy the Christmas letters that American astronomers would send but which were now undeliverable.[3]

For the last few years of his life, he was obsessed with a book that would harmonize quantum physics and relativity, a "theory of everything." In the summer of 1944, as he filled dozens of pages with his ideas, he began to feel unwell with stomach pains. He told no one, not his sister and not even C. J. A. Trimble, who visited him in October. Trimble noticed only that Eddington was consumed with work on his book, turning out draft after draft in his small and immaculate handwriting, as if he expected each version to be the last. World War II was filling English hospitals with a multitude of casualties that demanded immediate medical attention. Doctors were overextended. When Eddington finally let it be known that he was in considerable pain and needed to see a physician, there were no openings for him. He waited several weeks for just an X-ray.

A large cancerous tumor was discovered, critically advanced. This meant an emergency operation. He went home to prepare for the surgery the next morning. Douglas described that day: "All summer and autumn he had been working long hours to complete the sixth draft of his book. But two chapters and perhaps other sections remained incomplete when he had to lay down his pen on the night of 6 November."[4]

The next morning, Eddington left the observatory home he had known for over thirty years. He was driven to the hospital at Evelyn Nursing Home, in Cambridge. The operation was difficult, his condition now so serious that only Winifred was allowed into his room. He would linger for two more weeks, and in the facility's unsatisfactory conditions. Winifred wrote to a friend that her brother's care was not good enough, the nursing home being subject to wartime constraints. "My hands are so cold so do excuse the writing—we have to save fuel!" This need to ration resources was unfortunate, given that her brother would leave behind forty-seven thousand pounds, a considerable sum in those days. He would not live to see the end of this war, as he had the previous one. The founder of astrophysics died on November 22, 1944, at the age of sixty-two. His body was cremated at Cambridge Crematorium.

A memorial service was held five days later in the Trinity College Chapel. Along with the most prestigious representatives of Cambridge and other universities, in attendance were sister Winifred, a cousin John Eddington, and the loyal friend for so many years, Trimble. "Eddington had the true mystic insight to feel a deep significance behind everything in nature," Douglas would write of him. "He was moved by a harebell waving in the breeze and by the mighty rush of water over Niagara." Eddington's ashes were then buried in his mother's grave, on the hillside behind the observatory and near a row of cypress trees. He would be in good company. Since 1869, the earth there had taken in hordes of dignitaries, famous scholars and philosophers, poets and politicians. Sister Winifred would join her mother and brother a decade later, her own ashes sifted into the same grave. The Eddington family's journey, which had begun in the headmaster's house in northern England—a house declared unfit because of the typhoid fever that took their father away—was now over.

PERRINE: THOSE SOUTHERN STARS

William Campbell had recommended that Santa Clara University award Charles Dillon Perrine an honorary doctorate, qualifying him for the directorship of the Argentine National Observatory. Thus, Perrine became the forerunner in bringing astrophysics to that country. The great telescope he wanted built, the one he had hoped James Mulvey would live to enjoy, was still not completed when Perrine and his family visited California in 1923. His children at this time spoke limited English, having all been born and raised in Argentina. But Bell Perrine had never adapted to the different culture. In fourteen years of living with her husband at the observatory, she had learned only a few words in Spanish. The climate in the Southern Hemisphere was also not favorable for Bell's health. Charles, on the other hand, had learned to speak Spanish. He could read and write the language proficiently. After this visit to California, Charles returned to Córdoba without his family. Given what lay ahead, Bell was wise to remain where she was.[5]

Perrine wanted to see the great telescope built before he retired. As happened with the 1914 eclipse expedition, world events again held power over personal ambitions. Despite his contributions to astronomy, Perrine had nationalist enemies who wanted him gone. Argentina had been in the throes of political and financial agitation for some time. By 1930, the world was suffering its worst years of the Great Depression. The country would begin what is known as "the Infamous Decade," when a pro-Fascist military regime succeeded in a coup against the current president. Bombs exploded in the streets, and gunfire erupted. One evening, as Perrine sat at his office desk, a sniper fired a bullet through the window. Luckily, it missed the astronomer and hit the brick wall behind him. This was 1931, and Perrine had already been verbally attacked by members of the Argentine Congress. As he wrote to George Hale, he had become "the English Gringo," at least to the Fascists and despite his enormous accomplishments that were lauded by astronomers around the world. Two years later, the congress finally passed legislation that demoted him from director to senior officer. Now subject to political control, he clung tenaciously to his post, still struggling to find funding that would finish the great telescope.

Had Perrine been determined to retreat to California and leave the chaos behind, there remained the eternal curse of money. The Argentine government would honor his pension only if he stayed *in the country*. The government, in perpetual upheaval, wanted no currency to leave its borders. When the astronomer's daughter, Mary Perrine, graduated from the University of California in 1935, she returned to Córdoba to act as her father's secretary. Despite his asthma attacks and failing health, Perrine continued to publish numerous papers and books. In 1936, the stress heaped on him became too much, and he retired. He was already sixty-nine years old, his right hand now crippled from arthritis. He purchased a house north of Córdoba, in Villa del Totoral. A nurse came to live there and take care of him.

Perrine would witness the installation of the great telescope, finished in 1942, the largest in South America. Over the next few years, he remained prolific, writing papers on the solar system, comets, eclipses, and even earthquakes.[6] Early in 1951, six months before he passed away, he sent a letter to Harold Spencer Jones, the astronomer royal since Dyson's retirement. Perrine expressed his displeasure that the royal observatory had been moved from Greenwich to Sussex a few years earlier. As was the case with Campbell, Dyson, and Eddington, the world he knew was changing too rapidly with technology advancing daily. The time finally came to discuss with daughter Mary where he wanted his personal items to go, a gold watch to one son, a sextant to another, the silver objects to his wife and daughters. Some of these were souvenirs that had been given him as he traveled the world to observe total eclipses. As a final request, his loyal nurse should be allowed to live on in the house. At her death, it would go to his five children.

Charles Dillon Perrine, the man who liked to say that ham paid for his education, died of a heart attack on June 21, 1951. Since he had given his plot in the Cemetery of Dissidents to bury James Mulvey, his nurse made plans for Perrine to be laid to rest with her family. He had no friends among the Protestant clergy, but had kept a steady friendship with the Catholic priests. Since they were not allowed in the Protestant Cemetery, they had come to his bedroom the night before to give him last rites. The local papers took note of the famed astronomer's death with captions like "Dr. Carlos Dillon Perrine, North American in Origin, Who Lived for 42 Years in His New Country," and "Report of the Death of an American Citizen." These were the same newspapers whose editorials had attacked him years earlier. His pallbearers were the aging astronomers he had known in his career and the director who had replaced him. "Then the coffin was lowered into the grave," daughter Mary wrote, "and Papa was at last resting as he hadn't done in his whole busy life." Jones would write an obituary for the RAS. The man who had given up the stars over Mount Hamilton for the "untouched southern sky" was now at peace.

FREUNDLICH: THE DEVOTED COLLABORATOR

Einstein and his first believer, Erwin Finley-Freundlich, continued a strained relationship, with money often being the cause. Later, Freundlich would insist that Einstein had given him the original draft of the general theory of relativity. Einstein, and especially the enterprising Elsa, demanded it back. After much squabbling, the papers ended up at the Hebrew University of Jerusalem, in 1925, Einstein being one of the founders. But Freundlich was still not done with his mentor's theory. He had been unsuccessful at observing the 1914 eclipse, thanks to the war, although clouds would have done him in if Russia hadn't. In 1922, along with the Greenwich expedition, he had spent ten days aboard a ship being tossed about in a storm off Christmas Island, in the Indian Ocean. That venture was also unsuccessful. In 1926, in Sumatra, he managed to make his first observation but ended up photographing through haze. He returned to Sumatra for the 1929 eclipse in January and this time determined that the deflection was 2.2 arc seconds, even greater than what Einstein had calculated. A later measurement showed that the weighting system Freundlich used had been defective.

In 1933, with Hitler in power and anti-Semitism on the rise in Nazi Germany, Freundlich saw the danger in remaining in his homeland. Käte, his wife, was Jewish. They now had a family to worry about. When Käte's sister died that same year, they had adopted her young son and daughter. Freundlich was forced to leave the post he had worked so hard to achieve at Potsdam, where he had designed the solar telescope for the Einstein Tower. Einstein was in the same predicament in Germany. That autumn, he left Berlin with Elsa and would eventually accept a position at the Institute of Advanced Study, at Princeton, New Jersey.

Freundlich and his family went first to the observatory at Istanbul, where he was soon involved in modernizing the equipment. In 1937, they moved to Prague, where he accepted a position at Charles University. By 1939, Prague had also become unsafe under the Nazi occupation, so Freundlich moved again, this time to Holland, which soon suffered the same fate. With a recommendation from Eddington, the German astronomer found a stable home at the University of St. Andrews, in Scotland, where he was to oversee the building of a modern observatory. He was later appointed its first Napier Professor of Astronomy.

While at St. Andrews, Freundlich shared some Einstein stories with colleagues. Walter Ledermann had been born in Berlin in 1911, the same year Freundlich and Einstein sent their first letters back and forth. Ledermann was a young mathematics teacher at St. Andrews when he came to know and respect Freundlich. The older man told him about the day that Einstein was expounding on his general relativity theory. "He suggested that geometrical relations in space would not follow the Euclidean pattern and that, for example, the shortest distance between two points would not be the straight line joining them but a certain curve." Freundlich responded to this assertion: "Professor Einstein, what you have described is known to mathematicians as Riemannian geometry. It was discovered more than fifty years ago by Bernhard Riemann and developed by him." At this comeback, Einstein was so astonished that he called Freundlich a liar. To end the matter, Freundlich went to the library and signed out a copy of Riemann's original paper.[7]

In those early days, Freundlich had seen himself as more of a blackboard that Einstein used to simplify his mathematics than he saw himself a collaborator. Einstein had said that if a pupil of Felix Klein's couldn't understand his equations, then who could?

Ledermann grew close to the older man, who was in his midfifties when he arrived in Scotland. When the two men walked from the observatory into town, locals would see the tall and impressive Freundlich lumbering next to the shorter Ledermann and joke, "Here comes the sun and the moon." Freundlich's students knew him as "Herr Professor," and his colleagues considered him kind and generous with his time and knowledge. An able cellist, he enjoyed playing chamber music with Ledermann and others. Käte, on the other hand, was more fastidious and never fit into the less formal Scottish life. She was happy to return to Germany when they left Scotland in 1959. Freundlich was appointed honorary professor at the University of Mainz. Over half a century after he sent a letter to Perrine, mentioning a "Prof. Einstein," and signing his name with a confident and stylish E. *Finlay Freundlich*, he died at his home in Wiesbaden. It was 1964, and Freundlich was seventy-nine years old. His adopted daughter brought his cremated ashes back to Glasgow and scattered them in the Jewish cemetery.

DAVIDSON: INTO A MODERN WORLD

Charles Rundle Davidson had barely turned fifteen when he applied for a position at the Royal Observatory. William Christie was then the astronomer royal. The teenager learned the instrumentation so well that by 1896, he was an established computer. When Frank Dyson became astronomer royal in 1910, he soon relied on Davidson as an expert on the design and care of the instruments, and as the man who would have a match when his pipe went out. Davidson could have no way of knowing that he would outlive his good friend and colleague by over three decades. Elected as a Fellow of the Royal Society in 1931, he would be the last man standing in the group of dedicated astronomers who accepted Einstein's challenge to photograph the stars.

Davidson had gone on eight solar eclipse expeditions in his forty-year career at the Royal Observatory: Portugal in 1900; Tunisia in 1904; Brazil in 1912; Russia in 1914; Brazil in 1919; Sumatra in 1926; England in 1927; and, finally, Canada in 1932, his last. As was the case with all the other astronomers, his life had been part hard work and part exciting adventure. In his later years at the observatory, he developed an interest in stellar color photometry, a subject on which he would publish and copublish many papers. He retired in August 1937, after a career that lasted nearly fifty years. In a tribute to his loyal service, Harold Spencer Jones, who had stood next to him in the villa's garden north of Minsk in 1914, wrote a tribute. Jones reminded readers how the design and care of the instruments used in eclipse expeditions had often been placed under Davidson's supervision. "It is not possible to understand how much the Observatory has owed to Mr. Davidson," Jones added.

Davidson and his family moved out of London during World War II to a little English village called Ridlington, in the county of Rutland. When the children were older, Davidson's wife, Eliza, went back to school to become an elementary teacher. For years, the Davidsons had escaped during summer holidays to the seacoast with their four children, often to Cliftonville, on the North Sea. With money saved from teaching, Eliza bought the family a home there, where they could spend the entire summer. Eliza died in 1944 and was buried in the churchyard at Ridlington.

After Eliza's death, Charles kept busy writing and publishing papers. He later lost vision in his right eye and had a cataract removed from the left. He went to live with an unmarried daughter, a schoolteacher. But he wanted to return to Cliftonville where he had so many happy family memories. His daughter Stella had married by then, with three daughters of her own. She was living not far from the house that Eliza had bought. Charles would spend the last years of the 1960s with Stella and her family.

He knew the area well. He walked the beach daily for exercise and helped his granddaughters build sandcastles. He treated them from a tin of sweets he kept by his living room chair. He often visited the local shops and, unlike Campbell, loved gardening. "We were fascinated with his eyeglasses," his granddaughter recalled. "Because he had only one eye, they had two different lenses, one for close up and one for distance. They were hinged on both sides so that he appeared to turn them upside down and inside out to change between lenses." Charles spoke very little of his life as an astronomer. Still, the girls plied him for stories. They had heard from their mother of the exotic parrots and miniature monkeys that had come from far-off Brazil. "I was paid to enjoy my hobby" was all he would tell them.

He kept track of time by the bells he had grown used to as a schoolboy, especially for teatime. "Is it time for our seven-beller?" he would ask Stella. Bells on ship were rung every half hour, so this meant his 3:30 cup of tea. Approaching ninety years old in the mid-1960s, he had begun to grow deaf, and now the world of sound was also slipping away. But it was a world far different from the one that had seen him to manhood. Davidson remembered the floor-length Victorian dresses worn by his mother and the women of his teenage years. No wonder he was shocked to see miniskirts on the streets of London. "Do you think they forgot to put on their skirts?" he would ask a smiling granddaughter. He had gone from horse and carriage to man on the moon. Perhaps this was why, when his family gathered around a grainy television set in the summer of 1969, he doubted what they were telling him. The American astronaut Neil Armstrong had just stepped upon the surface of the moon. "It's not real," Davidson declared, when he read the story the next day in the *Daily Telegraph*. They couldn't persuade him otherwise.

Davidson was the man who had developed the famous photograph of the eclipsed sun in Brazil, taken by the four-inch telescope Cortie had been wise enough to send as a backup. When he slipped and fell in the street one day in 1970, having just turned ninety-five, he was taken to the local hospital. Muddled and unable to remember what had happened, he had difficulty expressing himself. He no longer recognized his family, those granddaughters with whom he had played cards, and checkers and chess. When he died, the granddaughters dressed in bright colors for his funeral. "We wanted to celebrate his extraordinary life. It was so long since he retired, no one was there from the astronomical world." Everyone who knew him in those early days of expeditionary travel were long gone, none left to write his official obituary. It was undertaken instead by then astronomer royal, Sir Richard Woolley, who had been chief assistant at the observatory for four years before Davidson left. "Although Davidson retired from the Royal Observatory staff more than thirty years ago," Woolley wrote, "there are still some Greenwich men who remember him well, not only because of his extreme skills with instruments but because he held strong principles yet was sensible enough to compromise."

The day of Charles Rundle Davidson's funeral marked the end of the Victorian-born astronomers who had once boarded fancy steamers and traveled around the world.[8]

EINSTEIN: THE FABRIC OF SPACE-TIME

When Allie V. Douglas, was working on her biography of Eddington after his death, she mailed a letter to Einstein, asking if he might contribute by offering his opinion on Eddington's contribution to the general theory. Two weeks later she had a reply, giving her permission to quote it:

> Eddington's main achievement, in my opinion, is his theory of the stars. His creative achievement in the field of relativity and the theory of matter did not carry conviction for me. But this may be my fault. The German physicist and philosopher Lichtenberg once said: "If a head and a book collide and it sounds hollow—this is not necessarily caused by the book!"

> Yours sincerely, A. Einstein

After fleeing Nazi Germany, Albert Einstein would remain at the Institute for Advanced Study in Princeton from 1933 until his death in 1955, a prominent symbol of the scientists who were compelled to leave under Hitler's regime. Unlike the astronomers who had traveled far and wide when a total eclipse beckoned them to test for light deflection, he did not slip between the pages of history. From the start of his fame, launched in 1919 by the meeting at Burlington House, until the day he died, celebrity pursued him. Hundreds of books have been written about him. Films and documentaries have struggled to define his genius. His detractors would continue to disparage his work, but unsuccessfully. When a book was published in 1931, in Germany, that listed one hundred "professors" speaking out against Einstein, it could have rattled him, except that it was a last sigh from the old guard, most of them too unqualified to speak in the first place. Einstein kept to his famous sense of humor instead. As he told the sculptor Jacob Epstein, who was struggling to create a bust of the physicist through a thick cloud of pipe smoke, "If I were wrong, one professor would have been quite enough."[9]

Einstein. To this day, his last name standing alone denotes genius.

In 2015, a century after the theory of general relativity was conceived, scientists announced that they had detected and recorded an unusual sound coming from the vastness of space. The fabric of space-time that Einstein had predicted was now rippling as two black holes collided a billion light-years away. It was the language of the universe, as if someone had held a giant seashell to the cosmos. The German physicist had been right. Time and space were interwoven, warped, able to stretch and shrink. Einstein would have been pleased, but not surprised. Those Victorian-born astronomers whose daring expeditions had challenged the universe of *their* era—Campbell, Perrine, Freundlich, Dyson, Eddington, Crommelin, and Davidson—would be gratified to learn that another great door to the mystery had been kicked wide open.

ACKNOWLEDGMENTS

We first thank those who generously provided family documents, personal letters, and photos to help with our research:

Bob Kelly-Thomas and Diana Merlo Perrine, the grandchildren of Charles Dillon Perrine, who sent many of their grandfather's letters, papers, and photos, and helped considerably with research.

Rosemary Cecily Hannibal and Catherine Stella Court, of England, the granddaughters of Charles Rundle Davidson. They answered dozens of questions so that we could better know their grandfather.

Peggy Campbell Rhoads, granddaughter of William Campbell, and Alison Campbell Clinch, great-granddaughter of William Campbell, for the archival help and encouragement (and in memory of their father and grandfather, Kenneth Campbell).

Christopher Wise, for the journal that his grandfather, Daniel Wise, kept during his stay in Sobral in 1919 with the English expedition.

Santiago Paolantonio, at Córdoba, for Herculean labors in sending his research materials on Charles Dillon Perrine's life in Argentina and on Perrine's expeditionary travels.

Jack Page, for all the research, reading, and encouragement.

Dr. José Maria de Castro Abreu Jr., for answering numerous questions about Brazil and medicine in the 1900s.

Our agent, Paul Lucas, at Janklow & Nesbit; our editor, Colleen Lawrie, and the team at PublicAffairs, including Melissa Veronesi, Amy Quinn, Patricia Boyd, Jocelynn Pedro, and Miguel Cervantes for their hard work.

And special thanks to Tom Viorikic, Dianna Abney, Delilah E. A. Gates, Sylvester J. Gates III, Stephon Alexander, Charles Kankelborg, Theodore Jacobson, David Spergel, Michael Turner, Clifford Will, Nicolas Yunes, Brian Keating, Kory Stiffler, and Mary Sutton.

A small army of archivists, librarians, writers, researchers, and professors in countries around the world were gracious enough to send letters, photos, papers, and even good wishes. And to those of you who read various chapters and drafts of the manuscript, your assistance and advice has been invaluable.

UNITED STATES: Shaun Hardy, Librarian, and Tina McDowell, Editor, Carnegie Institution for Science, Washington, DC; Sofia L. Birden, Associate Director, and Debra Durkin, Library Specialist, Blake Library, University of Maine at Fort Kent; Emily Walhout, Houghton Library, Harvard University; Jami Frazier Tracy, Curator of Collections, Wichita-Sedgwick County Historical Museum, Kansas; Jan Vetrhus, Jefferson County Historical Society in Madison, Indiana; Maria C. McEachern, John G. Wolbach Library, Harvard-Smithsonian Center for Astrophysics, Cambridge; Tony Misch, Director, Lick Observatory Historical Collection; the Mary Lea Shane Archives, Lick Observatory; Alix Norton and Maureen Carey, Special Collections and Archives, and Frank Gravier, Librarian, McHenry Library, University of California at Santa Cruz; Brian Kiss, Front Line Services Sterling & Bass Libraries, Yale University Library; Michael L. Strauss and his Fredericksburg Pennsylvania History website, for the information on James Lick; Allyssa Bruce, at the Kansas State University Libraries Help Desk, and the Richard L. D. and Marjorie J. Morse Department of Special Collections; Loma Karklins, Archivist for Reader Services, Archives & Special Collections, California Institute of Technology, Pasadena; the Bancroft Library, University of California, Berkley; Brenda Corbin, Librarian Emeritus, US Naval Observatory, Washington, DC; Dr. Valerie Rapson, Outreach, and Angela Matyi, Archives Assistant, Dudley Observatory, Schenectady, New York; Bill Kramer, for his information on astronomer David Todd and his website www.eclipse-chasers.com; Christina E. Barber, Deputy Archivist, Amherst College;

Luisa Haddad, Special Collections & Archives, University of California Santa Cruz; and Louis Pelletier III, Governor Janet T. Mills, Dr. Dora Mills, Tony Buxton, Karessa Grenier, Deborah Hodgkins, Maureen Connelly, Jan Greico, Don Zolotorofe, Angel Dionne, Randy Ford, Becky Hobbs, Kathleen Wallace King, Larry Wells, Larry Berz, James H. Page, Allen Jackson, Edgar Rothschild, Beurmond Banville, Nelson Eddy, Nancy Henderson, Michael Zeiler, Rick McLaughlin, David P. Ferguson, Lana Pelletier, Alistair Sponsel, and Colleen McLaughlin; Diana Kormos Buckwald at the Einstein Papers Project; Regina Filomeno, in Chicago; Daniel Meyer, Director, Special Collections Research Center, University of Chicago Library; Jacki Tolley, Science Source; Jacob Daugherty, at Bridgeman Images; Darrell McBreairty; Marsha Luttrell, at Mercer University Press; and Emma Grace Pelletier.

ENGLAND: Adam C. Green, Assistant Archivist and Manuscript Cataloguer, Trinity College Library, Cambridge; Graham Dolan, for his extensive information on the Royal Observatory and the early astronomers; Eric Franklin, and the Thrapston Town Council for the Cottingham information; Helen Chapman, and the Antiquarian Horological Society London, for Cottingham's photo; Neil Busby, at Thrapston Heritage; Mike Bevan, Archivist, and Martin Salmon, Archivist, Royal Museums Greenwich, National Maritime Museum; Mark Hurn, Departmental Librarian, Institute of Astronomy, University of Cambridge; Frank Bowles, Manuscripts Reading Room, Cambridge University Library; Dr. Sian Prosser, Librarian and Archivist, Royal Astronomical Society; Natalie Conboy, Collections and Archive Assistant, and Rachel Mollitor, Executive Assistant to the Chief Executive, Old Royal Naval College, Greenwich; James Kirwan, Digitisation Services Manager, Trinity College Library; Dr. Emma Saunders, Royal Observatory at Greenwich Archivist, Department of Manuscripts and University Archives, Cambridge University Library; John R. Fletcher F.R.A.S., Mount Tuffley Observatory, Gloucester; Dr. Margaret Penston, Institute of Astronomy, University of Cambridge; Professor Emeritus Derek Jones, Royal Observatory at Greenwich; Neil Rhodes, at Image-Restore, for the information on early photographic plates, Bordon, Hampshire; David Ball and Betty Ball, Ringstead; Wendy Page, for her help with the life of John Jepson "Atky"

Atkinson; Tom Leimdorfer, Claverham Meeting House, Bristol, for Eddington ancestry; Dr. Rebekah Higgitt, Senior Lecturer in History of Science, Rutherford College, University of Kent, Canterbury; Alison Moulds, Communications Officer, British Society for the History of Science; Beverley Larner, Executive Secretary, Cambridge Philosophical Society; Robson and Helena, at Selfridge's & Co., in London; Anne Gleave, Curator of Photographic Collections, Archives National Museums Liverpool, Merseyside Maritime Museum, Albert Dock; Rosemary Kingsland; Alexander von Weber, Süddeutsche Zeitung Photo; Martina Landmann, at ullstein bild; Joanna Adamczewska, Bonnier Media.

SCOTLAND: Professor Edmund F. Robertson, Mathematical Institute, University of St. Andrews; Rachel Nordstrom, Photographic Collection Manager, University of St. Andrews.

BRAZIL: Dr. Dr. José Maria de Castro Abreu Jr.; Ildeu de Castro Moreira, Institute of Physics at the Federal University of Rio de Janeiro; Kátia Teixeira dos Santos, the National Observatory, Rio de Janeiro; Emerson Ferreira de Almeida, Technical Scientific Director, Museo do Sobral; the folks at City Hall of Fortaleza, Brazil, for finding the seminary where Davidson and Crommelin stayed during their visit to Fortaleza after the eclipse (Prefeitura de Fortaleza).

AUSTRALIA: Linley Janssen, Sue Hunter, and Carol Smith, at the State Library of Western Australia, for the information and the many files of photographs; Roderick W. Home, Professor Emeritus, University of Melbourne, Australia.

CANADA: Heather Home, Private Records Archivist, Queen's University Archives, Kingston, Ontario; Sarina King, Assistant Archivist at Greyroots Museum in Owen Sound, Ontario, for information on David Thomson; Mylene Pinard, Liaison Librarian, McGill Library, McGill University, Montreal; Isaac L. Stewart, of Stratford, Prince Edward Island, peihistoryguy.com; the Royal Astronomical Society of Canada; the Arctic Institute of North America, University of Calgary, Alberta; Rita Shellard; and Callahan Levi Flipo, of Montreal.

GERMANY: Professor Klaus Hentschel, University of Stuttgart, Institute of History; Johanna Wolter, in Dresden, for help with translating; Michael Feist, CEO, Witte-Verlag Publishing, Hamburg; Anke Vollersen,

Hamburg Observatory Library, Department of Physics, University of Hamburg; Regina von Berlepsch, Head of the Library Department, the Leibniz Institute for Astrophysics, Potsdam; Rebecca Young, Heritage Centre Archivist, Royal Hospital School; Ellen Embleton, Picture Curator, The Royal Society, London.

ARGENTINA: Ing. Santiago Paolantonio, Magister Administración Educacional Área Historia, Enseñanza y Difusión de la Astronomía, Observatorio Astronómico Córdoba (Engineer and Master of Education, Astronomical Observatory Museum, National University at Córdoba).

PORTUGAL: José C. N. Duque Silva, Library, the Lisbon Geographic Society (Biblioteca da Sociedade de Geografia de Lisboa) for such patience in shipping us a needed book; Duarte Pape, author of *As roças de São Tomé e Príncipe*, for the amazing history and images of the roças; Joana Latas, for needed information; Paula Santos, Researcher and Curator, Astronomical Observatory of Lisbon, Portugal; Isabel Cruz, Lieutenant Commander, Hydrographic Institute, Navy, Lisbon; Alexandra Canha, Director of Biblioteca Municipal do Funchal (Madeira) for finding the 1919 newspaper interviews Eddington gave in 1919, before leaving for Príncipe; Chris Blandy, in Funchal, whose family business stored Eddington's equipment, in 1919.

ROMANIA: Carmen Zeries, Emanuel Zeries, and Eva Zeries, for the needed technical help.

PRÍNCIPE: Many thanks to Rita Alves.

RUSSIA: Elizaveta E. Kozlova, Library Director, Pulkovo Astronomical Observatory Library, Pulkovo.

And last, but certainly not least, a special thank you to GOOGLE TRANSLATE for working free of charge and without complaint while translating many emails and letters from around the world.

NOTES

CHAPTER 1: A PATH MADE OF MAGIC

1. James Fenimore Cooper might have missed the eclipse described in the epigraph had Yale not expelled him a year earlier for youthful mischief. He either brought a donkey to class or set off an explosion that blew the door off a fellow student's room. Whatever he did, Yale called it "misconduct" and sent young Cooper packing. The 1806 eclipse entered North America from the southwest, which was then Spanish controlled, and moved northeast toward New England. Fenimore's hometown, the Village of Otsego, New York, lay in the path of totality. His short story, "The Eclipse," beautifully evokes the reactions of the animals, as well as the townsfolk, and ranks among the most literary descriptions of a total eclipse.

2. Littmann, Espenak, and Willcox, *Totality: Eclipses of the Sun.*

3. In 1720, Edmond Halley succeeded John Flamsteed as Britain's second astronomer royal. It's believed that he was the first to draw the path of an eclipse as seen from above, looking down on the earth's surface. The eclipse of 1715—totality lasted for three minutes and thirty-three seconds—became known as Halley's Eclipse. But he is most famous for the comet named for him and for having predicted its return in 1759, which he did not live to see.

4. The incipient country's *first* expedition, and the first for Samuel Williams, was not for a solar eclipse. In 1716, Edmond Halley published a paper challenging scientists around the world to view the very rare transit of Venus, which would occur in 1761, some two decades after his death. With Benjamin Franklin's help, plans were made to send John Winthrop to Canada. Winthrop was teaching at Harvard then and took with him two students. Samuel Williams and Isaac Rand, chosen because of their mathematical skills, would miss their graduation because of the expedition. With the French and Indian War two years from its conclusion, they traveled under a flag of truce, with permission from France. The province of Massachusetts equipped Winthrop with a sloop that would carry them from Boston to "the savage coast of Labrador," a journey of thirteen days. They pitched their tents on what is today Kenmount Hill and were soon faced with a problem they hadn't foreseen, "the infinite swarms of insects, that were in possession of the hill." Despite this torment by the blackfly, from the family *Simuliidae*, Winthrop and his students had a front row seat for the transit on June 6.

5. Authors' note: An occasional reference is cited in an endnote to emphasize a specific source. But to avoid overwhelming the reader with excessive notation, unless otherwise noted, sources for the observations and quotations are listed in the Bibliography.

6. For a more complete story of this 1780 expedition by Samuel Williams, see Rothschild, *Two Brides for Apollo*.

7. The apparent eastward motion is because the speed of the moon's revolution toward the east is about twice the rate of the rotation of the earth from west to east. Because of the spherical shape of the earth, the umbra is slowest in the middle of its path.

8. The Reverend Bacon met up with his friend, the "kinematographer" John Nevil Maskelyne, another colorful BAA member famous for inventing the pay toilet, which required a penny to open the locked door. "Spend a penny" had thus become the euphemism of the day in Great Britain for needing a trip to the bathroom. Maskelyne was even more famous for co-inventing a card-playing automaton named "Psycho," which had gained fame with its hundreds of performances at London's Egyptian Hall. A magician, illusionist, and "cardsharp" who exposed cheaters and fake mediums, Maskelyne had come to Wadesboro hoping to be the first person to *film* a solar eclipse.

9. Walter Maunder, though a member of the Royal Astronomical Society for many years, had been a driving force behind the British Astronomical Association's goal to provide membership to amateur astronomers, regardless of class, education, or gender. After his wife's death five years earlier, he had married Annie Russell. Because married women were not allowed to hold public positions, Russell was obliged to quit her job as a "lady computer" at the Royal Observatory when she married Maunder. But her marital status would hardly end her career as an astronomer. She was so skilled at chronology that Arthur Eddington, years later, would ask Annie to "date the Nativity."

10. Maunder, *The Total Solar Eclipse 1900*.

CHAPTER 2: EINSTEIN'S VISIONARY YEARS

1. Einstein's biographer and colleague, the theoretical physicist Abraham Pais, who wrote *Subtle Is the Lord*, withheld the 1994 publication of his second book, *Einstein Lived Here*, while the editorial staff traveled to Yugoslavia and Hungary in an unsuccessful search for answers.

2. Granddaughter Evelyn Einstein—her father was the first son, Hans Albert—found the packet of letters in 1986. When Mileva died in 1948, the letters went to her son's first wife, Frieda Einstein, who brought them to the United States and put them on a shelf of her husband's closet. After Frieda's death, Evelyn was sorting through her mother's things when she found the letters her grandparents had written (Zackheim, *Einstein's Daughter*). When one considers the state of Mileva's emotional and financial situation during her final years, it's quite remarkable that she didn't sell them.

3. No researcher has spent more years and travel while seeking an answer to Lieserl's disappearance than Michele Zackheim. Visiting Serbia several times, twice during the Yugoslav Wars in the 1990s, Zackheim interviewed over a hundred people related to Marić's life. She believes she has deciphered the death of the unfortunate Lieserl. "And finally, by understanding the procedures of the postal system between Switzerland and the Austro-Hungarian Empire, I was able to determine the day that Lieserl died—September 21, 1903, the day of a solar eclipse, the day when the sun disappeared from the sky" (Zackheim, *Einstein's Daughter*).

4. This paper on photoelectricity earned Einstein the Nobel Prize in 1921. That photons carry discrete packets of energy had been anticipated by physicist Max Planck, who described them as "quanta." Planck's work was the beginning of the quantum theory. It's thus ironic that Einstein's Nobel Prize was awarded for work that was one of the cornerstones of quantum mechanics, an idea he would never accept in his lifetime.

5. The *concept* of an atom, and the word itself, was first discussed by fifth-century Greek philosophers. Isaac Newton left behind in various publications his thoughts that matter constituted particles in motion through a vacuum. In 1827, botanist Robert Brown studied pollen grains through a microscope and noticed that they wiggled about erratically, in what became known as the *Brownian motion*. Not until 1905 did Einstein prove the existence of the molecules responsible for this action when he observed the motion of particles smaller than pollen grains floating in the air. He could then connect the motion of the particles to the actual *size* of atoms.

6. By connecting energy and matter, $E = mc^2$ also led us to a true understanding of the energy source of the stars. Though this equation and its fullest application is rightly attributed to Einstein, he was not the first to write it. Several physicists, including Oliver Heaviside, Wilhelm Wien, and Max Abraham, wrote incorrect versions previously. The first physicist to write it correctly was Henri Poincaré, but he did not realize that it applied to real masses, not just to electricity and magnetism. The world would more fully understand this abstract idea when two atomic bombs were dropped on Japan, ending World War II.

7. Arc minutes and seconds have been used in astronomy since antiquity to measure positions and orbits of objects in the solar system. There are sixty arc minutes in one degree. Each arc minute is divided into sixty arc seconds. The size of an arc second would be comparable to the size of a dime at a distance of 1.3 miles.

8. Johann Georg von Soldner (1766–1833) remained better known for those works named for him, such as the Ramanujan-Soldner constant and the Soldner coordinate system. In an unpublished manuscript written around 1784—it was inspired by John Michell's paper on the escape of light from a massive body, published in 1783—Henry Cavendish (1731–1810) appears to have come to basically the same result that Soldner had.

CHAPTER 3: THE TWO ECLIPSES OF 1912

1. Leo Wenzel Pollak was then at the Institute for Cosmological Physics. On August 24, 1911, he wrote to the Berlin Observatory on Einstein's behalf. In a letter written on September 21, 1911, Einstein assured Freundlich: "You can keep the reprint, of course."

2. When atoms of different elements are in a high-temperature environment, each atom produces a rainbow-like array of colors, its *spectrum*. Hence a spectroscopist is a scientist expert in the study of this phenomenon.

3. This ambitious astronomical project was conceived and launched in 1887 by the Paris Observatory and the Observatory at the Cape of Good Hope. The objective was to "map the sky" by photographing and cataloging the positions of millions of stars, some as faint as eleventh and twelfth magnitude. Eighteen more observatories from around the globe, including the one at Córdoba, signed on. It would be Dyson's first project when he joined the Royal Observatory. Although the project was never finished, given the vastness of its reach and the enormous expense, international committees met over the upcoming decades to discuss the progress and confront problems.

4. Guglielmo Marconi had been offered free passage on the *Titanic* but, three days earlier, had chosen the *Lusitania* because he had work to do and preferred the public stenographer on that ship. Thanks to a transatlantic radiotelegraph service that had been put in place between Ireland and Glace Bay, Canada, in 1907, Marconi's invention was instrumental in rescuing the *Titanic* survivors.

5. The Southern Hemisphere has a continuing role in astronomy to this very day. The center of the Milky Way galaxy is visible from there. Therefore, if one wants to study the black hole at the center of our own galaxy, the best way would be to travel south of the equator; Chile plays a particularly prominent role.

6. The RMS *Arlanza* was one of the luxury passenger ships built by the Royal Mail Steam Packet Company. The company had become so famous that a decade earlier, Rudyard Kipling's *Just So Stories*, which ranked among Arthur Eddington's favorite "light literature," sang their praises in a popular poem of the day: *"I've never sailed the Amazon, / I've never reached Brazil; / But the Don and Magdalena, / They can go there when they will! / Yes, weekly from Southampton, / Great steamers, white and gold, / Go rolling down to Rio . . . And I'd like to roll to Rio / Some day before I'm old!"* These steamers were the largest machines on earth.

7. Argentina's second team would go even farther inland than the others, to Alfenas. With Morize going to Passa Quatro, the second Brazilian team would go to Cruzeiro, just a few miles to the south. Milan Rastislav Štefánik, as the only member of the French team, would stay at Passa Quatro with Morize and Eddington.

8. The president of Brazil was not interested in just the eclipse. He had been keen on using this natural event and a visit to Minas Gerais to let it be known that he hoped Pinheiro Machado would succeed him as president. Three years later almost to the day, Machado would be stabbed to death in the lobby of the Hotel dos Estrangeiros in Rio, where the English expedition had been lodged.

9. Astronomers the world over relied on this telegraph system, adapted and overseen by Pickering. Acting as a kind of communications center since 1898—its European counterpart was in Kiel, Germany, and included fifty observatories—the *Harvard College Observatory Bulletins* received information about the discovery and follow-up of astronomical phenomena and circulated it to astronomer members in the Western Hemisphere.

CHAPTER 4: EINSTEIN'S ENTREATY

1. János Bolyai (1802–1860); Carl Frederick Gauss (1777–1855); Nikolai Lobachevsky (1792–1856); and Bernard Riemann (1826–1866).

2. The International Union for Cooperation in Solar Research, formed by George Ellery Hale to broaden stellar research, met from July 30 to August 5, 1913. After a few days of sightseeing and having dinner on the Rhine, the eighty or so members got down to meetings and lectures. But mostly, what was written about and commented on were the personal friendships and camaraderie that existed in the scientific community. The union's publication later made this statement: "It is quite evident that if war and peace depended upon the feelings of those who partake in international conferences the era of peace would not be far away." After the outbreak of war, the union was discontinued. Its successor, the International Astronomical Union, formed after the war in 1919, had as two of its first vice presidents William Campbell and Frank Dyson.

3. As his earlier correspondence between them noted, Campbell would be happy to send on to Freundlich any successful photographic plates he might obtain during the eclipse. However, there were terms to this agreement: "Any results obtained by you might be announced by you in a preliminary way in the Nachrichten, or otherwise, and your full paper on the subject be published as a Lick Observatory Bulletin" (Crelinsten, *Einstein's Jury*, 76).

4. *The Swiss Years: Correspondence, 1902–1914.*

5. Mabel Loomis Todd wrote *Total Eclipses of the Sun*, which was widely read. She had carried on a long affair with Emily Dickinson's older brother, Austin, and became the poet's first editor, albeit posthumously. Percival Lowell had sent the Todds to South America to observe Mars and possibly identify the Martian canals that he believed were evidence of intelligent life. They were with him in Tripoli for the May 28, 1900, eclipse when he received a cable sent by his assistant, A. E. Douglas, back in tiny Washington, Georgia. "The race to beat the eclipse" was a current idea that a telegram would travel east faster than the eclipse, thanks to promotion by James Gordon Bennett and the *New York Herald*. If the 1900 eclipse could achieve it, it would mean that astronomers who saw a comet, or other phenomena, could develop their photographs quickly and cable the info ahead to astronomers in the east to be on the lookout when the eclipse reached them. The attempt failed in 1900, as it had in previous eclipses. The *New York Times* referred to David Todd's plan to fly after the eclipse in Russia as "chasing the sun" (see Bill Kramer, "Solar Eclipses Books," Eclipse Chasers website, September 15, 2017, www.eclipse-chasers.com/html/books.shtml).

6. The first director, Edward Holden, convinced wealthy San Francisco banker William H. Crocker to fund the Lick expeditions, which were then referred to as the Crocker expeditions." For clarity in this book, they will be referred to as the Lick expeditions. James Lick was a young carpenter and piano maker. When his sweetheart became pregnant, Lick asked her father for her hand in marriage. "Do you have a penny in your purse?" her father asked. Angry and insulted, Lick ended up in Argentina, where his pianos were in great demand. Wealthy two years later, he returned in victory to find his sweetheart happily married to another man. (Information on Lick is from Michael L. Strauss, Fredericksburg Pennsylvania History website: www.fredpah.com.)

7. The Pic Du Midi Observatory, in the French Pyrenees, was the first mountaintop observatory but was not staffed year-round, because of snowfall, until long after the completion of Lick. When forest fires occasionally broke out, the Lick Observatory staff became firefighters. See Holden, "Forest Fires on Mount Hamilton, July 29 to August 3, 1894"; and Holden, "Forest Fires at Mount Hamilton, July, 1891."

8. In the summer of 1894, Campbell had concluded that spectroscopes of the day were incapable of detecting water vapor or oxygen in the atmosphere on Mars. Therefore, he boldly sided with Maunder and rejected the assertion of the wealthy and influential Percival Lowell that life existed on that planet, an idea upheld by many famous astronomers and a favorite topic of conversation with the public. In his review of Lowell's book *Mars* (Boston: Houghton, Mifflin & Co., 1895), Campbell expounded on Lowell's assertion that the world would not be ready to accept a different new life: "Here Mr. Lowell is certainly wrong. In my opinion he has taken the popular side of the most popular scientific question afloat. The world at large is anxious for the discovery of intelligent life

on Mars, and every advocate gets an instant and large audience" (Campbell, "*Mars*, by Percival Lowell").

9. By "aberrations," Campbell is referring to a lens defect that prevents light rays from being sharply focused on a point, thereby causing any image formed to be distorted and blurred.

10. Frank Bequette Rodolph (1843–1923) was born in Wisconsin and traveled overland to California with his family in 1850. He was a successful commercial photographer in the Oakland area during the 1870s and 1880s. His impressive collection of work, which touches deeply on the familiar lives of the local people, is archived at the Bancroft Library, University of California, Berkeley.

11. Bertha's grandfather Alfred Krupp and other family members were known for their anti-Semitism. Alfred Krupp hoped that someone would "come and start a counterrevolution against Jews, socialists and liberals" and considered appointing himself for the job. During World War I, which was just around the corner, the Krupp company would manufacture the German army's heavy siege guns, including the ninety-four-ton howitzer Big Bertha, named for Bertha Krupp. Early in the war, their engineers went quickly to the drawing board to improve Krupp's long-range artillery.

12. The flash spectrum is observable for a few seconds as a total eclipse begins and just before it ends, when layers of the sun's atmosphere "flash" into prominence, producing bright lines caused by the hot, luminous gas. Charles Augustus Young (1834–1908), one of America's foremost solar spectroscopists, was first to observe it during the eclipse of 1870. Young viewed the 1900 eclipse in North Carolina with the English members of the BAA, Bacon, his daughter, and Maskelyne.

13. In 1773, Catherine the Great, as a patron of learning, welcomed to Russia all Jesuits expelled from other countries in Europe. Twenty-five years after her death in 1796 and under pressure from the Russian Orthodox Church, Tsar Alexander I exiled the Jesuits from Russia. In 1887, Father S. J. Perry, also of Stonyhurst, had been refused and then given a permit for an eclipse expedition. Cortie hoped the same would happen for him. The *Stonyhurst Magazine* speculated that the selection of Kiev might have been the problem, given that the city was then the highest governing body of the Russian Orthodox Church. See "Stonyhurst Astronomers at Hernösand."

14. Crelinsten, *Einstein's Jury*, 28.

15. The BAAS attendees were mostly from the British Isles but also included a Canadian, ten Americans, three South Africans, a Russian, a Pole, two Italians, a Swede, five Danes, and three scientists from India. There were also seven Germans and a few women. This was not Eddington's first BAAS meeting. He had been in Canada, in 1909, arriving in Quebec province. He then visited Niagara Falls before taking a train across the Canadian prairies to Winnipeg, where the meeting was held.

16. Gustav and Bertha Krupp bought Archduke Ferdinand's villa after his assassination.

17. Charles Francis Brush Jr. was the son of Charles Brush, a pioneer in electricity and the inventor of the arc lamp. The younger Brush would enlist in the US Army in 1917 and serve as a first lieutenant. He died in 1927, at age thirty-three, of complications during a blood transfusion for his seven-year-old daughter, Jane, who had died a few days earlier.

18. Isaacson, *Einstein: His Life and Universe*.

CHAPTER 5: THE 1914 ECLIPSE

1. To this day, astronomy research has been assisted by amateurs the world over. But the involvement of skilled amateurs has now grown to include science generally. In the 1990s, the term *citizen science* was coined independently by the American ornithologist Rick Bonney and the British sociologist Alan Irwin. In opening up science to the public, Irwin saw the relationship as twofold: science must be responsive to the needs of citizens, who in turn can produce dependable scientific knowledge. Bonney was unaware of Irwin's writing when he also defined the term to cover the contributions of nonscientists (such as amateur birdwatchers) to the overall body of scientific research. The *Oxford English Dictionary* now carries the terms *citizen science* and *citizen scientists* as of June 2014.

2. Lywood, "Our Riviera, Coast of Health."

3. Livadia Palace was a lavish retreat for Nicholas II and his doomed family. Yalta was also where Anton Chekhov, suffering from tuberculosis, had built a home a few years before his death in 1904. In 1945, during the Yalta Conference, Franklin Delano Roosevelt and members of the American delegation were housed at the palace. The Russian delegation was lodged at Yusupov Palace, nearby in the town of Koreiz. The British took up residence in Vorontsov Palace, at the foot of the Crimean Mountains some five miles away. Livadia Palace is home to a museum today but is still used occasionally for international summits.

4. The English expedition from the Imperial College London, headed by Alfred Fowler, would be the most short-lived of the 1914 expeditions. Traveling through the Kiel Canal, the three members arrived in Riga on August 1, hours before Germany declared war on Russia. Their equipment had not been sent on to Kiev as planned since the military had taken over the railroad. The city was in an uproar. They decided to join Cortie, their banned team member, in Sweden. After waiting days for a steamer to Stockholm, they boarded the ship only to disembark when their permits did not arrive. On August 6, when England entered the war against Germany, the British embassy in Saint Petersburg recalled astronomer and soldier Major Edmond H. Grove-Hills, back to England for active duty. Fowler knew then that the expedition was hopeless. The team booked passage on a steamer to Copenhagen and was back in England by mid-August.

5. In this group with Penck was the Austrian anthropologist Bronislaw Malinowski. His is the most astonishing story. When Malinowski was detained in Australia, the government gave him permission to carry on his work and even provided funding. He chose the Trobriand Islands, off the coast of New Guinea and spent several years there. He published *Argonauts of the Western Pacific*, which was based on his work there and which established his career.

6. The RMS *Arlanza* was the same steamer that Eddington, Davidson, and Atkinson had taken from Southampton to Rio for the eclipse of 1912. The ship was off the coast of Brazil on August 16, 1914, returning to Southampton with over one thousand passengers, when it was intercepted by the cruiser *Kaiser Wilhelm der Grosse*, which issued an order to stop or the cruiser would open fire. The *Arlanza*'s captain was ordered to dismantle the radio aerials and throw them overboard. When asked how many women and children were aboard, the captain replied that there were 335 women and 97 children. At this, the German cruiser ordered the *Arlanza* to proceed. Thus, the rumor that Perrine read was just that, a rumor. A few months later, the *Arlanza* was requisitioned as an armed merchant cruiser. It would go back into civilian service again in 1920, and Perrine would take

the *Arlanza* again to Southampton in 1922. The vessel was sold for scrap in 1938 after a cheering crowd in Bueno Aires bade her farewell as the ship sailed back to Southampton for the last time.

7. Peter the Great founded the city and named it Sankt-Peterburg, to reflect his appreciation of all things Dutch. With the outbreak of war, the imperial government changed it to Petrograd on September 1, 1914, replacing the German *burg* with *grad*. The city later became Leningrad in 1924 and was changed back to Saint Petersburg in 1991.

8. The folks at Amherst, and their friends, were worried. But David Todd and wife Mabel had skedaddled home through Scandinavia, as had most of the astronomers, despite newspapers declaring that the "Amherst savant" was "lost in Russia" and urging William Jennings Bryant to help find him (Julie Dobrow, "The Star-Crossed Astronomer," *Amherst Magazine*, July 28, 2017).

9. Boney, "The Summer of 1914," describes a personal diary written by F. O. Bower, a botanist at the University of Glasgow. With insight and humor, Bower details his journey to and from Australia for the 84th meeting of the BAAS, as well as events there. His departing ship was the *Orvieto*, which left from Tillbury, near the mouth of the Thames River. Also on the ship was Sir Oliver Lodge, the previous BAAS president; his family; physicist Henry Moseley; and Moseley's mother. While the trip south was pleasant, Bower reflects on some of his fellow passengers with splendid comic remarks. As the ship came through the Great Australian Bight, passengers had been warned that it would be a stormy passage. When things went smoothly, Bower remarked that "the Bight is not as bad as the Bark," a comment later stolen by Lodge and used as his own in a speech he gave the same night. Bower also records the quick demise of the "scientific brotherhood" expected in Adelaide. Bower, Lodge, and many others cut their visit short with the outbreak of war. From Brisbane, they caught two trains south, to Sydney. There, they booked passage on the SS *Morea*, "a noisy and creaky ship." Two of the German scientists, Albrecht Penck and Otto Maas, had been sent back from Australia also on the *Morea*, much to Lodge's disapproval. First, Lodge accused Penck of having "spy" documents. (These were the notes and drawings that Penck, as a geographer, had made of the coast near Sydney.) Lodge wanted Penck confined to his cabin and taken to England as a prisoner. He believed that the two Germans scientists were sending secret codes to enemy boats. Lodge also reported a woman for smoking on deck at night. A crack of thunder was once thought to be a torpedo and caused havoc aboard. A "Russian geologist" who kept two life belts strapped around his waist heard a loud bang on deck and bolted in terror from the smoking room. Suffering from what Bower called "war hysteria," Lodge worked to get Penck and Maas taken off the ship at Malta. Determined, he wrote a letter to the governor in the pretense that it came from all the BAAS members. Bower referred to Lodge as "a man over his head and ears in conceit." In trying to obtain a full copy of Bower's diary (these notes refer to the very detailed fifteen pages written by Boney), we learned that it had burned in the 1990s in a fire at the archives.

CHAPTER 6: A MAGIC CARPET MADE OF SPACE-TIME

1. In 1675, King Charles II commissioned Christopher Wren to build an observatory and living quarters for England's first astronomer royal, John Flamsteed (1646–1719). A study of the moon by Flamsteed and other astronomers would assist sailors with navigation at sea and possibly prevent shipwrecks, which were costly in lives and cargo. Since

funding was a problem, Wren used the brick and stone remnants from an old Tudor fort to build the Octagon Room and Flamsteed's living quarters just below it. When Dyson arrived in Greenwich, six of his eight children had already been born. Additional rooms were built in the basement to accommodate the family. The observatory is today a museum, with graffiti scrawled by some of Dyson's children still visible on the walls.

2. With its roots dating back to the seventeenth century, the Royal Arsenal munitions factory, in Woolwich, covered thirteen hundred acres and employed a labor force of eighty thousand. In 1915, the government built nearly thirteen hundred homes to accommodate the more senior and skilled employees. By 1994, it was no longer a military organization.

3. The concept of a time ball is now outdated, and time balls are used today mostly for historical reference or as tourist attractions. But the time ball was once an important mechanism around the world for keeping time. It was a large ball, usually wooden or metal, that would drop at a scheduled time to let people know the time (for many decades, only the rich could afford clocks and reliable timepieces). More importantly, it was meant to aid offshore ships in verifying the settings of their chronometers and maintaining accurate longitude. The time ball at Greenwich, installed in 1833, was red and built atop the observatory, where it was easily seen by ships in the Thames River. It dropped each day at 1:00 p.m., as it still does for tourists, although its use has long been replaced by electronic time signals.

4. The onion dome at the Royal Observatory was built to house the Great Equatorial Telescope, the twenty-eight-inch refracting telescope that is seventh-largest in the world. The onion dome was built specifically to house it, as were other observatory domes with large telescopes. Since aluminum was still too expensive to be practical, Christie decided to use the same paper-and-wood framing used in boat building. Thus, in the nineteenth and twentieth centuries, many observatories had papier maché panels over a wood or, sometimes, an iron framework. Being lightweight, the domes could rotate more easily so that astronomers could position telescope openings in any direction of the sky. The onion dome at Greenwich was damaged by an air raid during World War II. A fiberglass replica replaced the original dome in 1971.

5. Dyson told a humorous story about his Scottish chief assistant, John Jackson. "I found only one man who really enjoyed air raids," Dyson told the RAS Club one night at a dinner. "He was an old night watchman who calmly lighted his pipe and watched the show as if it had been a display of fireworks. The most amusing incident I remember on these occasions is that I said to Jackson one night in the front court: "Let's get out of this and go under cover," to which he replied indignantly in his native tongue, 'the Obsairvatory was built to obsairve the Moon, and I am going to obsairve the Moon'" (Wilson, *Ninth Astronomer Royal*).

6. Eddington is known for "the Eddington number," which, in astrophysics, is the number of protons in the universe that can be observed from earth. However, the Eddington number in *cycling* was a system he devised for tracking a cyclist's long-distance rides and thus his lifetime progress. Eddington's own Eddington number was 84. Not long before he died, he was still cycling an occasional 80 miles in a day. His longest ride during his lifetime was from Doncaster to Cambridge, a distance of 122 miles.

7. Along the lines of fighting in Northern France, the shooting stopped on Christmas Eve. Both German and British soldiers climbed out of the trenches to sing carols, exchange small presents, and even engage in a game of soccer. Graham Williams, Fifth

London Rifle Brigade, recalled the scene: "First the Germans would sing one of their carols and then we would sing one of ours, until when we started up 'O Come, All Ye Faithful' the Germans immediately joined in singing the same hymn to the Latin words Adeste Fideles. And I thought, well, this is really a most extraordinary thing—two nations both singing the same carol in the middle of a war" (Williams, quoted in Vinciguerra, "The Truce of Christmas, 1914"). This would be the only truce to occur during World War I after Allied commanders gave orders to shoot any soldiers who violated the no-fraternization rule.

8. In 1909, Perrine was offered a yearly salary of 12,000 pesos by the Argentine government, or $5,400 in gold (about $150,000 today). Campbell's salary at Lick Observatory in 1909, as director, was $4,000. The drawback to this offer, which promised Perrine an increase in equipment over time, was that his salary would not be increased at all from 1909 until his retirement, twenty-seven years later. The last letter Perrine wrote, asking to be reimbursed for the money spent in Russia, was written in 1934, two full decades after the expedition. There is no indication that he was ever reimbursed. (Santiago Paolantonio, Engineer and Master of Education, Museum Astronomical Observatory, National University of Cordoba, Argentine, email correspondence with authors, July–August 2018; and Bob Kelly-Thomas, grandson of Charles Perrine, email correspondence with authors, August–September 2018.)

9. Fritz Haber would win the Nobel Prize for chemistry in 1918; Otto Hahn, who also took part in the hopeful Christmas truce of 1914, would win it for chemistry in 1944; James Franck would win for physics in 1925; and Gustav Hertz for physics in 1925.

10. Clara Immerwahr Haber (1870–1915) was the first woman in Germany to earn a PhD in chemistry. Like her friend Mileva Marić, she, too, felt oppressed as a woman scientist by the demands of an imperfect marriage and household duties. The story arose that Fritz had attended a celebration party for the successful chlorine gas attack and that Clara killed herself the same night in opposition to her husband's work. The suicide notes she left behind did not survive. She died hours later in the arms of their only child, thirteen-year-old Hermann. After his wife's death, Fritz Haber returned immediately to the front to oversee the second gas release.

11. Charles Perrine, letter to George Hale, May 31, 1915. Museo Astronómico, Córdoba.

12. William Wordsworth, *The Prelude, Or Growth of a Poet's Mind, An Autobiographical Poem* (London: Edward Moxon, 1850). While a student, Wordsworth (1770–1850) could see Newton's statue from his room at nearby St. John's College.

13. Richard Feynman, one of the twentieth century's greatest physicists, would say this of Maxwell's achievement: "From a long view of the history of mankind, seen from, say, ten thousand years from now, there can be little doubt that the most significant event of the 19th century will be judged as Maxwell's discovery of the laws of electrodynamics. The American Civil War will pale into provincial insignificance in comparison with this important scientific event of the same decade" (Feynman, Leighton, and Sands, *Feynman Lectures on Physics*).

14. Albert Einstein to Hans Albert Einstein, November 4, 1915, *Collected Papers of Albert Einstein*, vol. 8, doc. 134.

15. Einstein letter to Mileva, February 6, 1916. See also Isaacson, *Einstein: His Life and Universe*.

16. Toward the end of the nineteenth century, physicists were searching for a mysterious medium through which light waves passed, what came to be called ether. In 1887, in the basement of a dormitory, physicist Albert Michelson (1852–1931), of Case School of Applied Science, and chemist Edward Morley (1838–1923), of Western Reserve University (now Case Western Reserve University), conducted what is known as the Michelson-Morley experiment. In an attempt to detect the relative motion of matter through the stationary luminiferous ether, it compared the speed of light in perpendicular directions. They found no substantial difference between the speed of light in the direction of movement through the presumed ether and the speed at right angles. This negative result is considered the first evidence against the prevailing ether theory so strongly embraced by Lodge and most scientists of the day. While twenty-one-year-old Einstein once had his professor at the Swizz Polytechnic Institute reject his proposal to do an experiment that would measure how fast the earth moved through ether for having been done too many times before, his 1905 special theory would invalidate its existence. Einstein later wrote of the famous 1887 experiment, "If the Michelson-Morley experiment had not brought us into serious embarrassment, no one would have regarded the relativity theory as a (halfway) redemption." Lodge, however, believing deeply in "the theory of spiritual evolution," would continue to reject this repudiation of ether by insisting that both the universe and the afterworld were filled with it. On September 8, 1913, the *New York Times* published a dispatch from Birmingham, England, about an upcoming BAAS meeting in which "Sir Oliver Lodge, in Presidential Address, Will Combat the 'Theory of Relativity.'" The dispatch mentioned "Prof. Einstein of the University of Zurich" and stated that his theory argued against ether. This was the first time Einstein's name appeared in the *New York Times*.

17. See Bruton, "Sacrifice of a Genius"; and British Association for the Advancement of Science, *Report of the Eighty-Fourth*.

18. Nobel Prize–winning physicist Ernest Rutherford, later known as the "father of nuclear physics," had been Moseley's supervisor at the University of Manchester. Rutherford was so distraught over this loss that he lobbied the British government until a policy was introduced that would no longer allow promising and prominent scientists to volunteer for combat duty. The American physicist and Nobel Prize winner Robert Millikan wrote eloquently of Moseley's death: "In a research which is destined to rank as one of the dozen most brilliant in conception, skillful in execution, and illuminating in results in the history of science, a young man twenty-six years old threw open the windows through which we can glimpse the sub-atomic world with a definiteness and certainty never dreamed of before. Had the European War had no other result than the snuffing out of this young life, that alone would make it one of the most hideous and most irreparable crimes in history" (Millikan, "Radiation and Atomic Structure," 195). Isaac Asimov would write years later, "In view of what he might still have accomplished, his death might well have been the most costly single death of the War to mankind generally" (Asimov's *Biographical Encyclopedia of Science and Technology*, 921).

19. Erwin Planck, a politician and resistance fighter, would be executed by the Gestapo in January 1945 for participating in a failed attempt on Adolph Hitler's life several months earlier. Sadly, both of Planck's daughters also died young, during childbirth. Many years later, a letter was discovered in which Max Planck, at eighty-seven years old, begged Hitler for his son's life: "As the gratitude of the German people for my life's work, which has become an everlasting intellectual wealth of Germany, I am pleading for my

son's life." The plea fell on deaf ears. Planck died two years later. (Max Planck, letter to Adolph Hitler, October 25, 1944, from the personal collection of Graham Farmelo.)

20. In 1923, the Marić family received a letter from Milos that he was not only alive, but on his way home to Serbia. The family's joy—Mileva rushed to Vojvodina to meet her brother—soon turned to disappointment since Milos changed his mind. He sent his wife a postcard that she should consider herself a free woman. He was staying in Russia, and he did so, becoming a well-respected professor of histology. His wife never saw him again; nor did Mileva ever see her brother.

CHAPTER 7: UNRIDDLING THE UNIVERSE

1. What is more amazing is that Schwarzschild's 1900 paper also preceded by over two decades the work by the Belgian mathematician Georges Lemaître (1894–1966), a Roman Catholic priest; the Russian physicist and mathematician Alexander Friedmann (1888–1925), who died young of typhoid fever contracted in Crimea; and the American mathematician and physicist Howard P. Robertson (1903–1961), who taught at both the California Institute of Technology and Princeton. Schwarzschild's paper may have been the first modern work on the large-scale structure of space-time, what we now call cosmology.

2. This was the first time conscription went into place for England, Scotland, and Wales. With Irish republicans launching a rebellion against British rule that resulted in the Easter Rising in April of that year, the Irish were exempt from conscription, although many nonetheless volunteered to fight.

3. Eddington told one biographer that he added this postscript about peeling potatoes to the letter (Chandrasekhar, *Eddington: The Most Distinguished Astrophysicist of His Time*). Luminaries at Cambridge then included such names as Newall, their first professor of astrophysics who had stood beside Perrine in the Crimean vineyard for the 1914 eclipse. There was also Sir Joseph Larmor, the Lucasian Professor, who seemed to perfectly represent the gap between the classical physics of Newton and the new physics of quantum theory and relativity. Larmor also adopted a conservative view on many non-scientific topics, perhaps in keeping up his persona. In 1920, he spoke out in opposition to installing baths at Cambridge. "We've done without them for four hundred years," he was quoted as saying. "Why begin now?" He soon became a devoted visitor to the baths, however, turning up each morning in a mackintosh and cap, which he had never worn before.

4. The celebrated senior wrangler was considered the most prestigious intellectual achievement in Great Britain, awarded to the top mathematics undergraduate among the other "wranglers." Many senior wranglers, such as Arthur Eddington, would go on to direct observatories. Others would achieve fame in various fields beyond physics and mathematics.

5. Sir Gilbert Thomas Walker (1868–1958) was a British mathematician and meteorologist. As director general of observatories in the Indian Meteorological Service, he focused on the importance of the monsoon and conducted extensive statistical studies of worldwide correlations.

6. Braithwaite, "Friends' School in Kendal."

7. Biographies of Eddington mention only his father's death by typhoid fever. It seems unlikely that even Eddington himself knew about the *Leeds Mercury* newspaper account in 1885, or the Quaker magazine's follow-up. His early memories of Kendal, as

he revealed to his biographer Subrahmanyan Chandrasekhar, contain a somewhat ironic note: "The traditions of Kendal have been woven into my earliest memories as the home of the brief married life of my parents . . . Kendal has an earlier association with science, that great chemist, perhaps the greatest of all chemists, who was headmaster of Stramongate school, the same school of which a century later my father became headmaster and where I was born. From John Dalton we had the atom. And now I have become an atom chaser myself. John Dalton must have left some germ behind him which lingered in the walls of Stramongate" (Chandrasekhar, *Eddington*). Dalton was headmaster at Stramongate from 1781 to 1793.

8. Wilson, *Ninth Astronomer Royal*, writes that her father was one of the first to recognize Eddington's "mathematical genius." Dyson's relationship with his assistants and computers appeared to be congenial and professional. Wilson recalled the major difference between life in Edinburgh for the astronomer royal and life in Greenwich. In Greenwich, only the astronomer royal lived within the observatory grounds. "Dyson fell into the habit of bringing first one assistant and then another in to the mid-day meal. He would greet his wife with:— 'Here's Chapman, Carrie,' or 'Can you find some dinner for Eddington?' and Mrs. Dyson, smiling hospitably on the visitor, would wonder anxiously whether there was enough food to go around. These meals must have been rather an ordeal for the shy bachelor assistants, surrounded by school-girls in gym tunics." (Dyson's mention of "Chapman" is to Sydney Chapman [1888–1970], a mathematician and geophysicist who died in Boulder, Colorado, where he was involved at the High Altitude Observatory.)

9. Eddington's biographer, Allie Vibert Douglas (his postdoctoral student at Cambridge in 1921 and the first Canadian woman to become an astrophysicist), described him as a loner: "His need for relaxation after long periods of intensive mental concentration found its outlet only rarely in social intercourse and then to a limited extent. He never married nor did he ever wish to marry. Apart from his mother and sister who were the homemakers, his interest in women was simply and solely as acquaintances or, in the case of the very few women astronomers in various countries, as friendly colleagues." Douglas, a devoted former student, wrote the biography at the request of Eddington's sister, Winifred, after his death. There is only one name mentioned in the dedication: "To Arthur Stanley Eddington's friend C. J. A. Trimble," the man with whom Eddington maintained a close forty-year friendship. A modern British documentary, *Einstein and Eddington*, directed by Philip Martin, written by Peter Moffat (BBC Two, 2008), depicts Eddington in love with a young man who died at Ypres during the war. This romantic presentation is fiction. Trimble was to write Eddington's biography, but he had suffered over the years from mental health problems. Eventually, Douglas became the author.

10. Although many scientists, from ancient Greece onward, had estimated the distance over the centuries and although Halley believed observing those rare transits of Venus would reveal an answer, it was Arthur Hinks, a British astronomer, geographer, and cartographer who determined the astronomical unit. For this accomplishment, he was awarded the Gold Medal of the RAS and elected a fellow of the Royal Society.

11. Crelinsten, *Einstein's Jury*, 22. The letter is dated March 8, 1914.

12. Original copies of Hinks's and the secretary-general's correspondence are in Sociedade De Geografia De Lisboa, *Comemoracoes do 90° Aniversario*.

13. See Campbell, "The Lick Observatory Community in War Service."

14. A hundred pounds needed for the instruments in 1917 would be equivalent to $8,462 in 2017. The thousand pounds needed for travel would be ten times greater.

15. In the summer of 1917, A. G. Freeman, a California businessman, traveled to Petrograd on business. Campbell and the University of California had given Freeman permission and the credentials to represent them to the Imperial Academy of Science on behalf of the instruments. Freeman secured the deal, and the shipment left Pulkovo on August 15, 1917. It arrived in Vladivostok in mid-December. Freeman cabled Vladivostok, urging the port to promptly ship the cargo.

16. The Friends Ambulance Unit was formed within days of the outbreak of the war by a group of young Quakers. They were convinced that current ambulance services as they were would be inadequate and that, by offering such services themselves, they could save lives. The Friends Ambulance Unit would also allow those conscientious objectors to make a nonviolent contribution to the war effort. With conscription still not in place in 1914, this consideration wasn't necessary on their part. It was a commendable effort.

17. "All Ready for Solar Eclipse at Stations," *Glendale (WA) Sentinel*, June 6, 1918, http://gld.stparchive.com/Archive/GLD/GLD06061918p01.php.

18. These details come from Campbell, *Life on Mount Hamilton*, in which author Kenneth Campbell recalls the details in a conversation with C. Donald Shane. Shane was one of the Mount Hamilton boys who had gone to the war effort. Kenneth Campbell recalls the night before the eclipse when word came that his brother Douglas had been shot down. Campbell says that his parents didn't know if Doug was alive or dead. Shane is not listed on the official report of volunteers, but since he was stationed at Tacoma, he had come to Goldendale to be present and take part unofficially. In 1945, Shane became Lick's director.

19. Campbell's dispatch was the third time Einstein's name was mentioned in the *New York Times*. The first had been in 1913, in a dispatch from Birmingham, England, announcing that Lodge would argue for ether and that the theory by "Prof. Einstein of the University of Zurich" argued against ether. Einstein was mentioned the second time on November 30, 1917, in a discussion of Georg Friedrich Nicolai's book *The Biology of War*. Nicolai was one the three signers of "A Manifesto to Europeans," a call for unity, not war, and a response to "Manifesto to the Civilized World," which had been signed by ninety-three German intellectuals. The article listed the two pacifist signers as "professors Albert Einstein and Wilhelm Forster." By the time Nicolai's book was released in the United States, the book's translator added this update to the preface: "Berne, Switzerland, June 1918, Postscript—Since writing the above the world outside Germany has been gratified to learn that Professor Nicolai has escaped from Germany in a German aeroplane and has reached Denmark" (Nicolai, *The Biology of War*, iii).

20. It was wise that Cortie did not go to Brazil. His ailing predecessor, Father Walter Sidgreaves (1837–1919), had been director of the observatory since 1889. Sidgreaves died on June 12, 1919, over two months before Crommelin and Davidson returned to England. His own predecessor had been another priest-astronomer, Father Stephen Perry, who died while on an eclipse expedition to French Guiana in December 1889 and was buried at sea.

CHAPTER 8: THE RMS *ANSELM* SETS SAIL

1. Quotation from Wilson, *Ninth Astronomer Royal*. Wilson was Frank Dyson's second-oldest daughter and was still living at Flamsteed House in 1919; she remembered

the visit by Eddington and Cottingham on those two nights. The "suicide" reference that she used is not repeated in other books that describe this scene. For example, Douglas, *The Life of Arthur Stanley Eddington*, has Dyson saying, "Eddington will go mad and you will have to come home alone." Chandrasekhar, *Eddington*, also writes this milder version, being told it by Eddington. We prefer Wilson's version and have used it here. It's possible the reference to suicide was deliberately left out by Eddington and others over the years for being too off-color to tell in polite gatherings and public lectures. Wilson notes that it was her father and Cottingham who first told the story at an RAS dinner later that year.

2. The idea of a grand tour for young upper-class elites arose from Richard Lassels, *Voyage or a Complete Journey through Italy* (Paris, 1670). Lassels (c. 1603–1668) was a priest and travel writer who tutored members of the English nobility. After having traveled to Italy five times in the employ of aristocrats, he wrote the book, in which he implored "young lords," armed with letters of introduction, to experience the world for themselves. "Indeed, the coral-tree is neither hard nor red, till taken out of the Sea, its native home." Without travel to France and Italy, he feared, "the country lord" would think "the Lands-End to be the World's-end." His introduction asserted that the benefits of this tour would be "intellectual," "social," "political," and "ethical." With the book's popularity among the upper classes, many young wealthy elites, both men and women, toured the European continent with tutors, sometimes gone for several years. A tourist industry soon grew up in the wake of the book's publication, with tour guides and maps.

3. The city called Pará is actually Belém, which is Portuguese for "Bethlehem." The original name was "City of Saint Mary of Belém of the Great State of Pará." So it's easy to see why non-Brazilians of the nineteenth and early twentieth centuries referred to it on maps, postcards, and in conversation as simply Pará. The city's name was later changed to Belém, the capital of Pará, a state in northern Brazil.

4. See Einstein, "Elementary Theory of Water Waves and of Flight."

5. "It is correct that I committed adultery. I have been living together with my cousin, the widow Elsa Einstein, divorced Löwenthal, for about 4½ years and have been continuing these intimate relations since then. My wife, the Plaintiff, has known since the (spring) summer of 1914 that intimate relations exist between me and my cousin. She has made her displeasure known to me" (Albert Einstein's Deposition in Divorce, December 23, 1918, Collected Papers of Albert Einstein). The later decree (Divorce Decree, Feb. 14, 1919, Collected Papers of Albert Einstein, 9:6) came with the provision that Einstein would not remarry for a period of two years.

6. British Summer Time had first been proposed in 1895 by George Hudson, a New Zealander born in London. Hudson was an astronomer and explorer, but also an entomologist who wanted more daylight time to collect insects. His proposal was unbeknownst to Willet, a builder who in 1905 had taken an early ride through London before his breakfast and was not pleased to realize that Londoners were asleep during daylight hours. (An avid outdoorsman and sportsman, he found that the standard time, as it then stood, also shortened Willet's golf game.) Thus, in 1907, he published a pamphlet titled *The Waste of Daylight*, which proposed advancing the clock during summer months. It was not a popular concept. In 1912, Willet visited Dyson at the Royal Observatory on a campaign to enlist supporters. "Astronomers, however, do not take kindly to any juggling with time, and when Willet referred to his scheme, the response was unenthusiastic. He was vindicated, however, when the war of 1914 broke out" (Wilson, *Ninth Astronomer Royal*).

In early 1916, in an effort to conserve coal, Germany and Austria-Hungary adopted the daylight-saving plan, calling it *Sommerzeit*. England adopted the plan a few weeks later. Russia and other countries followed suit in 1917, and the United States in 1918.

7. Manuel Peres (1888–1968), who was then director of the Portuguese observatory on Mozambique, had hoped to join Eddington's expedition to Príncipe. He wrote to Director Oom in November 1918, asking about the possibility, and again in March 1919. He noted the problems with funding and securing passage on a steamer (Mota, Crawford, and Simões, "Einstein in Portugal"). It would appear that Peres didn't find the same help and hospitality that had been shown to the British.

8. Wilson, *Ninth Astronomer Royal*.

9. Wilson, *Ninth Astronomer Royal*.

10. In September, the funicular would blow up, killing several passengers. On his way to Brazil in 1912, Eddington had ridden the toboggan 3,300-feet down the mountain to Funchal.

11. The French gunboat *Surprise* lost thirty-four crew members in this attack. A second vessel was the elderly *Dacia*, a British cable-laying ship. The third was the SS *Kanguroo*, a French ship built to transport submarines before the war. Max Valentiner (1883–1949) had been branded a war criminal in 1915 for sinking the British passenger ship *Persia* off the coast of Crete while its passengers were enjoying lunch. In an act strictly against the rules of war, Valentiner gave the liner no warning that a torpedo was headed its way. The ship sank in a few minutes, killing 343 of the 519 people on board.

12. Alexandra Canha, director of Biblioteca Municipal do Funchal found the newspaper article that Eddington mentions in a letter home. The two Madeira papers in 1919 were the *Diário de Noticias* and *the Diário da Madeira*. In the *Diário de Noticias*, appearing April 8, a day after Eddington met the editor, Cyriaco de Brito Nóbrega, was a short description of the expeditions. On April 9, the *Diário de Noticias* informed readers that the total eclipse would not be visible in Funchal, only the partial, and times to watch. The governor of São Tomé and Príncipe at this time was João Gregório Duarte Ferreira, who would leave office in June, a month before the Englishmen left Principe. The Bella Vista Hotel where they stayed is now a religious school.

13. British historian Robin Furneaux perhaps best summarized the intense gaudiness in the jungle: "No extravagance, however absurd, deterred them . . . If one rubber baron bought a vast yacht, another would install a tame lion in his villa, and a third would water his horse on champagne" (Furneaux, *The Amazon: The Story of a Great River*).

14. Henry Wickham (1846–1928) was a British explorer now considered a biopirate, although there was no law against smuggling seeds from Brazil in 1876. (He might have misrepresented what his cargo contained.) The transport of the material was no easy task, considering that seventy thousand para tree seeds, each seed three-quarters of an inch long, would weigh about a half ton. It took more than thirty years for the trees to grow on plantations in Asia, even though they were planted much closer together than in the wild, thus facilitating production. Wickham was much praised for his contribution to the rubber industry, but he paid a great price. He and several of his family members became seriously ill during the rigorous trips to the jungle, and several family members died there in 1872. For further reading on Wickham and the rubber boom, see Jackson, *Thief at the End of the World*.

15. See Grandin, *Fordlandia*.

16. The railway taken several times by all the astronomers during their time in northern Brazil, between Camocim and Sobral, was the Estrada de Ferro Viação Cearense.

17. Crispino and Lima, "Expediçâo norte-americana"; Mattos, "A Febre Amarella no Norte."

CHAPTER 9: PRÍNCIPE AND SOBRAL

1. Ludwig Boltzmann hanged himself on September 5, 1906, in his hotel room near Trieste (then Austria) while his wife and daughter were swimming. He was sixty-two years old. While some scholars attribute his futile attempts to prove his theories as the cause of his suicide—two of his biggest opponents were Ernst Mach and Wilhelm Ostwald—others point out that Boltzmann suffered from asthma attacks and severe depression all his life. His physical health and eyesight were also in decline, although the academic refutations certainly did not help. He did not leave behind a suicide note.

2. During his fourth visit to the New World (1502–1504), Christopher Columbus saw his first cocoa beans in what would be Nicaragua today. By this time, local peoples such as the Olmecs and Mayans had been using the cocoa bean as a monetary unit, and to make a favorited drink, for over a thousand years. The story goes that the bean's use as a drink dates back to when the Olmecs got the idea from watching rats devour the beans with a ravenous gusto. Later, the Mayans built what can be considered the first cocoa plantations, around 600 AD. Of the indigenous peoples he encountered on this journey, Columbus would write of their regard for the beans: "They seemed to hold these almonds at a great price, for when they were brought on board ship together with their goods, I observed that when any of these almonds fell, they all stooped to pick it up, as if an eye had fallen."

3. The journalist Henry Woodd Nevinson would later write in his book, *A Modern Slavery*, that the two islands were "the islands where slaves die." Referring to the repatriation fund, whereby three-fifths of a laborer's monthly wages went to insure free passage back to the person's homeland after five years of labor, Nevinson wrote, "A more ingenious trick for reducing the price of labor has never been invented, but, for very shame the Repatriation Fund has ceased to exist, if it ever existed. Ask any honest man who knows the country well. Ask any Scottish engineer upon the Portuguese steamers that convey the 'servicaes' to the islands, and he will tell you they never return. The islands are their graves."

4. John, *Life and Times of Henry W. Nevinson*, 1.

5. Nevinson, *A Modern Slavery*, 179, writes of a heartrending scene as they boarded the ship. First-class passengers were watching from the upper deck as a young slave mother struggled to board the ship from below, her newborn in her arms as the waves knocked against her. "At last she reached the top, bruised and bleeding, soaked with water, her blanket lost, most of her gaudy clothing torn off or hanging in strips. On her back the little baby, still crumpled and almost pink from the womb, squeaked feebly like a blind kitten. But swinging it round to her breast, the woman walked modestly and without complaint to her place in the row with the others. I have heard many terrible sounds, but never anything so hellish as the outbursts of laughter with which the ladies and gentlemen of the first class watched that slave woman's struggle up to the deck. A few days later, a slave leaped overboard. A boat quickly captured him, and when he was returned to the ship, beaten and bruised, the passengers yelled, "Flog him! Flog him! A

good flogging!" If a slave died during passage to the islands, they were thrown overboard while the passengers were having a first-class breakfast, 'so that the feelings of the passengers might not be harrowed.'"

6. See Swan, *Slavery of To-Day*.

7. Nevinson, *A Modern Slavery*, 24.

8. Nevinson, *A Modern Slavery*, 188.

9. Since automobiles were owned only by the wealthy in those days, and since the majority of the population could not drive anyway, it was common for the vehicles to be rented with a driver. Or, for the wealthier, the cars were purchased and stored in private garages around the city of Rio, with chauffeurs as part of the household staff. It's not clear whether this driver, whose name was Antonio Rodrigues de Carvalho, was employed by Studebaker or if he was a chauffeur for hire in the city.

10. See Marengo, Torres, and Alves, "Drought in Northeast Brazil."

11. Here is an excerpt from the Davidson and Crommelin article: "The theory is of too mathematical a nature, to make it possible to attempt to explain it in popular language. It will suffice to say that it is a four dimensional system, adding time as a fourth dimension to the familiar length, breadth and height; it includes some speculations as to a possible curvature of space and its re-entry into itself, so that if one travelled far enough one would return to the starting point" (Mourão, *Einstein: de Sobral para o mundo*).

12. Brito, "Repercussões. Novidades Scientificas."

13. Eddington and others refer to the cacao trees on Príncipe as "cocoa trees"; thus the references in this book. The flowers of the cacao tree (*Theobroma cacao*) are not pollinated by bees, but are pollinated by tiny flies (*Forcipomyia*), which are biting midges. The trees begin to bear fruit when they are approximately four years old. Each tree bears twenty to thirty pods, even though it may have over six thousand flowers. Each pod weighs about a pound, is six to eight inches long, and holds twenty to forty seeds, which are allowed to ferment in heaps and then are raked out and dried to become the cocoa bean. The pods on Príncipe were harvested in June and December. The workers sliced the low-hanging pods with a sharp blade to cut the stalk. For high-growing pods, the cutting tool would be placed on the end of a long pole. It requires the yearly crop of a single tree to yield one pound of chocolate.

14. The chemical process of coating glass with a reflective substance is called *silvering*. When glass mirrors first came into great demand in Europe during the sixteenth century, most had been silvered with a mixture of mercury and tin. By the 1800s, mirrors were generally made through a process by which silver is coated onto a glass surface.

15. Just as Wickham took the rubber tree seeds to Asia, José Ferreira Gomes brought the cacao tree to Príncipe. And the seeds of the cashew tree were taken by the Portuguese to India in around 1560 (Jostock, "Cashew Industry").

16. Details of the first car ride, May 10, 1919, are from the journal of Daniel M. Wise, of the Carnegie Institution, courtesy of his grandson Christian Wise. The other experiences with the car are from Morize's diary of the 1919 eclipse, in Mourão, *Einstein: de Sobral para o mundo*.

17. Sometimes, historic leavings, written and published quickly or with a motive in mind, can leave an ambiguous or downright inaccurate picture for posterity. A write-up under "Notes," *Observatory* 42, no. 541 (July 1919): 294–295, contained some interesting

social information from the expedition on Príncipe. Apparently, a letter dated May 9 had arrived from Edwin Cottingham. It described the beauty of Príncipe, "a delightful tropical island," that was filled with many trees and "plumaged birds and gorgeous butterflies." According to excerpts of his letter, their unventilated hut was too hot for developing plates, so he and Eddington "use our bedroom." In uncorking a small bottle of developer, a piece of glass broke from the bottom and "trickled down my new thin pyjamas." The letter goes on to say that "they looked poor things in the morning, but I managed to get a good deal of the stain out by fixing them in the hypo. Eddington says they were exposed, developed and fixed, but after final washing there was no image." Cottingham added that the sleeping sickness was nearly over, and there were few mosquitos, so very little malaria. He drank to the health of those who would be attending the RAS meeting (where the letter was likely read) with "Eno's Fruit Salt," the only effervescent liquid available. Eno's Fruit Salt had been invented in Newcastle a half century earlier and was popular with sailors for their health on long sea voyages. Its cures listed everything from "Biliousness" to "Constipation," from "Giddiness" to "Gouty Poison." It would cleanse and invigorate the entire digestive tract by driving out disease germs and clearing the intestines. If Cottingham's intestine needed "clearing" in Príncipe, circa 1919, a bottle would have had to come from England with him. But what is more interesting about the letter is that the wording sounds typically Eddington, and as will be shown later in chapter 12, Dyson had been careful to keep news of the expeditions before the public. It would have been unlikely that Eddington intended to develop inside an outdoor hut when he had Roca Sundy at his disposal and a controlled environment. May 9 is ten days earlier than when Eddington wrote that they developed the check plates. And that they shared a bedroom at the accommodating Roca Sundy is also questionable. Also in this issue is a report that Father Cortie was in Brazil to conduct an "examination of the Aurora Borealis" during the eclipse and had telegrammed (from Stonyhurst) that it would be a "very anxious five minutes." Of course, Father Cortie remained at Stonyhurst given that his predecessor was expected to pass away, and Crommelin went in his place. As a matter of fact, Walter Sidgreaves's obituary is printed four pages earlier in the same issue.

CHAPTER 10: "THROUGH CLOUD, HOPEFUL"

1. Morize, "*O Eclipse de 29 de Maio de 1919.*" It is logical that Crommelin and Davidson were sent to Brazil, as had been planned for Father Cortie, because of their shared Catholicism with the majority of the residents.

2. See Basso, "Observatório Nacional reúne as Placas Fotográficas"; and Morize's journal in Mourão, *Einstein: de Sobral para o mundo.*

3. The part of French West Africa touched by the shadow now belongs to the Ivory Coast. The British Gold Coast is now Ghana.

4. Sobral and Príncipe times used in this chapter are local times. However, Eddington, in his letter home and his paper, says the eclipse would be at 2:15 Universal Time.

5. Sodium sulfite (Na_2SO_3) is more commonly known as sodium thiosulfate. It's a white powered or crystalline compound used in developing photographs and silvering mirrors. It is also used in the making of dyes and to preserve certain foods. Formalin (formaldehyde) is still used in low concentrations for some photographic developing needs. Rodrigues, "Entre Telescópios e Potes de Barro," criticizes the British for

transporting formalin in their baggage. Formalin can be a severe skin and eye irritant and is a suspected human carcinogen.

6. Dyson, Eddington, and Davidson, "A Determination of the Deflection of Light," cites nineteen plates. Thomson, "Joint Eclipse Meeting," cites *eighteen* plates. Often overlooked today, astronomers like Eddington, Crommelin, Davidson, Campbell, and Perrine also had to know photography and darkroom skills. Amateur astronomers, for the most part, introduced astrophotography as a scientific tool in the mid-1800s. The brands of glass plates Eddington had ordered for the two expeditions were Ilford Special Rapid and Ilford Empress, and Imperial Sovereign and Imperial Special Sensitive. The wet collodion process used to develop early plates was replaced by gelatin dry plates in the late nineteenth century. These glass plates began to lose favor in the consumer market in the early years of the 1900s as less fragile films replaced them. In the late 1800s, a business grew up when junk collectors would buy the old, unwanted glass plates from photographers and observatories. They would then scrape away the emulsion and resell the glass to dry-plate manufacturers and even to greenhouse manufacturers. Glass was superior to film because the plates were stable and the imaging less likely to distort or bend. Thus, astronomers continued to use glass plates as late as the 1990s.

7. Wilson, *Ninth Astronomer Royal*, says that the telegram came to her father on June 3, 1919. Eddington says he developed the twelve plates for "six nights." If he began on the night of May 29, he would have finished on the night of June 3. It's likely that Carneiro or Gragera sent the cablegram for him from Santo Antonio. During the planning meeting, it had been decided to use a code in conveying news back to Dyson by telegram. "Splendid," which Crommelin had wired from Brazil, meant success. Eddington, not yet certain, had chosen "Hopeful." Our research did not find what the code for "failure" would have been.

8. William Campbell letters, June 2, 1919, Mary Lea Shane Archives, University of California, Lick Observatory.

9. For a detailed account of how Curtis did his measurements, see Crelinsten, *Einstein's Jury*, 131–140.

10. The IAU, formed on July 28, 1919, would have as two of its first four vice presidents Campbell and Dyson. This organization would carry on the work of the Carte du Ciel committee; the International Union for Cooperation in Solar Research (formed by George Ellery Hale), which had been discontinued with the outbreak of war in 1914; and other organizations (Fowler, "International Astronomical Union, Formed").

11. For records of the speeches given at this meeting, see *Observatory* 26, no. 542 (August 1919). In order of speaking, the American astronomers were: L. A. Bauer (1865–1932) of the Carnegie Institution, also a geophysicist and magnetician, and whose colleagues Wise and Thomson had just left Sobral (Bauer was on his way home from Liberia, where he had observed the total eclipse minutes before it reached Príncipe); Walter Sydney Adams (1876–1956) of the Mount Wilson Observatory; Benjamin Boss (1880–1970) of the Dudley Observatory; Frank Schlesinger (1871–1943) of the Allegheny Observatory; Samuel A. Mitchell (1874–1960) of the Leander McCormick Observatory; Frederick H. Seares (1873–1964) of the Mount Wilson Observatory; Charles E. St. John (1857–1935) of the Mount Wilson Observatory; and Joel Stebbins (1878–1966) of the University of Illinois Observatory. When the November 6, 1919, results were read (Dyson, Eddington,

and Davidson, "A Determination of the Deflection of Light"), L. A. Bauer was referenced: "A station at Cape Palmas did not seem desirable from the meteorological reports, though, as the event proved, the eclipse was observed in a cloudless sky by Prof. Bauer, who was there on an expedition to observe magnetic effects."

12. Turner was referring to the fourth conference of the International Union for Co-operation on Solar Research, which was held August 29 to September 3, 1910, at the Mount Wilson Observatory. The union was Hale's brainchild. He, Turner, and Arthur Schuster put the idea together when they met at an international congress held during the St. Louis World's Fair, in 1904. When planning began for this 1910 conference, the union was urged to expand its scope beyond solar research to include all of spectroscopic astrophysics. Hale knew that bringing Pickering aboard was essential. Not only was Pickering the most influential astronomer of the day, but his years of work in developing systems of spectral classification and stellar magnitudes was unparalleled. The members of the union would meet again at Bonn, in 1913, after which Campbell traveled on to Berlin to visit Freundlich. The union was disbanded at the outbreak of World War I, thus canceling proposed meetings in Rome in 1916 and Cambridge in 1919. The newly formed IAU, formed at the Brussels meeting in July 1919, took up its work anew.

13. Wise had left by train on June 11 for Novas Russas, "New Russia," to the south of Sobral, for more observations. Thomson, Morize, and the Rio entourage had taken the special train back north to Camocim, where Wise joined them on the afternoon of June 14. After a flurry of telegrams to people representing the Yellow Fever Prophylaxis Commission, asking where to store the automobile, Morize was finally instructed to drop it off at a company in Camocim. In his diary, Morize is grateful for electric fans and fewer mosquitos than he encountered in the warmer climate of Sobral. On June 25, his team left Camocim on the steamer *Prudente de Morais*, named for a Brazilian president. In strong sea winds, the entire team quickly suffered from seasickness. Thomson and Wise eventually sailed home from Pará. They would be back at the Carnegie Institution in Washington, D.C., on August 25.

14. Personal details are from Rosemary Cecily Hannibal and Catherine Stella Court (granddaughters of Charles Rundle Davidson), correspondence with authors, September 2018 to May 2019. They knew their grandfather well before his death in 1970. He lived with their family in his last years. "We were told one of the parrots lived to old age," Hannibal wrote. Crommelin, "The Expedition to Sobral," described the return trip to England. He is obviously mistaken in writing that they went first to Pará (now Belém) and then back to Maranhao (now São Luís), where they caught the *Polycarp*. This itinerary would mean backtracking hundreds of miles. Plus, the *Polycarp*'s route was Pará to Liverpool. They had already visited Pará in the beginning, and they would be eager to get home with the important plates. Crommelin probably mixed up the order of the cities when he came to write the report. A corrected version of the route is also in Mourao, *Einstein: De Sobral Para O Mundo*.

15. Erwin Freundlich, *Die Grundlagen der Einsteinschen Gravitationstheorie*, was the first book on general relativity written by anyone other than Einstein. In the foreword, Einstein singled Freundlich out as being "the first among fellow-scientists who has taken pains to put the theory [of general relativity] to the test." Freundlich's editor was Arnold Berliner (1862–1942), a German physicist who had obtained a degree in physics from the

University of Breslau. He was the founder and first editor of a new scientific magazine, *Naturwissenschaften* (Natural sciences), which debuted in 1913 and published many of Einstein's papers and articles. Einstein published "Max Planck ala Forscher" (Max Planck as Scientist) in its first issue. The magazine was also published by Springer Books. Berliner was editor from 1913 to 1935, before he was removed by the Nazi government for being non-Aryan. Berliner took his own life on March 22, 1942, one day before being deported to an extermination camp by an "evacuation order."

16. The telegram Einstein received from Lorentz is dated September 22, 1919. The letter Einstein sent to his mother, citing "Some good news today!" is dated September 27, 1919.

CHAPTER 11: GREEK DRAMA

1. In 1664, a wealthy lawyer, architect, and poet name Sir John Denham built the house as his private mansion. Four years later, he sold it to Richard Boyle, the first Earl of Burlington, for whom the building is named. Famous guests down the years included Jonathan Swift and Alexander Pope. Composer George Frideric Handel lived at Burlington House from 1714 to 1717 at the invitation of Boyle. Since 1854, when it was sold to the British government and expanded, it has housed the Royal Academy of Arts and five prestigious societies: the Royal Astronomical Society, the Geological Society of London, the Linnean Society of London, the Society of Antiquaries of London, and the Royal Society of Chemistry. Founded in 1660, the Royal Society was at Burlington House from 1873 until its move to Carlton House Terrace, in 1967. According to Wootton, "Brief History of Facts," from its inception, the Royal Society was "dedicated to experimental knowledge and declared that it would concern itself with 'facts not explanations.' 'Facts' became part of a modern vocabulary for discussing knowledge—also including theories, hypotheses, evidence and experiments—which emerged in the 17th century. All these words existed before, but with different meanings: 'experiment,' for example, simply meant 'experience.'" For the November 6, 1919, meeting discussions referred to in this chapter, see Thomson, "Joint Eclipse Meeting of the Royal Society."

2. The four-inch telescope at Sobral had found a 1.98 arc seconds displacement, and the larger, problematic astrograph measured 0.9. At Príncipe, Eddington had measured 1.61 arc seconds. An averaging of these three amounts would reveal a displacement of 1.64. Because of the malfunction of the Greenwich astrograph (measuring 0.9), the researchers decided to throw out the suspect measurement and average the two remaining amounts. The result gave Einstein's full deflection of 1.75 arc seconds. The month before, Eddington had written to Dyson: "Dear Dyson, I was very glad to have your letter & measures. I am glad the Cortie plates gave the full deflection not only because of theory, but because I had been worrying over the Principe plates and could not see any possible way of reconciling them with the half deflection. I thought perhaps I had been rash in adopting my scale from few measures" (A. S. Eddington, letter to Sir Frank Dyson, October 1919. Royal Observatory/Cambridge University Library).

3. A. S. Eddington, letter to Walter Sydney Adams, January 1918. Adams (1876–1956) was an American astronomer whose field of study was stellar spectra and solar spectroscopy. Adams became the second director of the Mount Wilson Observatory in 1921, following its founder, George Ellery Hale.

4. Newton never wrote the equations in the form they were presented on the blackboard. These forms could have only come from the work of two other men, either Joseph-Louis Lagrange (1736–1813), an Italian mathematician and astronomer, or Sir William Hamilton (1805–1865), an Irish mathematician.

5. David Hilbert (1862–1943), a brilliant German mathematician, had published his paper "The Foundations of Physics" at almost the same time that Einstein's final paper on general relativity was published. Hilbert credited Einstein as the theory's originator, and the two men were never in dispute personally about this issue. For a more complete understanding of this interaction with Einstein and Hilbert, see Isaacson, *Einstein: His Life and Universe*, 212–222.

6. For months after the special meeting, astrophysicist Newall would continue to put forth his speculation of a refracted atmosphere around the sun, the properties of which could explain the deflection of starlight. At the November 6 meeting, after Newall spoke, Frederick Lindemann stood to address Newall's conjecture by stating that if this were the case, comets would slow down as they passed near the sun. "Have we not evidence from the motion of comets passing near the Sun that matter outside the Sun is distinctly diffuse? The comets suffer no noticeable check in their paths." It was Lindemann's tests that had made Freundlich believe early on that photographing stars during the daytime might be possible.

7. As late as 1936, Ludwik Silberstein, in "Two-Centers Solution of the Gravitational Field Equations, and the Need for a Reformed Theory of Matter," *Physical Review* 49, no. 5 (February 1, 1936), was claiming that Einstein's theory was flawed and needed revising. This assertion compelled Einstein and Nathan Rosen to respond to the editor (Einstein and Rosen "Two-Body Problem in General Relativity Theory," *Physical Review* 49, no. 5 [March 1, 1936]). Silberstein, who had been invited to lecture at the University of Toronto, then engaged the popular press by going public. On March 7, 1936, he published "Fatal Blow to Relativity Issued Here," in the *Toronto Evening Telegram*. He would, of course, be proven wrong. For more on Silberstein's refute of the general relativity theory, see Crelinsten, *Einstein's Jury*, 232–235. A story that has been written many times and is most likely apocryphal—despite Subrahmanyan Chandrasekhar's stating that Eddington told it to him personally—is that Silberstein had once approached Eddington to discuss the general theory. The still-skeptical Silberstein noted that "only three men in the world could understand it," and he had been told Eddington was one of them. When Eddington didn't respond, Silberstein said, "Don't be modest, Eddington!" to which Eddington replied, "On the contrary, I'm just wondering who the third might be." Whether truth or fiction, Silberstein earned his place as the butt of the joke. The story has been told in various versions, often with a journalist taking the place of Silberstein.

8. Whitehead, *Science and the Modern World*, described the philosopher's interpretation of the events at the meeting he attended, in a chapter in which he discussed the effects of Greek dramatic literature. Like many other authors who write about Einstein, Frank, *Einstein: His Life and Times*, mentions this paragraph by Whitehead.

9. The RAS Club has continued as a social group that still meets for dinner and talk after organized scientific meetings. According to Wilson, *Ninth Astronomer Royal*, these lively dinners were, in her father's day, filled with songs and libations. Her father missed very few monthly meetings. When he was appointed astronomer royal for Scotland, he

was celebrated at a dinner on December 8, 1905, with haggis and champagne on the menu and a piper for entertainment: "According to the records of the club he began his appointment as a Scotchman by plying the piper with so much neat whisky that he nearly killed him! The RAS Club was always on the lookout for occasions to celebrate." The club came up with bestowing on members the title of "Centurion" for those who had eaten one hundred dinners at the club. By 1920, Dyson had become a Centurion. Even during his stay in Scotland, as his daughter writes, he still attended 25 percent of the club's monthly dinners, and close to 100 percent when he lived in Greenwich. He was elected president of the RAS Club in 1936. Wilson recalls the toasts given that night, the Cottingham story, and Turner's full song, as written in this chapter. She quotes her father as saying, "Einstein wanted 1.74, but Cottingham wanted to double this amount."

10. According to Douglas, *The Life of Arthur Stanley Eddington*, 45, the published paper was Arthur Eddington, "On a Formula for Correcting Statistics for the Effects of a Known Probable Error of Observation," *Monthly Notices of the Royal Astronomical Society* 73, no. 5 (March 14, 1913): 359–360.

11. In this play of words, Turner's uses *Einstein*, since *stein* is "stone" or "rock" in German: "Nor left Einstein unturned, Sir." The full song was first published in the *Observatory* following the joint meeting. It can be found in H. H. Turner, "From an Oxford Note-book," *Observatory* no. 546 (1919): 25; and Wilson, *Ninth Astronomer Royal*.

CHAPTER 12: THE SEARCH FOR ACCURACY

1. In a remarkable sleuthing job, Sponsel, "Constructing a Revolution in Science," writes that it was noted zoologist Peter Chalmers Mitchell (1864–1945), a Royal Society Fellow since 1906 and then secretary of the Zoological Society of London, who sent the *Times of London* his report on the meeting. Why a zoologist? "This much is clear," Sponsel writes of Mitchell, "he had a strong connection to the proprietor of *The Times*; as a Fellow of the Royal Society he had a right to attend the joint meeting; and he was a dedicated internationalist, so he had reason to share Eddington's desire to publicize and celebrate the British confirmation of a German theory." Working in conjunction with Nicholas Mays, then deputy archivist for News International, Sponsel has pieced together an amazing estimation of how news about the eclipse expeditions found its way to the press over the months preceding the special meeting.

2. Sponsel, "Constructing a Revolution in Science," reviews Marshall Missner's analysis of Einstein's fame: "The link between newspaper reporting and Einstein's fame is well established in the secondary literature. Focusing on Einstein's renown in the USA, Marshall Missner has argued that 'the American press was the instrument that made Einstein into a celebrity'" (See Missner, "Why Einstein Became Famous in America"). "Missner's analysis of the role of newspapers in the spread of Einstein's fame is drawn largely from comparative readings of the journals themselves. Missner does not mention that there are two secondary accounts of how the *New York Times* got hold of the joint-meeting story from London" (See Hillier Krieghbaum, "American Newspaper Reporting of Science News," *Kansas State College Bulletin* 25 [August 15, 1941], 1–73; and Berger, *Story of the New York Times*).

3. Van Anda still maintained the hours he had kept as a night editor at the *New York Sun*, from 10:00 p.m. to dawn. He was at his desk at 1:20 a.m. on April 15, 1912, when the rope attached to the wooden box that carried bulletins down from the wire room to

the news floor below began to rattle. An Associated Press bulletin had arrived from Cape Race, Newfoundland. A Morse code distress signal, a CQD, was received from the *Titanic*. The ship had hit an iceberg off the coast of Newfoundland and needed assistance. Van Anda acted immediately. While other editors at competing papers were reluctant to write that the "unsinkable" ship had sunk, Van Anda thought differently. He ordered that the lead story for that day be torn out to make room for the bulletin and a brief history of the *Titanic*'s maiden journey. The headline would read TITANIC SINKING IN MID-OCEAN; HIT GREAT ICEBERG. Relying on information from the *Times* correspondent in Halifax—the re-write desk reporters also contacted any White Star Line officials they could reach at that hour—Van Anda learned that a half hour after the first cable, the *Titanic* had sent an SOS that the ship was going down. Then, nothing. He was now certain. When dawn broke and newspapers flew across the city, he had beaten the world's press to one of the greatest disaster stories of the times. By learning hieroglyphics, he would later make the 1923 opening of King Tut's tomb into a news story that would grip the public's imaginations for months and influence fashion and architecture. See Berger, *Story of the New York Times*.

4. For this first *New York Times* article on November 9, Crommelin also noted (as had Charles Davidson the day before in the *Times of London*) that two of the tests, the perihelion of Mercury and light deflection, "might now be looked on as established, at least with great probability." Crommelin was also careful to mention the third essential test, "a shift of the lines in the spectrum toward the red in a strong gravitational field." He stated that the idea of revising Newton was so "fundamentally important that consideration was already being given to the next total eclipse in September, 1922, visible in the Maldive Islands and Australia."

5. There are mistakes in Meyer Berger's account of Henry Charles Crouch and his cabled article on the meeting. Berger writes that the meeting was November 8 and that the *Times of London* ran it first on November 9. But he was two days off the facts in both cases. He also writes that between the meeting and when the story was printed, "the astronomers had by then examined photographs of stars" taken during the eclipse. He has Crouch telephoning Eddington, the most famous of the expeditionary foursome, to get a full report. It's more likely Crouch phoned Crommelin at the Royal Observatory, since Crommelin is directly quoted and Eddington's name is not even mentioned. Berger holds Van Anda in high esteem because Van Anda caught a mistake made by the brilliant L. P. Eisenhart, of Princeton, who had meant to give the *diameter* of the universe, but gave the *radius* instead. Van Anda apparently insisted that the translated Einstein lectures done for the paper by a local professor had an incorrect equation. When the professor insisted, "No, that is what Einstein said," the paper demurred, given Van Anda's brilliance in the subject matter. Einstein, also at Princeton by then, was telephoned. According to Berger, Einstein scanned his notes. "Yes, Mr. Van Anda is right. I made a mistake in transcribing an equation on the blackboard." See Berger, *The Story of the New York Times*.

6. The "12 wise men" notion would have longevity, even to today. When Albert and Elsa Einstein made their first visit to America, in March 1921, Albert was confronted with the notion. According to Isaacson, *Einstein: His Life and Universe*, "by the time Einstein reached Chicago, where he gave three lectures and played violin at a dinner party, he had become more adept at answering irksome questions, particularly the most frequent one, which was sparked by the fanciful *New York Times* headline after the 1919 eclipse that only twelve people could understand his theory. 'Is it true only twelve great

minds can understand your theory?' the reporter from the *Chicago Herald* and *Examiner* asked. 'No, no,' Einstein replied with a smile. 'I think the majority of scientific men who have studied it can understand it.'"

7. It's amusing that these two major newspapers of the world had Einstein age five years in two days, with the *Times of London* putting his age at forty-five and the *New York Times*, at fifty. His physics concepts from the relativity theory point to a possibility for how this fast aging could actually happen. Einstein could remain on the planet earth while all other inhabitants would take a round-trip journey on a spaceship. This vehicle needed to leave earth, traveling near the speed of light, and then return. The so-called time dilation effect could have implied that Einstein had aged five years in what would have only been experienced as two days for all those on the spaceship. In truth, he was actually only forty years old. But unfortunately, not even his theories of relativity allow for travel backward in time.

8. In a 1921 letter, Campbell wrote: "The fact is that we should not have attempted any observations on that subject with the imperfect and untested lenses which we borrowed only one month before the date of the 1918 eclipse." In a lecture delivered in San Francisco on January 19, 1923, and later reprinted that same year by the Astronomical Society of the Pacific, Campbell gave his opinion on the British results: "Professor Eddington was inclined to assign considerable weight to the African determination, but, as the few images on his small number of astrographic plates were not so good as those on the astrographic plates secured in Brazil, and the results from the latter were given almost negligible weight, the logic of the situation does not seem entirely clear." The British observers were the first to say that, in view of the fundamental importance of the general subject, confirmation should be sought at the eclipse of September 21, 1922.

9. In 1904, Paul Ehrenfest married Tatyana Alexeyevna Afanasyeva. Of their four children, one daughter became a mathematician, and the other an illustrator and author of books for children. The older son became a physicist. Wassik, the one Einstein called "little Crawlikins," had been born with Down syndrome in 1918, when Tatyana was forty-two years old. Paul Ehrenfest's career took an unfortunate spiral in the following decade, during which he began an affair with a young artist. He also suffered bouts of depression. Einstein wrote to the University of Leiden's board, asking that his friend's workload be reduced. On September 25, 1933, after writing farewell letters, Ehrenfest traveled to the Amsterdam institute where fifteen-year-old Wassik, known as a cheerful boy, was being kept. Ehrenfest was carrying a pistol hidden in his jacket pocket. In the waiting room, he shot Wassik first and then himself, dying immediately. Wassik lived for a few hours. Paul Ehrenfest wrote that he did not want to kill himself and leave his other children responsible for Wassik's care.

10. A. S. Eddington to Albert Einstein, December 1, 1919; Albert Einstein to A. S. Eddington, December 15, 1919; Erwin Freundlich to Albert Einstein, December 6, 1919, Einstein, *Collected Papers of Albert Einstein*, vol. 9, *The Berlin Years: Correspondence, January 1919–April 1920*.

11. In the paper they submitted a week before the November 6 meeting (and also during the meeting), both Dyson and Eddington pointed out that the gravitational redshift, the third test that Einstein said was necessary to prove his theory, did not affect his law of gravity. In the autumn of 1923, John Evershed, who had been unsuccessful at the eclipse in Australia a year earlier because of technical problems, would make a startling

announcement. Having once disagreed with Einstein's prediction of the gravitational redshift, and after years of work on the problem, Evershed changed his mind. From a modern perspective, we know he was wrong. It was not until 1971 that the sought-after effect was observed, and not by looking at light from the sun. Instead, it was light from the star Sirius B, together with a completely correct mathematical understanding pioneered by Robert Marshak and Subramanyan Chandrasekhar, that finally allowed observation of Einstein's redshift in starlight as his theory predicted. This is why the work of Karl Popper (1902–1994), inspired in part by the results of the theory of general relativity, is so important. Science has built into its formal structure and practice a very interesting self-protection mechanism for how it deals with accuracy, errors, and mistakes. A scientific theory is never proven correct, precisely because a later observation using more advanced technology that yields increased accuracy might show its mathematical predictions are no longer supported. For this observation, Popper continues to be recognized as one of the greatest philosophers of science of the twentieth century.

12. Campbell had disagreed about location with Arthur Hinks, the Englishman who had been so diligent in steering Eddington away from clouds on Príncipe. Hinks thought that Eighty Mile Beach on the west coast of Australia was inaccessible, with its twenty-six-foot surf and no land transportation to speak of. Campbell believed this spot was nothing compared with his expedition to Flint Island in 1907–1908. He had originally planned to take just Trumpler as his team member. In Fremantle, they would board a schooner that would be loaded with supplies for the sheep ranches near Wallal. The schooner would return with tons of wool from the sheared sheep. When the Australian government offered to put a naval vessel at his disposal, his team members grew. Campbell credited the Australian government with showing him and his team the most generous hospitality of his career. In his 1922 published paper, he referred to the coastal area where his team came ashore as Ninety Mile Beach, which it was then called. In 1946, the name was changed to Eighty Mile Beach to avoid confusion with the Ninety Mile Beach in the state of Victoria, on the southeast coast. For a detailed account of the Lick expedition and others in Australia in 1922, see Crelinsten, *Einstein's Jury*, chap. 8; and Campbell, "The Total Eclipse of the Sun, September 21, 1922." See also Burman and Jeffery, "History of Australian Astronomy."

EPILOGUE: MEN MADE ON DREAMS

1. Historian David Ball has an interesting observation about how Cottington's work affected the students at Trinity: "There had been a tradition for students on the day of the Matriculation Dinner to try to run around the Great Court of Trinity, some 401 yards, while the clock struck 12 o'clock (in fact it struck 24 times). This event was made famous worldwide by the film *Chariots of Fire*, in which the future Olympic athletes Liddell and Abrahams raced each other. The "At Random" column in *The Observer* later noted: 'To change the pace of a public clock is akin to the sin of removing one's neighbour's landmark and the famous horologist who has just died, Mr Edwin Cottingham, played at least a small part in deranging records. In tending the clock of Trinity Church, Cambridge, he speeded it up slightly, spoiling the sport of the undergraduates . . .' In fact, no-one was then able to beat the chimes until Lord Burghley in 1927. That is why we must assume that E. T. Cottingham's careful work was cursed by a generation of students" (Ball, "Cottingham, Edwin Turner").

2. The face of the wall clock at St. James Church was stolen in 1984 and likely resides in a private collection. The Thrapston Town Council voted to have the mechanism for the chimes in the bell tower repaired in commemoration of Queen Elizabeth's eightieth birthday. Several years ago, local historian David Ball went in search of the building that once had the Cottingham name proudly displayed over the large front window. "None knew where the shop had been," Ball writes. "In fact, none of them had heard of Edwin T. Cottingham. Eventually a man did confirm that he had vaguely heard of him and he might have had the last shop before the mini roundabout." Ball discovered the very shop where Cottingham had made such beautiful clocks. It had become an Indian restaurant. Eric Franklin, of the Thrapston Town Council and also a local historian, adds that "Cottingham Way is a small area of relatively new industrial units and, although well off the beaten track and of no real merit, at least commemorates one of the unsung heroes of our town." (Ball, "Cottingham, Edwin Turner").

3. One of Eddington's fellow scientists who stayed in touch during these last years was none other than Subrahmanyan (Chandra) Chandrasekhar, who was then at the Ballistic Research Laboratories in Maryland. Chandra sent Eddington parcels of rice to help make up for food rationing. This gift seems a subtle, unconscious symbol, showing again the chasm that lay between the two men culturally. *Rice*, not English sausage or tea. Eddington's public RAS ridicule of the younger man's work is well known among physicists and astronomers. Chandra had shown Eddington some calculations he had written that would be the first irrefutable proof that black holes existed. Thinking Eddington was in support of his work, he was shocked when the famous astronomer instead humiliated him and his concept in front of the members. After winning the Nobel Prize in 1983, Chandra was asked his opinion on why Eddington had not eventually relented, as many scientists have done when they realize that they are mistaken. Chandra, who saw the attack as "racially motivated," replied, "Eddington worked on the universe." See Miller, *Empire of the Stars*.

4. For the last years of his life, Eddington was obsessed with a book that would harmonize quantum physics and relativity, a "theory of everything." When it was published after his death, in 1946, it was given the title *Fundamental Theory*. Eddington usually destroyed his personal letters over the years, even those to other scientists, once the business in the letters was concluded. Perhaps he knew that death was near, for he set about destroying any personal correspondence that remained. Winnifred Eddington, once her brother was gone, finished the job. But, luckily for posterity, spared was a large file cabinet stuffed with papers that he bequeathed in his will to the RAS. The cabinet went to Frederick Stratton, known as "Colonel Stratton" for his war service. Stratton had been with Newall in the Crimean vineyard where Perrine and Freundlich had also set up, before being recalled when the war broke out. It was Stratton who, as president of the RAS in 1935, had called Chandra to the podium that fateful January day when Eddington so harshly criticized the young man's work. Stratton sorted through the file cabinet of Eddington's papers and then threw them away for being "merely of biographical interest."

5. Much of the information about Charles Perrine's private and professional life comes from Bob Kelly-Thomas and Diana Merlo Perrine, grandchildren of Charles Perrine, email correspondence with authors, June 2017 to May 2019.

6. Some papers written and published by these accomplished astronomers and men of science didn't always pertain to their work at hand. Two papers Perrine published in 1914 were about lightning and a chicken. When lightning hit the dome of his observatory in January, before he and Mulvey left for the Crimea, Perrine commented on how well the metal dome fared as he stood outside the shop and watched the flashes. "Mr. Mulvey was in the underground optical shop at the time and thought there had been an explosion." Another paper began, "On December 15, 1914, a chicken was hatched out of our settings which had four legs. It lived from one evening to the next noon when it was stepped upon by the mother-hen and killed. It seemed to be normal in every other respect, eating and walking about like the others" (Charles Perrine, "Effect of Lightning on a Reinforced Concrete and Steel Dome," *Science* 40, no. 1032 [October 9, 1914]; and Charles Perrine, "A Chicken with Four Legs," *Science* 42, no. 1072 [July 16, 1916]). In the spring of 1925, Einstein traveled to Argentina and also visited Córdoba, although he did not visit the observatory. Charles Perrine—the first man to journey thousands of miles in a failed attempt to test for light deflection—was in California until that September. Albert Einstein never visited Mount Hamilton either, although he was invited over the years. The famed physicist's schedule was pressing when in the United States. In 1931, he and William Campbell finally met in person, but at Mount Wilson, near Pasadena, California.

7. An oral history interview with Lilo Leyendecker, the German housekeeper who came to Scotland after answering a newspaper ad placed by Freundlich's brother, is quoted in Ledermann's archive: "Freundlich often spoke of his friend Einstein, but by 1953 he had lost touch with him, something which he rather regretted. Einstein used to visit Freundlich's house in Potsdam, Germany. After they had eaten dinner one evening, Einstein and Freundlich used the table cloth to write mathematics on. This greatly annoyed Mrs. Freundlich who was cross that her best table cloth was ruined. . . . After Freundlich came to St Andrews he sometimes went to London to act as an interpreter for Einstein, who did not speak English very well at that time, when he came to meetings. On one occasion Einstein, after listening to a speaker at a meeting, said, 'That man is someone who can speak English well—I can understand what he says.' The man in question spoke English with a very heavy German accent!" (John O'Connor and Edmund F. Robertson, "Walter Ledermann," MacTutor History of Mathematics Archive, School of Mathematics and Statistics, University of St. Andrews, Scotland, www-history.mcs.st-andrews.ac.uk/index .html). See also Forbes and Ross, *Dictionary of Scientific Biography*.

8. The personal memories of the astronomer are from Rosemary Cecily Hannibal and Catherine Stella Court (Davidson's granddaughters), correspondence with the authors, September 2018 to May 2019. In Woolley's obituary, he has Davidson being educated at Christ's Hospital, which had no record of his attending. His granddaughters are certain that his education was at the Royal Hospital School in Greenwich, which is confirmed by Graham Dolan's research.

9. Israel, Ruckhaber, and Weinmann, *Hundert Autoren gegen Einstein*; and Epstein, *Let There Be Sculpture*.

BIBLIOGRAPHY

PAPERS

Adams, John C. "The Observations of the Total Solar Eclipse of July 28, 1851." *Memoirs of the Royal Astronomical Society* 21 (1852): 101–107.

Aiken, R. G. "The Lick Observatory-Crocker Eclipse Expedition to Fryeburg, Maine." *Publications of the Astronomical Society of the Pacific*, September 15, 1932.

Airy, George B. "Observations of the Total Solar Eclipse of 1842, July 7 (July 8, civil reckoning)." *Memoirs of the Royal Astronomical Society* 15 (1846): 9.

Baily, Francis. "Some Remarks on the Total Eclipse of the Sun, on July 8th, 1842." *Monthly Notices of the Royal Astronomical Society* 5, no. 25 (November 11, 1842): 208–220. https://doi.org/10.1093/mnras/5.25.208a.

Ball, David. "Cottingham, Edwin Turner (1869–1940): Modern Times." In *Ringstead People: Biographies of Ringstead (Northamptonshire) People*, October 14, 2010. http://ringstead.squarespace.com/ringstead-people/2010/10/14/cottingham-edwin-turner-1869-1940-modern-times.html.

Barboza, Christina Helena da Motta. "Science and Nature in the Astronomical Expeditions to Brazil (1850–1920)." *Boletim do Museu Paraense Emílio Goeldi. Ciências Humanas*, 2010. www.oalib.com/paper/2943129#.XIvw0ShKiUk.

Boney, A. D. "The Summer of 1914: Diary of a Botanist." *Notes and Records of the Royal Society of London* 52, no. 2 (July 1998): 323–338. www.jstor.org/stable/531864.

Bracher, Kate. "A Fateful Eclipse." Chapter 1 in *Stars for All: A Centennial History of the Astronomical Society of the Pacific*. San Francisco: Astronomical Society of the Pacific, 1989. https://www.astrosociety.org/about/history/ch01.pdf.

Braithwaite, Chas. Ll. "Friends' School in Kendal: [Letter] to Westmorland Quarterly Meeting." *British Friend* 43, nos. 1–12 (February 2, 1885): 33.

Brito, Paulino de A. "Repercussões. Novidades Scientificas." Estado do Pará, Belém, April 22, 1919. Cited in "Expedição norte-americana e iconografia inédita de Sobral em 1919" [North-American expedition and unpublished iconography from Sobral in 1919], by Luís Carlos Bassalo Crispino and Marcelo Costa de Lima. *Revista Brasileira de Ensino de Física*, June 2, 2017. http://dx.doi.org/10.1590/1806-9126-rbef-2017-0092.

Burgess, Anika. "The 1922 Eclipse Adventure That Sought to Confirm the Theory of Relativity." Eclipse Madness series. *Atlas Obscura*, August 11, 2017. www.atlas obscura.com/articles/the-1922-eclipse-expedition-to-remote-western-australia.

Burman, R. R., and P. M. Jeffery. "History of Australian Astronomy, Wallal: The 1922 Eclipse Expedition." *Proceedings of the Astronomical Society of Australia* 8, no. 3 (1990): 312–313. http://articles.adsabs.harvard.edu/cgi-bin/nph-iarticle_query ?1992PASAu..10..168B&data_type=PDF_HIGH&whole_paper=YES &type=PRINTER&filetype=.pdf.

Campbell, Kenneth (with comments by C. Donald Shane). "Life on Mount Hamilton 1899–1913." Interviews conducted and edited by Elizabeth Speddling Calciano. Santa Cruz, 1971. https://archive.org/details/lifemounthamilton00camprich and https://escholarship.org/uc/item/1z90c0c2.

Campbell, W. W. "The Crocker Eclipse Expedition from the Lick Observatory June 8, 1918." *Publications of the Astronomical Society of the Pacific* 30, no. 176 (August 1918): 218–249. http://iopscience.iop.org/article/10.1086/122736/pdf.

———. "International Meetings of Astronomers in Germany." *Publications of the Astronomical Society of the Pacific* 25, no. 150 (October 1913).

———. "The Lick Observatory Community in War Service." *Publications of the Astronomical Society of the Pacific* 30, no. 178 (December 1918): 353–357.

———. "Mars, by Percival Lowell" (book review). *Publications of the Astronomical Society of the Pacific* 8 (August 1896). www.jstor.org/stable/10.2307/40667612.

———. "The Nature of an Astronomer's Work." *Publications of the Astronomical Society of the Pacific* 10, no. 122 (October 10, 1908).

———. "The Total Eclipse of the Sun, September 21, 1922." *Publications of the Astronomical Society of the Pacific* 35, no. 203 (February 1923): 11–44.

Campbell, W. W., and Heber D. Curtis. "The Lick Observatory-Crocker Eclipse Expedition to Brovary, Russia." *Publications of the Astronomical Society of the Pacific* 26, no. 155 (December 1914): 224–237.

Cooke, W. E. "The Total Eclipse of 1922 September 21, Observations by the Sydney Observatory Expedition at Goondiwindi, Queensland." *Monthly Notices of the Royal Astronomical Society* 83, no. 9 (July 14, 1923): 511–515. https://academic.oup.com /mnras/article/83/9/511/1302868.

Cortie, A. L. "Preliminary Report: Sun, Eclipse of, 1914 August 20–21:- Expedition of the Joint Permanent Eclipse Committee to Hernösand, Sweden (Plates 9–12)." *Monthly Notices of the Royal Astronomical Society* 75, no. 3 (January 8, 1915): 105–117. https:// academic.oup.com/mnras/article/75/3/105/1016068.

———. "Stonyhurst Astronomers at Hernösand." *Stonyhurst Magazine* 13, no. 195 (July 1914): 929–930. www.worldwar1schoolarchives.org/wp-content/uploads/2013/11/1914 _07.pdf.

Crispino, Luis C. B., and Marcelo Costa de Lima. "Amazonia Introduced to General Relativity: The May 29, 1919, Solar Eclipse from a North-Brazilian Point of View." *Physics in Prospective* 18 (December 2016): 379–394. https://link.springer.com /article/10.1007/s00016-016-0190-3.

———. "Expediçâo norte-americana e iconografia inédita de Sobral em 1919" [North-American expedition and unpublished iconography from Sobral in

1919]. *Revista Brasileira de Ensino de Física* 40, no. 1 (June 2, 2017). http://dx.doi .org/10.1590/1806-9126-rbef-2017-0092.

Crommelin, A. C. D. "The Eclipse of the Sun on May 29." *Nature* 102 (1918–1919): 444–446.

———. "The Expedition to Sobral." *Observatory* 42, no. 544 (October 1919): 368– 371. http://articles.adsabs.harvard.edu/cgi-bin/nph-iarticle_query?1919Obs....42 ..368C&data_type=PDF_HIGH&whole_paper=YES&type =PRINTER&filetype=.pdf.

Davidson, C. "Observation of the Einstein Displacement in Eclipse of the Sun." *Observatory* 45, no. 578 (July 1922): 224–225. http://adsabs.harvard.edu/abs/1922 Obs....45..224D.

Davidson, C. R., and J. Jackson. "Report of the Expedition from the Royal Observatory, Greenwich, to Observe the Total Eclipse of the Sun on 1932 August 31." *Monthly Notices of the Royal Astronomical Society* 93, no. 1 (November 11, 1932): 3–14. https:// academic.oup.com/mnras/article/93/1/3/1022914.

Davis, C. D. P., J. L. E. Dreyer, A. Fowler, G. J. Newbegin, A. A. Rambaut, W. Sidgreaves, A. L. Cortie, and J. H. Worthington. "Summary of Observations of the Solar Eclipse of 1912 April 16–17." *Monthly Notices of the Royal Astronomical Society* 72, no. 7 (May 17, 1912): 542–546.

de la Baume Puluvinel, A. "L'Eclipse Total de Soleil du 21 Août 1914." *L'Astronomie* 28 (1914): 113–117. http://adsabs.harvard.edu/abs/1914LAstr..28..113D.

de Sitter, Willem. "Errata in Professor De Sitter's Third Paper on 'Einstein's Theory of Gravitation.'" *Monthly Notices of the Royal Astronomical Society* 78, no. 4 (February 8, 1918): 341. https://academic.oup.com/mnras/article/78/4/341/1053430.

———."On Einstein's Theory of Gravitation and Its Astronomical Consequences. Second Paper." *Monthly Notices of the Royal Astronomical Society* 77, no. 2 (December 8, 1916): 155–184. https://academic.oup.com/mnras/article/77/2/155/979347.

———. "Space, Time, and Gravitation." *Observatory* 39, no. 505 (October 1916): 412–419. http://articles.adsabs.harvard.edu/pdf/1916Obs. . . . 39..412D.

DeVorkin, David H. "The International Union for Cooperation in Solar Research: Prelude to the IAU." Smithsonian Institution, Washington, DC, 1995, 117–118. www.cambridge.org/core/services/aop-cambridge-core/content/view/ADB31B7FC C210E914B8DBF6213659FB5/S1539299600010467a.pdf/international_union_for _cooperation_in_solar_research_prelude_to_the_iau.pdf.

Dietrich, Sarah, and Stephen Klassen. "Physics Comes to Winnipeg: The 1909 Meeting of the British Association for the Advancement of Science." *Proceedings of the Second International Conference on Story in Science Teaching.* Munich, July 2008. www .researchgate.net/publication/225415295.

Dobrow, Julie. "The Star-Crossed Astronomer." *Amherst Magazine*, July 28, 2017.

Duerbeck, H. W., D. E. Osterbrock, L. H. Barrera S., and R. Leiva G. "Halfway from La Silla to Paranal—in 1909." *European Southern Observatory*, March 1999. www.eso .org/sci/publications/messenger/archive/no.95-mar99/messenger-no95-34-37.pdf.

Dyson, F. W. "On the Opportunity Afforded by the Eclipse of 1919 May 29 of Verifying Einstein's Theory of Gravitation." *Monthly Notices of the Royal Astronomical Society* 77, no. 5 (March 9, 1917). https://academic.oup.com/mnras/article/77/5/445 /1272333.

————. "Report of the Expedition from the Royal Observatory, Greenwich to Observe the Total Solar Eclipse of 1927 June 29 (Plates 8–12)." *Monthly Notices of the Royal Astronomical Society* 87, no. 9 (July 14, 1927): 657–665. https://academic.oup.com/mnras/article/87/9/657/1006288

————. "W. W. Campbell." *Obituaries of the Royal Astronomical Society* 2, no. 7 (January 1939).

Dyson, F., A. S. Eddington, and C. R. Davidson. "A Determination of the Deflection of Light by the Sun's Gravitational Field from Observations Made at the Eclipse of May 29, 1919." *Observatory* 42, no. 545 (November 1919): 291–315.

Eddington, A. S. "The Deflection of Light during a Solar Eclipse." *Nature* 104 (1919): 372.

————. "Gravitation and Principle of Relativity." *Nature* 98 (1916–1917): 328–330.

————. "Joint Eclipse Meeting of the Royal Society and the Royal Astronomical Society." *Observatory* 42, no. 545 (November 1919): 392–394. http://articles.adsabs.harvard.edu/cgi-bin/nph-iarticle_query?1919Obs....42..389.&data_type=PDF_HIGH&whole_paper=YES&type=PRINTER&filetype=.pdf.

————. "Notes on Some Points Connected with Recent Progress of Astronomy." *Monthly Notices of the Royal Astronomical Society* 77, no. 4 (February 9, 1917): 377–382. https://academic.oup.com/mnras/article/77/4/352/1048770.

————. "Sir David Gill." Obituary in "Report to the Council to the Ninety-Fifth General Meeting." *Monthly Notices of the Royal Astronomical Society* 75 (February 1915): 236–247.

————. "Sir Frank Watson Dyson, 1868–1939." *Royal Society Obituary Notices*, January 1, 1940, 159–162. https://doi.org/10.1098/rsbm.1940.0015.

————. "Some Problems of Astronomy, XIX Gravitation." *Observatory* 38 (February 1915): 93–98. http://articles.adsabs.harvard.edu/cgi-bin/nph-iarticle_query?1913Obs....36..142E&data_type=PDF_HIGH&whole_paper=YES&type=PRINTER&filetype=.pdf.

————. "The Total Eclipse of 1919, May 29, and the Influence of Gravitation on Light." *Observatory* 42, no. 537 (March 1919): 119–122. http://articles.adsabs.harvard.edu/cgi-bin/nph-iarticle_query?1919Obs....42..119E&data_type=PDF_HIGH&whole_paper=YES&type=PRINTER&filetype=.pdf.

Eddington, A. S., and C. R. Davidson. "Total Eclipse of the Sun, 1912 October 10: Report on an Expedition to Passa Quatro, Minas Geraes, Brazil." *Monthly Notices of the Royal Astronomical Society* 73, no. 5 (March 14, 1913): 386–390. https://academic.oup.com/mnras/article/73/5/386/972834.

Einstein, Albert. "Elementary Theory of Water Waves and of Flight." *Naturwissenschaften* 4 (1916): 509–510.

————. "On the Influence of Gravitation on the Propagation of Light." *Annalen der Physik*, June 1911.

————. "Principles of Research" (speech delivered at Max Planck's sixtieth birthday celebration, 1918). In *Mein Weltbild* (Amsterdam: Querido Verlag, 1934); reprinted in *Ideas and Opinions* (New York: Crown, 1954), 224–227, 235.

Einstein, Albert, and Nathan Rosen. "Two-Body Problem in General Relativity Theory." *Physical Review* 49, no. 5 (March 1, 1936).

Evershed, J. "Report of the Indian Eclipse Expedition to Wallal, West Australia." *Kodaikanal Observatory Bulletin* 61 (February 26, 1923).

Favaro, G. A. "Le Osservazioni Dell' Eclisse Solare Del 21 Agosto 1914, Fatto a Osservatorio Astrofisico di Catania" [The observations of the solar eclipse of August 21, 1914, made at the R. Astrophysical Observatory of Catania]. John Wolbach Library, Smithsonian Center for Astrophysics, Cambridge, MA.

Flewelling, Lindsey. "William Cadbury, Chocolate, and Slavery in Portuguese West Africa." *Isles Abroad* (blog of British and Irish Global History), May 11, 2016. https://britishandirishhistory.wordpress.com/2016/05/11/william-cadbury-chocolate-and-slavery-in-portuguese-west-africa.

Flower, A., E. H. Hills, and W. E. Curtis. "The Total Eclipse of the Sun, 1914 August 21.—Report on the Kiev Expedition." *Monthly Notices of the Royal Astronomical Society* 75, no. 3 (January 8, 1915): 117–125. https://academic.oup.com/mnras/article/75/3/117/1016078.

Fowler, Alfred. "The International Astronomical Union, Formed on July 28, 1919." *Popular Astronomy*, November 1919.

Halley, E. "A Description of the Passage of the Shadow of the Moon Over England As It Was Observed in the Late Total Eclipse of the Sun April 22, 1715." London: J. Senex, 1715. Digital image. University of Cambridge, Images from the Institute of Astronomy Library. www.repository.cam.ac.uk/handle/1810/221308.

Headrick, Daniel R. "Sleeping Sickness Epidemics and Colonial Responses in East and Central Africa, 1900–1940." *PLOS: Neglected Tropical Diseases* 8, no. 4 (April 24, 2014). https://journals.plos.org/plosntds/article?id=10.1371/journal.pntd.0002772.

Heffernan, M., and Jons, H. "Research Travel and Disciplinary Identities in the University of Cambridge, 1885–1955." *British Journal for the History of Science* 46, no. 2 (June 2013).

Henroteau, F. "The International Astronomical Union at Cambridge." Royal Astronomical Society of Canada, Dominion Observatory, Ottawa, 1925.

Hentschel, Klaus. "Erwin Finlay Freundlich and Testing Einstein's Theory of Relativity." OPUS—Publication Server of the University of Stuttgart, 1994. https://elib.uni-stuttgart.de/bitstream/11682/7171/1/hen16.pdf.

Hilbert, David. "Die Grundlagen der Physik" [Foundations of physics]. *Nachrichten von der Gesellschaft der Wissenschaften zu Göttingen – Mathematisch-Physikalische Klasse* 3 (1915): 395–407.

Holden, Edward S. "Forest Fires at Mount Hamilton, July, 1891." *Publications of the Astronomical Society of the Pacific* 3 (1891): 292.

———. "Forest Fires on Mount Hamilton, July 29 to August 3, 1894." *Publications of the Astronomical Society of the Pacific* 6 (1894): 240.

Israel, Hans, Erich Ruckhaber, and Rudolf Weinmann, eds. *Hundert Autoren gegen Einstein*. Leipzig: Voigtländer, 1931.

Janssen, Michel, and Jurgen Renn. "Arch and Scaffold: How Einstein Found His Field Equations." *Physics Today*, November 1, 2015. https://physicstoday.scitation.org/doi/10.1063/PT.3.2979?journalCode=pto&.

Jones, Bessie Zaban, and Lyle Gifford Boyd. "E. C. Pickering: The Early Years." In *The Harvard College Observatory: The First Four Directorships, 1839–1919*, 193–198. Cambridge, MA: Belknap Press of Harvard University Press, 1971.

Jones, H. Spencer. "1939 February 10 Annual General Meeting." *Monthly Notices of*

the Royal Astronomical Society 99, no. 4 (February 1, 1939): 279–281. https://doi
.org/10.1093/mnras/99.4.279.

———. "The Royal Observatory Eclipse Expedition to Minsk, Russia." *Observatory* 479,
no. 37 (October 1914): 379–384. http://articles.adsabs.harvard.edu/cgi-bin/nph
-iarticle_query?1914Obs....37..379S&data_type=PDF_HIGH&whole
_paper=YES&type=PRINTER&filetype=.pdf.

Jones, H. Spencer, and C. R. Davidson. "Sun, Eclipse of, 1914 August 20–21: Preliminary
Account of the Observations Made at Minsk, Russia." *Monthly Notices of the Royal
Astronomical Society* 75, no. 3 (January 8, 1915): 125–132. https://academic.oup.com
/mnras/article/75/3/125/1016085.

Jostock, Carolyn. "Cashew Industry." In *Encyclopedia of Latin American History and Culture*, 2:5. New York: Charles Scribner's Sons, 1996.

Kennefick, Daniel. "Not Only Because of Theory: Dyson, Eddington and the Competing Myths of the 1919 Eclipse Expedition." In *Proceedings of the 7th Conference on
the History of General Relativity, Tenerife, 2005*. September 5, 2007. https://arxiv.org
/abs/0709.0685.

Lywood, William George. "Our Riviera, Coast of Health: Environment, Medicine, and
Resort Life in Fin-de-Siècle Crimea." PhD diss., Ohio State University, 2012.

Marengo, Jose A., Roger Rodrigues Torres, and Lincoln Muniz Alves. "Drought in
Northeast Brazil—Past, Present, and Future." *Theoretical and Applied Climatology*
129, no. 3–4 (August 2017): 1189–1200. https://link.springer.com/article/10.1007
/s00704-016-1840-8.

Mauchly, S. J. "Section of Terrestrial Electricity." Carnegie Institution of Washington,
Department of Terrestrial Magnetism, Year Book 19, 1920, pp. 307–320.

Mercer, Frank. "Mr. E. T. Cottingham." *Nature* 3678 (April 27, 1940): 653.

Millikan, R. A. "Radiation and Atomic Structure." *Proceedings of the American Physical
Society* 10, no. 2 [1917]: 195.

Minniti, E. R., and S. Paolantonio S. "Córdoba Estelar: Historia del Observatorio Nacional Argentino. Observatorio Astronómico de la Universidad Nacional de Córdoba."
Córdoba: Editorial de la Universidad, 2009. www.cordobaestelar.oac.uncor.edu.

Missner, Marshall. "Why Einstein Became Famous in America." *Social Studies of Science*
15, no. 2 (1985): 267–291.

Mota, Elsa, Paulo Crawford, and Ana Simões. "Einstein in Portugal: Eddington's Expedition to Principe and the Reactions of Portuguese Astronomers (1917–25)." *British
Journal for the History of Science*, 2008. doi:10.1017/S0007087408001568.

Newall, H. F. "The Eclipse Expedition to Theodosia (Crimea)." *Observatory* 37, no. 479
(October 1914): 384–387. http://articles.adsabs.harvard.edu/cgi-bin/nph-iarticle
_query?1914Obs....37..384N&data_type=PDF_HIGH&whole_paper
=YES&type=PRINTER&filetype=.pdf.

———. "Sun, Eclipse of, 1914 August 20–21: Report on an Expedition from the Solar
Physics Observatory, Cambridge, to Theodosia, Crimea." *Monthly Notices of the
Royal Astronomical Society* 75, no. 3 (January 8, 1915): 134–138. https://academic
.oup.com/mnras/article/75/3/134/1016104.

O'Connor, J. J., and E. F. Robertson. "Erwin Finlay Freundlich." MacTutor History of
Mathematics Archive, University of St Andrews, Scotland, October 2003. www
-history.mcs.st-andrews.ac.uk/Biographies/Freundlich.html.

Paolantonio, S. "A un siglo del primer intento de verificar la Teoría de la Relatividad. *Historia del Astronomía* (blog), 2012. https://historiadelaastronomia.wordpress.com /documentos/primerintento.

———. "De Córdoba al Mar Negro: Relatos de una aventura científica." *Historia del Astronomía* (blog), 2010. https://historiadelaastronomia.wordpress.com/documentos /de-cordoba-al-mar-negro.

Paolantonio, S., and E. R. Minniti. "Argentinean Attempts to Prove the Theory of Relativity" [in Spanish]. *Boletín de la Asociación Argentina de Astronomía* 50 (2007): 359–362. https://historiadelaastronomia.files.wordpress.com/2008/12/2007baaa50359p .pdf.

———. "Attempts to Prove Einstein's Theory of Relativity." Translated by A. D. Slopes, E. E. Scorians, and M. E. Valotta. Córdoba Estelar: Historia del Observatorio Nacional Argentino [Córdoba estelar: history of the Argentine National Observatory]. 2013. http://sion.frm.utn.edu.ar/WDEAIII/wp-content/uploads/2018/09 /Relatividad-Eng-Final2.pdf.

———. "Historia de la astronomía Argentina y Latinoamericana." *Historia de la Astronomía* web page. https://historiadelaastronomia.wordpress.com/documentos.

———. "Historia del Observatorio Astronómico de Córdoba." *Historia de la Astronomía Argentina* (Asociación Argentina de Astronomía) (2009): 51–167. https://historia delaastronomia.files.wordpress.com/2008/12/historia-del-ona1.pdf.

Paolantonio, S., et al. "The Argentinean Attempts to Prove the Theory of General Relativity: The Total Solar Eclipses of 1912, 1914 and 1919." *Proceedings of the International Astronomical Union* 13, symposium 349 (2018): 516–519.

Pearson, John. "The Lick Observatory Solar Eclipse Expeditions," chap. 4 of "The Role of the 40 Foot Schaeberle Camera in the Lick Observatory Investigations of the Solar Corona." PhD diss., James Cook University, 2009. https://researchonline.jcu.edu .au/10407/3/03Chapter4.pdf.

Pearson, John, and Wayne Orchiston. "The 40-Foot Solar Eclipse Camera of the Lick Observatory." *Journal of Astronomical History and Heritage* 11, no. 1 (2008): 25–37.

Perrine, C. "Contribution to the History of Attempts to Test the Theory of Relativity by Means of Astronomical Observations." *Astronomische Nachrichten*, 219 (1923), 281–284.

———. "James Oliver Mulvey." Obituary in *Publications of the General Notes, Astronomical Society of the Pacific*, December 15, 1915, 94. https://archive.org/details /jstor-40691926/page/n1.

Rodrigues, Joyce Mota. "Entre Telescópios e Potes de Barro: As Commissões Scientíficas do Eclipse Solar em 1919—Sobral" [Between telescopes and clay pots: the scientological commissions of the solar eclipse in 1919—Sobral]. Universidade Federal do Ceará, 2012. https://core.ac.uk/display/44852351.

Scheckter, John. "Modern in Every Respect: The 1914 Conference of the British Association for the Advancement of Science." *Journal of the European Association for Studies of Australia* 5, no. 1 (2014).

Schwarzschild, Karl. "On the Permissible Numerical Value of the Curvature of Space" [in German]. *Vierteljahrsschrift der Astronomische Gesellschaft* [Quarterly of the Astronomical Society] 35 (1900): 337–347.

———. "Uber das Gravitationsfeld eines Massenpunktes nach der Einsteinschen Theorie." *Sitzungsberichets der Königlich-Preussischen Akademie der Wissenschaften* (Berlin) (January 1916): 189–196.

Shears, Jeremy. "The British Astronomical Association and the Great War of 1914–1918." *Journal of the British Astronomical Association* 124, no. 4 (2014): 187–197.

Silberstein, Ludwik. "Two-Centers Solution of the Gravitational Field Equations, and the Need for a Reformed Theory of Matter." *Physical Review* 49, no. 3 (February 1, 1936).

Soldner, Johann Georg von. "On the Deflection of a Light Ray from its Rectilinear Motion." Translated by Wikisource. Berlin, 1801. https://en.wikisource.org/wiki/Translation:On_the_Deflection_of_a_Light_Ray_from_its_Rectilinear_Motion.

Sponsel, Alistair. "Constructing a Revolution in Science: The Campaign to Promote a Favourable Reception for the 1919 Solar Eclipse Experiments." *British Journal for the History of Science*, December 2002.

Stanley, Matthew. "An Expedition to Heal the Wounds of War: The 1919 Eclipse and Eddington as Quaker Adventure." *Isis, A Journal of the History of Science Society* 94, no. 1 (March 2003): 57–89. www.journals.uchicago.edu/toc/isis/2003/94/1.

Thomson, Andrew. "Results of Atmospheric-Electric Observations Made at Sobral, Brazil, during the Total Solar Eclipse of May 29, 1919." *Journal of Geophysical Research Atmospheres*, January 1920.

Thomson, J. J. "Presidential Address." *Proceedings of the Royal Society of London* 96 (1919–1920): 316.

Thomson, Joseph. "Joint Eclipse Meeting of the Royal Society and the Royal Astronomical Society." *Observatory* 42, no. 545 (November 1919): 388–398. http://articles.adsabs.harvard.edu/cgi-bin/nph-iarticle_query?db_key=AST&bibcode=1919Obs....42..389.&letter=.&classic=YES&defaultprint=YES&whole_paper=YES&page=389&epage=389&send=Send+PDF&filetype=.pdf.

Treschman, Keith John. "Early Astronomical Tests of General Relativity: The Gravitational Deflection of Light." *Asian Journal of Physics* 23, nos. 1 and 2 (2014): 145–170. www.researchgate.net/publication/296183387_Early_astronomical_tests_of_General_Relativity_the_gravitational_deflection_of_light/download.

———. "General Relativity in Australian Newspapers: The 1919 and 1922 Solar Eclipse Expeditions." *Historical Records of Australian Science* 26 (2015): 150–163. www.researchgate.net/publication/283741047_General_Relativity_in_Australian_Newspapers_The_1919_and_1922_Solar_Eclipse_Expeditions.

[Turner, Herbert Hall]. "Obituary Notices: Fellows: John Jepson Atkinson." *Monthly Notices of the Royal Astronomical Society* 85 (February 1925): 305–308.

UK Meteorological Committee. "Monthly Weather Report of the Meteorological Office." Majesty's Stationary Office, London, March 1919, Vol. XXXVI, No. III.

Weinstein, Galina. "From the Berlin 'Entwurf' Field Equations to the Einstein Tensor I: October 1914 until Beginning of November 1915." January 25, 2012. www.researchgate.net/publication/221659835_From_the_Berlin_ENTWURF_Field_Equations_to_the_Einstein_Tensor_IOctober_1914_until_Beginning_of_November_1915/download.

Willard, Joseph. "A Memoir, Containing Observations of a Solar Eclipse, October 27, 1780, Made at Beverly: Also of a Lunar Eclipse, March 29, 1782: a Solar Eclipse,

April 12, and of the Transit of Mercury over the Sun's Disc, November 12, the Same Year, Made at the President's House in Cambridge." *Memoirs of the American Academy of Arts and Sciences* 1 (1783): 129–142. Cited in *College Men at War*, by John P. Monks (Boston: American Academy of Arts and Sciences, 1957).

Willet, William. *The Waste of Daylight* (brochure). July 1907.

Wootton, David. "A Brief History of Facts." *History Today*, February 13, 2017. www .historytoday.com/david-wootton/brief-history-facts.

Wright, W. H. "Biographical Memoir of William Wallace Campbell, 1862–1938." *National Academy of Sciences Biographical Memoirs* 25 (Third Memoir) (1947): 5–74. www.nasonline.org/publications/biographical-memoirs/memoir-pdfs/william -campbell.pdf.

ONLINE LECTURE

Bruton, Elizabeth. "Sacrifice of a Genius: Henry Moseley's Role as a Signals Officer in World War One" (audio lecture). Royal Society, London, October 11, 2013. https:// royalsociety.org/science-events-and-lectures/2013/henry-moseley.

BOOKS

Asimov, Isaac. *Asimov's Biographical Encyclopedia of Science and Technology: The Lives and Achievements of 1195 Great Scientists from Ancient Times to the Present, Chronologically Arranged.* New York: Doubleday, 1972.

Berger, Meyer. *The Story of the New York Times, 1851–1951.* New York: Simon and Schuster, 1951.

British Association for the Advancement of Science. *Report of the Eighty-Fourth Meeting of the British Association for the Advancement of Science, Australia: 1914, July 28–August 31.* London: John Murray, 1915. https://archive.org/details/reportofbritisha15adva.

Bryce, Viscount. *Report of the Committee on Alleged German Outrages, Appointed by His Britannic Majesty's Government.* New York: Macmillan and Company, 1914. https:// archive.org/details/reportofcommitte00ingrea/page/n1.

Cadbury, William. *Labour in Portuguese West Africa.* London: G. Routledge and Sons, 1910. New York: E. P. Dutton and Co., 1910.

Chandrasekhar, Subrahmanyan. *Eddington: The Most Distinguished Astrophysicist of His Time.* Cambridge: Cambridge University Press, 1983.

Crelinsten, Jeffrey. *Einstein's Jury, The Race to Test Relativity.* Princeton, NJ: Princeton University Press, 2006.

Crommelin, A. C. D., and Philip H. Cowell. *Investigation of the Motion of Halley's Comet, from 1759 to 1910.* Edinburgh: Neill & Co., 1910.

Douglas, Allie Vibert. *The Life of Arthur Stanley Eddington.* London: Thomas Nelson, 1951. Reprint, Kingston, Ontario: Queen's University, 2018.

Eddington, A. S. *Report on the Relativity of Gravitation.* London: Fleetway Press, 1918. Second edition, 1920.

———. *Space, Time and Gravitation: An Outline of the General Relativity Theory.* London: Cambridge University Press, 1920.

Epstein, Jacob. *Let There Be Sculpture.* New York: G. P. Putnam's Sons, 1940.

Farrow, John Pendleton. *The History of Islesborough Maine.* Bangor, ME: T. W. Burr, 1893.

Ferguson, David P. *Chasing the Stars: The Amazing Life of William Wallace Campbell.* Findlay, OH: David P. Ferguson, 2017.

Feynman, Richard P., Robert B. Leighton, and Matthew Sands. *The Feynman Lectures on Physics,* vol. 2 (Reading, MA: Addison-Wesley, 1964).

Frank, Philipp. *Einstein: His Life and Times.* New York: Knopf, 1953.

Freundlich, Erwin. *Die Grundlagen der Einsteinschen Gravitationstheorie.* Berlin: Springer (Verlag von Julius Springer), 1916.

Furneaux, Robin. *The Amazon: The Story of a Great River.* New York: Putnam, 1967.

Grandin, Greg. *Fordlandia: The Rise and Fall of Henry Ford's Forgotten Jungle City.* New York: Henry Holt and Company, 2010.

Higgs, Catherine. *Chocolate Islands: Cocoa, Slavery, and Colonial Africa.* Athens: Ohio University Press, 2013.

Isaacson, Walter. *Einstein: His Life and Universe.* New York: Simon & Schuster, 2007.

Jackson, Joe. *The Thief at the End of the World: Rubber, Power, and the Seeds of Empire.* New York: Viking, 2008.

Jaff, Fay. *They Came to South Africa.* Cape Town, South Africa: Howard Timmins, 1963.

John, Angela V. *War, Journalism and the Shaping of the Twentieth Century: The Life and Times of Henry W. Nevinson.* London: I.B Tauris & Co., 2006.

Jones, Clifford. *The Sea and the Sky: The History of the Royal Mathematical School of Christ's Hospital.* Horsham, West Sussex, 2015.

Littmann, Mark, Fred Espenak, and Ken Willcox. *Totality: Eclipses of the Sun,* 3rd ed. Oxford: Oxford University Press, 2008.

Malinowski, Bronislaw. *Argonauts of the Western Pacific.* London: George Routledge & Sons, 1922. American printing, New York: E. P. Dutton & Co., 1932.

Maunder, E. Walter, ed. *The Total Solar Eclipse 1900: Report of the Expeditions Organized by the British Astronomical Association to Observe the Total Solar Eclipse of 1900 May 28.* London: British Astronomical Association, 1901. https://archive.org/details /eclipstotalsolar00britrich/page/n3.

Miller, Arthur. *Empire of the Stars: Obsession, Friendship, and Betrayal in the Quest for Black Holes.* New York: Houghton Mifflin, 2005.

Mourão, Ronaldo Rogério de Freitas. *Einstein: de Sobral para o mundo.* Sobral, Brazil: Edições Universidade Estadual Vale do Acaraú, 2003.

Nicolai, Georg Friedrich. *The Biology of War.* New York: Century Co., 1918.

Nevinson, Henry W. *A Modern Slavery.* New York: Harper & Brothers, 1906.

Osterbrock, Donald E., John R. Gustafson, and W. J. Shiloh Unruh. *Eye in the Sky: Lick Observatory's First Century.* Berkeley: University of California Press, 1988.

Pais, Abraham. *Einstein Lived Here.* Oxford: Oxford University Press, 1994.

———. *Subtle Is the Lord: The Science and the Life of Albert Einstein.* Oxford: Oxford University Press, 1982. Reprinted 2005.

Pape, Duarte, and Rodrigo Rebelo de Andrade. *As roças de São Tomé e Príncipe.* Lisbon, Portugal: Edições Tinta-da-China, 2013.

Rothschild, Robert Friend. *Two Brides for Apollo: The Life of Samuel Williams (1743–1817).* Bloomington, IN: iUniverse, 2009.

Satre, Lowell J. *Chocolate on Trial: Slavery, Politics, and the Ethics of Business.* Athens: Ohio University Press, 2005.

Simonis, H., *The Street of Ink: An Intimate History of Journalism.* London: Cassell & Company, 1917.

Sociedade De Geografia De Lisboa. *Comemoracoes do 90° Aniversario da Expedicao Cientifica de Eddington a Ilha do Principe.* Lisbon, Portugal: Sociedade De Geografia De Lisboa, 2010.

Stanley, Matthew. *Practical Mystic: Religion, Science and A. S. Eddington.* Chicago: University of Chicago Press, 2007.

Swan, Charles A. *The Slavery of To-Day or, The Present Position of the Open Sore of Africa.* Glasgow: Pickering & Inglis, 1909.

Vance, John L., ed. *The Great Flood of 1884 in the Ohio Valley.* Gallipolis, OH: Bulletin Office, 1884.

Whitehead, A. N. *Science and the Modern World.* New York: Macmillan, 1925.

Wilson, Margaret Dyson. *Ninth Astronomer Royal: The Life of Frank Watson Dyson.* Cambridge: W. Heffer & Sons, 1951.

Zackheim, Michele. *Einstein's Daughter: The Search for Lieserl.* New York: Riverhead, 1999.

LETTERS AND ARCHIVES

A. Vibert Douglas Archives, Queen's University Archives, Kingston, Ontario, Canada.

Carey, Maureen, and Alix Notron. Elizabeth Ballard Campbell Family Papers. University of California, Santa Cruz, Online Archives of California. https://oac.cdlib.org/findaid/ark:/13030/c8vq3767.

Davis, John, Letters to Thomas Davis with Other Papers, 1779–1781 (MS Am 2688). Houghton Library, Harvard University.

Lick Observatory Records (Mary Lea Shane Archives), Special Collections, University Library, University of California Santa Cruz.

William Campbell Letters. Mary Lea Shane Archives, University of California, Lick Observatory.

EINSTEIN LETTERS

Einstein, Albert. *The Collected Papers of Albert Einstein.* Princeton, NJ: Princeton University Press. https://einsteinpapers.press.princeton.edu. The following volumes were used for quotations and other observations:

Vol. 1, *The Early Years: 1879–1902.* Document 127. https://einsteinpapers.press.princeton.edu/vol1-trans.

Vol. 3, *The Swiss Years: Writings, 1909–1911.* Document 23. https://einsteinpapers.press.princeton.edu/vol3-trans.

Vol. 4, *The Swiss Years: Writings, 1912–1914.* https://einsteinpapers.press.princeton.edu/vol4-trans.

Vol. 5, *The Swiss Years: Correspondence, 1902–1914.* Documents 33–36, 281, 287, 303–305, 336, 472, 477, 483. https://einsteinpapers.press.princeton.edu/vol5-trans.

Vol. 6, *The Berlin Years: Writings, 1914–1917.* Document 30. https://einsteinpapers.press.princeton.edu/vol6-trans.

Vol. 7, *The Berlin Years: Writings, 1918–1921.* Document 26. https://einsteinpapers.press.princeton.edu/vol7-trans.

Vol. 8, *The Berlin Years: Correspondence, 1914–1918.* Documents 53, 54, 59, 63, 243, 244,

272, 273, 290, 311, 312, 353, 402, 404, 433, 562, 676. https://einsteinpapers.press
.princeton.edu/vol8-trans.

Vol. 9, *The Berlin Years: Correspondence, January 1919–April 1920*. Documents 1, 2, 3, 4, 6, 8, 14, 15, 88, 89, 99, 101, 103, 105, 106, 110, 113, 116, 119, 121, 127, 128, 137, 140, 148, 149, 154, 155, 159, 160, 162–165, 168, 173, 175, 180, 184–187, 189, 194, 197, 203, 204, 206a, 208, 216, 233, 242, 261, 295, 271, 272, 293, 309, 311, 312, 315, 320, 325, 328, 330, 331, 332, 335, 348, 349, 351. https://einsteinpapers.press.princeton.edu/vol9-trans.

Vol. 10, *The Berlin Years: Correspondence, May–December 1920, and Supplementary Correspondence, 1909–1920*. Documents 101, 231, 309. https://einsteinpapers.press
.princeton.edu/vol10-trans.

Vol. 12, *The Berlin Years: Correspondence, January–December 1921*. https://einsteinpapers
.press.princeton.edu/vol12-trans.

Vol. 13, *The Berlin Years: Writings & Correspondence, January 1922–March 1923*. https://
einsteinpapers.press.princeton.edu/vol13-trans.

Vol. 14, *The Berlin Years: Writings & Correspondence, April 1923–May 1925*. Documents 14, 16. https://einsteinpapers.press.princeton.edu/vol14-trans.

NEWSPAPER ARTICLES

"Astronomer C.O. Given One Year Exemption TO VIEW AN ECLIPSE." *Cambridge Daily News, Cambridge Tribunal*, July 12, 1918, 3.

"The Eclipse Comes on Time—A Grand Sight." *Hickory Nut and Upson (GA) Vigil*, June 1, 1900. Thomaston-Upson Archives. Reprinted in *Thomaston (GA) Times*.

Mattos, Emygdio de. "A Febre Amarella no Norte: Partida da Missão Sanitaria; Quem é e o que Fará o Seu Chefe." *O Combate (São Paulo)*, April 12, 1919, 1.

Morize, Henrique. "O Eclipse de 29 de Maio de 1919." *Folha do Littoral, Camocim, Ceará*, March 23, 1919.

"The Ninety-Three Today," *New York Times*, March 2, 1921, 7.

"Professor of Astronomy and 'C.O.' Refused Exemption." *Cambridge Daily News, Appeal Tribunal*, June 14, 1918, 3.

Silberstein, Ludwik. "Fatal Blow to Relativity Issued Here." *Toronto Evening Telegram*, March 7, 1936.

Smith, Leon. "The Day Middle Georgia Was in the Dark." *Atlanta Journal and Constitution Magazine*, May 25, 1958.

"Vice-Chancellor on Government Demands on University DEPLETED COLLEGE STAFFS." *Cambridge Daily News, Borough Tribunal*, May 16, 1918, 4.

Vinciguerra, Thomas. "The Truce of Christmas, 1914." *New York Times*, Week in Review, December 25, 2005.

WEBSITES

Ball, David. "Edwin Turner Cottingham at the Science Museum." In *Ringstead People: Biographies of Ringstead (Northamptonshire) People*. May 24, 2014. https://ringstead
.squarespace.com/ringstead-people/2014/5/24/edwin-t-cottingham-at-the-science
-museum.html.

Basso, Paulo César Cristófano. "Observatório Nacional reúne as Placas Fotográficas do eclipse de Sobral que contribuíram para comprovar a Teoria da Relatividade Geral de Albert Einstein" [National Observatory gathers photographic plates of Sobral eclipse that contributed to prove the theory of Albert Einstein's general relativity]. *Tema Espetacular* (blog). November 18, 2015. http://paulosciencebasso.blogspot .com/2015/11/observatorio-nacional-reune-as-placas.html.

Dolan, Graham. "The Royal Observatory, Greenwich." www.royalobservatorygreenwich .org.

Fletcher, John. "'The Comet Man': A. C. D. Crommelin, B.A., D.Sc., F.R.A.S." In "Andrew Claude de la Cherois Crommelin" web page. Crommelin Family website, accessed April 12, 2019. www.crommelin.org/history/Biographies/1865Andrewdela Cherois-Crommelin/index.htm.

Kramer, Bill. "Chasing the Solar Eclipse." Eclipse Chaser website, updated April 18, 2015. www.eclipse-chasers.com.

Page, Wendy. "John Jepson Atkinson." *Cosgrove & Furtho History.* Accessed April 12, 2019. www.cosgrovehistory.co.uk/doc/people/at-01.html.

"Roped Climbers Plunge to Their Deaths." Crommelin Family website, accessed April 12, 2019. www.crommelin.org/history/Biographies/Cherois/ClimbingDeaths/Climbing Deaths.htm.

ILLUSTRATION CREDITS

INSERT, PAGE 1:

Sir Isaac Newton: Copyright © The Royal Society, London
Optiks: Public Domain
Soldner: Public Domain
Solar Corona: Public Domain

INSERT, PAGE 2:

Young Albert: Copyright © SZ Photo / Scherl / Bridgeman Images.
Pauline Koch Einstein: Public Domain.
Wedding photo: *The Collected Papers of Albert Einstein, Volume 5* (courtesy of Evelyn Einstein), Princeton University Press, 1993.
Mileva and Sons: *The Collected Papers of Albert Einstein, Volume 5* (courtesy of Evelyn Einstein), Princeton University Press, 1993.

INSERT, PAGE 3:

Mount Hamilton/Lick: Courtesy Special Collections, University Library, University of California Santa Cruz. [Mary Lea Shane Archives of the Lick Observatory].
W. W. Campbell: Courtesy Special Collections, University Library, University of California Santa Cruz. [Mary Lea Shane Archives of the Lick Observatory].
Charles Perrine: Courtesy Special Collections, University Library, University of California Santa Cruz. [Mary Lea Shane Archives of the Lick Observatory].

INSERT, PAGE 4:

Flamsteed House: Courtesy of Graham Dolan.
Eddington: Courtesy of the Master and Fellows of Trinity College, Cambridge.
Dyson: Royal Astronomical Society/Science Source.

INSERT, PAGE 5:

1912 Teams at Passa Quatro, Brazil: Courtesy of Observatório Nacional, Rio de Janeiro.
Freundlich: Walter Ledermann Collection, University of St. Andrews, Scotland.

Perrine Camp/Telescopes/1912: Archivo Histórico, Observatorio Astronómico de Córdoba, digitalized by S. Paolantonio.

INSERT, PAGE 6:

Henry Moseley: University of Oxford, Museum of the History of Science, courtesy of AIP Emilio Segrè Visual Archives, Physics Today Collection.

Eclipsed Sun 1914 Crimea: Archivo Histórico, Observatorio Astronómico de Córdoba, digitalized by S. Paolantonio.

Raymond Lodge: Frontispiece for *Raymond or Life and Death*, by Oliver Lodge, published by Methuen & Co. Ltd., London, 1916.

Oliver Lodge: The National Portrait Gallery, London.

INSERT, PAGE 7:

The 1914 Newspaper Graphic: Courtesy of Michael Zeiller.

INSERT, PAGE 8:

1918 Campbell Expedition: Courtesy Special Collections, University Library, University of California Santa Cruz. [Mary Lea Shane Archives of the Lick Observatory].

Three Pilots: Courtesy of Peggy Campbell Rhoads.

INSERT, PAGE 9:

RMS *Anselm*: Courtesy of Graham Dolan.

Map 1,000 Miles Amazon: From the Booth Line Brochure, published 1931.

INSERT, PAGE 10:

Saboya's House: Courtesy of the Carnegie Institution for Science, Department of Terrestrial Magnetism.

Tents/Racetrack: The Special Collections Research Center, University of Chicago Library.

INSERT, PAGE 11:

Sobral Equip: Courtesy of the Charles Rundle Davidson Family Collection.

1919 Seated/Stand Group: Courtesy of Observatório Nacional, Rio de Janeiro.

INSERT, PAGE 12:

Roça/Plantation: Public Domain.

Shackles in Tree: Reproduced from *The Slavery of To-day*, by Charles Swan, published in Glasgow by Pickering & Inglis, 1909.

Edwin Cottingham: Courtesy of the Antiquarian Horological Society, London.

INSERT, PAGE 13:

Café Einstein: State Library of Western Australia, 4131B/2/4.

Leaving Wallal Boats: Courtesy Special Collections, University Library, University of California Santa Cruz. [Mary Lea Shane Archives of the Lick Observatory].

INSERT PAGE, 14:

1922 Eclipse Plate/Stars: Courtesy Special Collections, University Library, University of California Santa Cruz. [Mary Lea Shane Archives of the Lick Observatory].

INSERT, PAGE 15:

Einstein & Elsa/Japan: Public Domain.

President Campbell: Courtesy of Peggy Campbell Rhoads.

Perrine at Desk: Archivo Histórico, Observatorio Astronómico de Córdoba, digitalized by S. Paolantonio.

INSERT, PAGE 16:

Crommelin: Royal Astronomical Society/Science Source.

Eddington: The National Portrait Gallery, London.

Davidson: Courtesy of Cecily Maud Davidson.

Dyson: The National Portrait Gallery, London.

INDEX

Credit: Allen Jackson

Cathie Pelletier is the critically acclaimed author of twelve novels, including *The Funeral Makers*, *The Weight of Winter*, and *The One-Way Bridge*. Several have been translated into numerous languages, and two have been made into films. She lives in Allagash, Maine, on the banks of the St. John River, in the house where she was born.

Sylvester James "Jim" Gates, Jr., is a theoretical physicist who is most known for his work on supersymmetry, supergravity, and superstring theory. He is the Brown Theoretical Physics Center Director, Ford Foundation Professor of Physics, Affiliate Professor of Mathematics, and Faculty Fellow, Watson Institute for International Studies & Public Affairs at Brown University. In 2013, he was awarded the National Medal of Science, the highest award in the United States given to scientists. He is former president of the National Society of Black Physicists, a Fellow of the American Association for the Advancement of Science and the Institute of Physics in the United Kingdom. Vice President of the American Physical Society, Professor Gates regularly gives lectures for general audiences and makes frequent appearances in documentaries about science.

Credit: John T. Consoli

PublicAffairs is a publishing house founded in 1997. It is a tribute to the standards, values, and flair of three persons who have served as mentors to countless reporters, writers, editors, and book people of all kinds, including me.

I. F. STONE, proprietor of *I. F. Stone's Weekly*, combined a commitment to the First Amendment with entrepreneurial zeal and reporting skill and became one of the great independent journalists in American history. At the age of eighty, Izzy published *The Trial of Socrates*, which was a national bestseller. He wrote the book after he taught himself ancient Greek.

BENJAMIN C. BRADLEE was for nearly thirty years the charismatic editorial leader of *The Washington Post*. It was Ben who gave the *Post* the range and courage to pursue such historic issues as Watergate. He supported his reporters with a tenacity that made them fearless and it is no accident that so many became authors of influential, best-selling books.

ROBERT L. BERNSTEIN, the chief executive of Random House for more than a quarter century, guided one of the nation's premier publishing houses. Bob was personally responsible for many books of political dissent and argument that challenged tyranny around the globe. He is also the founder and longtime chair of Human Rights Watch, one of the most respected human rights organizations in the world.

· · ·

For fifty years, the banner of PublicAffairs Press was carried by its owner Morris B. Schnapper, who published Gandhi, Nasser, Toynbee, Truman, and about 1,500 other authors. In 1983, Schnapper was described by *The Washington Post* as "a redoubtable gadfly." His legacy will endure in the books to come.

Peter Osnos, *Founder*